Perspectives on
Plant Competition

Perspectives on Plant Competition

Edited by

James B. Grace
Department of Botany
Louisiana State University
Baton Rouge, Louisiana

David Tilman
Department of Ecology
University of Minnesota
Minneapolis, Minnesota

Academic Press, Inc.
Harcourt Brace Jovanovich, Publishers
San Diego   New York   Berkeley   Boston
London   Sydney   Tokyo   Toronto

Front cover illustration by Kim Johnson

Academic Press, Inc.
San Diego, California 92101

United Kingdom Edition published by
Academic Press Limited
24–28 Oval Road, London NW1 7DX

Library of Congress Cataloging-in-Publication Data

Perspectives on plant competition / edited by James B. Grace,
    David Tilman.
        p.   cm.
    Includes index.
    ISBN 0-12-294452-6 (alk. paper).
    1. Plant competition. 2. Plant communities.
I. Grace, James B.
II. Tilman, David. Date.
QK911.P37   1990
581.5'247–dc20   89-6863
                                                               CIP

Printed in the United States of America
90 91 92 93   9 8 7 6 5 4 3 2 1

Contents

Part II
The Role of Competition in Community Structure

12. Plant–Plant Interactions in Successional Environments
F. A. Bazzaz

13. Competitive Hierarchies and Centrifugal Organization in Plant Communities
Paul A. Keddy

14. Disorderliness in Plant Communities: Comparisons, Causes, and Consequences
Norma L. Fowler

15. The Role of Competition in Structuring Pasture Communities
Roy Turkington and Loyal A. Mehrhoff

16. The Role of Competition in Agriculture
S. R. Radosevich and M. L. Rousch

Part III
The Impact of Herbivores, Parasites, and Symbionts on Competition

17. The Mediation of Competition by Mycorrhizae in Successional and Patchy Environments
Edith B. Allen and Michael F. Allen

18. The Impact of Parasitic and Mutualistic Fungi on Competitive Interactions among Plants
Keith Clay

19. Herbivore Influences on Plant Performance and Competitive Interactions

Svata M. Louda, Kathleen H. Keeler, and Robert D. Holt

20. Predation, Herbivory, and Plant Strategies along Gradients of Primary Productivity

Lauri Oksanen

Contributors

Numbers in parentheses indicate the pages on which the authors' contributions begin.

Edith B. Allen (367), Department of Biology and Systems Ecology Research Group, San Diego State University, San Diego, California 92182

Michael F. Allen (367), Department of Biology and Systems Ecology Research Group, San Diego State University, San Diego, California 92182

M. P. Austin (215), Commonwealth Scientific and Industrial Research Organization, Division of Wildlife and Ecology, Lyneham Australian Capital Territory 2602, Australia

F. A. Bazzaz (240), Department of Organismic Biology, Harvard University, Cambridge, Massachusetts 02138

Frank Berendse (93), Center for Agrobiological Research, NL-6700 AA Wageningen, The Netherlands

Keith Clay (391), Department of Biology, Indiana University, Bloomington, Indiana 47405

Joseph H. Connell (9), Department of Biological Sciences, University of California, Santa Barbara, California 93106

Wim Th. Elberse (93), Center for Agrobiological Research, NL-6700 AA Wageningen, The Netherlands

L. G. Firbank (165), Anglia Higher Education College, Cambridge, England

Norma L. Fowler (291), Department of Botany, University of Texas, Austin, Texas 78713

Deborah E. Goldberg (27), Department of Biology, University of Michigan, Ann Arbor, Michigan 48109

James Grace (3, 51), Department of Botany, Louisiana State University, Baton Rouge, Louisiana 70803

Robert D. Holt (414), Museum of Natural History, University of Kansas, Lawrence, Kansas 66045

Paul A. Keddy (266), Department of Biology, University of Ottawa, Ottawa, Ontario K1N 6N5, Canada

Kathleen H. Keeler (414), School of Biological Sciences, University of Nebraska, Lincoln, Nebraska 68588

Svăta M. Louda (414), School of Biological Sciences, University of Nebraska, Lincoln, Nebraska 68588

Loyal A. Mehrhoff (308), Department of Botany, University of British Columbia, Vancouver, British Columbia V6T 2B1, Canada

Lauri Oksanen (445), Umea Universitet, Institutionen for ekologisk botanik, 901 87 Umea, Sweden

Stephen W. Pacala (67), Department of Ecology and Evolutionary Biology, University of Connecticut, Storrs, Connecticut 06268

S. R. Radosevich (341), Department of Forest Science, Oregon State University, Corvallis, Oregon 97331

M. L. Rousch (341), Departments of Forest and Crop Science, Oregon State University, Corvallis, Oregon 97331

John A. Silander, Jr. (67), Department of Ecology and Evolutionary Biology, University of Connecticut, Storrs, Connecticut 06268

U. Sommer (193), Max Planck Institute of Limnology, Plön, Federal Republic of Germany

David Tilman (3, 117), Department of Ecology and Behavioral Biology, University of Minnesota, Minneapolis, Minnesota 55455

Roy Turkington (308), Department of Botany, University of British Columbia, Vancouver, British Columbia V6T 2B1, Canada

A. R. Watkinson (165), School of Biological Sciences, University of East Anglia, Norwich, England

G. Bruce Williamson (143), Department of Botany, Louisiana State University, Baton Rouge, Louisiana 70803

Preface

Ever since Darwin, competition has been considered to be one of the major forces shaping the morphology and life history of plants and the structure and dynamics of plant communities. Because of this central position, plant competition has been approached from a great variety of perspectives. This has led to the development of divergent conceptual frameworks, each with its own vocabulary, goals, theory, and empiricisms. Because comparable divergences led to a decade of often unproductive controversy in animal ecology (see Strong *et al.*, 1984), we feel it is time to explore various perspectives on plant competition and hope to avoid further controversy by clarifying the underlying definitions, goals, and concepts associated with each perspective. Furthermore, by providing a forum for such communication and clarification, we wish to encourage the synthesis of the unique strengths and insights of each perspective into new, more general approaches.

This book is designed to encourage such communication. Each of its 20 chapters was written, reviewed, and edited to present, in as unambiguous a manner as possible, the unique perspectives of its author or authors. As such, they are statements of the present status of knowledge in this field. However, this book was motivated by the belief that, once stripped of differences in vocabulary and once different goals were acknowledged, there were many commonalities in studies of plant competition. These commonalities are opportunities for syntheses that lead to the development of new theories. It is our hope that this book can provide the raw material and the impetus for such syntheses during the coming decade.

Parallel to the process of book preparation, a meeting of most authors was held at the University of Minnesota's Cedar Creek Natural History Area in October 1987. This stimulating meeting was marked by the free exchange of ideas and opinions. The main accomplishment of this meeting was for each of us to come to understand each other's perspectives. The process of manuscript review and revision did much to further this assimilation of contrasting ideas. In the summer of 1988, a second symposium was held at the annual AIBS meeting in Davis, California. The integration of perspectives evident at this meeting was a gratifying accomplishment. It is our hope that this same benefit can be gained by all who read this collection of views on plant competition.

We are grateful for the assistance provided by the Department of Botany at Louisiana State University and the Departments of Ecology and Behavioral Biology at the University of Minnesota. We thank the Graduate School of the University of Minnesota for funding the October 1987 meeting that initiated this book and the staff of Cedar Creek Natural History Area for their assistance with the logistics of that symposium. In addition, all the authors of chapters in this book and many others have made substantial contributions to the quality of this book through their reviews of chapters written by others. In addition to the authors, we thank the following for their assistance in manuscript review: Lonnie Aarssen, Donald N. Alstad, Ralph Boerner, Steven Carpenter, Norman Christensen, Carmen Cid-Benevento, Roger del Moral, Patrice A. Morrow, Philip Grime, David Hartnett, David Jones, Peter Jordan, Jane Lubchenco, James McGraw, Tom Miller, Matthew A. Parker, John Pastor, Robert Peet, William J. Platt, Richard Primack, David Read, Ruth Shaw, Steve Simmons, and Arthur J. Stewart.

JAMES B. GRACE
DAVID TILMAN

I

Perspectives on the Determinants of Competitive Success

1

Perspectives on Plant Competition: Some Introductory Remarks

James Grace   **David Tilman**

Competition among plants was likely discovered by the first farmers during neolithic agriculture. Some of the first scientific treatments of the subject can be found in the works of de Crescentiis (1305) and DeCandolle (1820). Inspired by the logic of Malthus (1798), Darwin (1859) wrote extensively about competition as an important selective agent for all types of organisms. Early botanists and vegetation ecologists considered interspecific competition to be an integral part of nature. Agricultural and forestry practices have long attempted to minimize the effects of undesired plants. One of the first exclusive treatments of the subject was published in 1929 by Clements *et al*. This seminal work contains a detailed description of the early history of plant competition as well as a wealth of empirical information. As with most areas of science, the literature dealing with competition has grown dramatically in recent decades.

A wide range of meanings has been ascribed to the word "competition." Definitions range from the narrow to the general, from operational to philosophical, and from phenomenological to mechanistic. This range of definitions has caused confusion and continues to cloud discussions of the substance of competition. Several authors have attempted to define competition as a precise term (e.g., Harper, 1961; Milne, 1961) but it seems unlikely that a narrow definition is possible for a term that has been used so broadly. Rather, a more profitable approach may be to define competition broadly, but to study specific kinds of competition, such as resource competition or interference competition (e.g., Tilman, 1982; Begon *et al.*, 1986). Because this is a book about contrasting perspectives, it is important to examine the definition used by each author, especially the operational definition of competition (i.e., how each measures competitive effect or competitive ability). In practice, far more confusion is generated through differences in operational definitions than through contrasting conceptual definitions because the latter are often quite general.

In reading this book, and in reading the plant competition literature in general, it is important to distinguish between two markedly different operational definitions of competition. The first stresses the total competitive effect of a species or of an entire community on another species. This definition, which has been used by Keddy, Grime, and others, is operationally based on the difference between the biomass that a target plant attains in the absence of some or all neighbors compared to its biomass in the presence of all neighbors. The additional biomass attained after the removal of competitors is the competitive effect of those plants on that species. This operational definition does not adjust for differences in the biomass of competitors removed, and thus does not measure the intensity of competition per unit neighbor biomass. Its use lies in its ability to demonstrate whether or not competition is occurring. If competition is occurring, it is expected, from first principles, that the magnitude of the total competitive effect would increase directly (though not necessarily linearly) with the amount of neighbor biomass removed. Thus, when all neighbors are removed, a target plant should be able to attain greater biomass in a productive community with an initial total biomass of 1000 g/m^2 than in an unproductive one with an initial biomass of 100 g/m^2. Because it depends on the amount of biomass removed, the total competitive effect, by itself, tells us little about the mechanism of competition, especially when comparing habitats that differ in productivity or standing crop.

A second operational definition of competition commonly used is the intensity of competition *per unit biomass*. Here the total competitive effect of all neighbors is divided by the amount of biomass removed to obtain a measure of the intensity or strength of competition. This measure can be used to make comparisons both within and among plant communities. Thus, it would be possible to ask if the intensity of competition per unit of neighbor biomass depended on the species of the neighbors, on the productivity of a habitat, or the disturbance rate of a habitat. This operational definition of the intensity of competition has been the standard for animal ecology, and is qualitatively consistent with the traditional definitions provided by the competition coefficient of the Lotka–Volterra equations and by the relative yields used to analyze replacement experiments, and with the terminology in the work of Tilman, Goldberg, Pacala, Silander, and others. A portion of the controversy between Grime and Tilman may have resulted from their different operational definitions of competition. Grime, in asserting that the "strength of competition" was greater in more productive habitats, was most likely referring to what we have termed the "total competitive effect." In contrast, Tilman, in asserting that strength of competition should be approximately equal across a productivity gradient was referring to the "intensity of competition per unit neighbor biomass." The discussion above suggests that they may both be correct, once their definitions are understood. This illustrates the importance of understanding an author's operational definitions.

The contents of this volume are organized around three main subdivisions: (1) Perspectives on the Determinants of Competitive Success, (2) The Role of Competition in Community Structure, and (3) The Impact of Herbivores, Parasites, and Symbionts on Competition. The first section of the book deals with the question of "What determines competitive success?" In addressing this question, several of the papers deal, by necessity, with definitions of competition and competitive success. The paper by Connell distinguishes the traditional definition of competition in which members of a pair of species inhibit each other through effects on resources or on one another's abiotic environment, from "apparent" competition in which indirect negative effects are mediated through additional species, often on other trophic levels. Goldberg discusses some of the ways that plants interact via limiting resources by pointing out the distinction between the effects of plants on resources versus the response of plants to resource depletion. Grace deals further with definitions and with the determinants of competitive success in a chapter that discusses the similarities and differences between the theories of Grime and Tilman. Silander and Pacala present spatial models of neighborhood competition and use them to analyze experimental studies of competition among neighboring plants. Berendse and Elberse focus on the determinants of competitive success for nutrient-limited heathland and grassland species. Tilman discusses several different models of the mechanisms of nutrient competition to illustrate how each of them "abstracts" reality and how each incorporates both a plant's response to and its effect on resources. Finally, Williamson addresses some of the long-standing questions concerning the way to study allelochemic interactions among plants. He evaluates proposed methodologies by focusing on the degree to which apparent allelopathy might be adaptive.

The second section of this book provides a contrasting set of views about the consequences of competitive interactions for plant community structure. It is a tacit assumption in all of these chapters that competition is one of, but by no means the only, force in natural communities. The section begins with Firbank and Watkinson, who address some of the methodological limitations associated with traditional additive and substitutive experiments and present a hybrid approach to studying the phenomenon of competition. Sommer analyzes the mechanisms of nutrient competition among phytoplanktonic algae and discusses how well such mechanisms can predict the patterns of species abundances in lakes. Both Austin and Keddy explore the role of competition in the distributions of species across habitats. Within-habitat effects of competition are examined by Bazzaz, who illustrates the complexity that exists in actual interactions among individuals as habitats change over time. The themes of complex explanations of pattern and even the lack of strong pattern within communities are addressed in the papers by Turkington and Mehrhoff and Fowler. Finally, Radosevich and Rousch examine the role of competition in agricultural systems.

The final section of the book places competition within the context of the interactions of plants with organisms on other trophic levels. These chapters demonstrate that the outcome of competition depends on the effects of herbivores, parasites, and symbionts. Allen and Allen discuss the role of mycorrhizae in altering competitive abilities, particularly during succession. Clay illustrates that nonmycorrhizal fungi can have strong effects on competitive interactions through both pathogenic and mutualistic mechanisms. The role of herbivores in both modifying competitive relations and in regulating populations are the themes addressed in the final two papers by Louda, Keller, and Holt and by Oksannen. All of the papers in this final section emphasize the need to integrate the mechanisms of competition into the framework of the entire foodweb. This was a major theme to come from our meetings, and is a fitting theme to end this book.

References

Begon, M., Harper, J. L., and Townsend, C. R. (1986). "Ecology." Sinauer, Sunderland, Massachusetts.

Clements, F. E., Weaver, J. E., and Hanson, H. C. (1929). "Plant Competition: An Analysis of Community Functions." Carnegie Institution, Washington, D.C.

de Crescentiis (1305). Cited in Clements *et al.*, 1929.

Darwin, C. (1859). "The Origin of Species," Harvard Facsimile 1st ed., reprinted in 1964. Harvard Univ. Press, Cambridge, Massachusetts.

DeCandolle, A. P. (1820). "Essai Elementaire de Geographie Botanique." Cited in Clements *et al.*, 1929.

Harper, J. L. (1961). Approaches to the study of plant competition. *Symp. Soc. Exp. Biol.* **15,** 1–39.

Malthus, T. R. (1798). "First Essay on Population," reprinted in 1927 for the Royal Economic Society. London, Macmillan.

Milne, A. (1961). Definition of competition among animals. *Symp. Soc. Exp. Biol.* **15,** 1–39.

Tilman, D. (1982). "Resource Competition and Community Structure." Princeton Univ. Press, Princeton, New Jersey.

2

Apparent versus "Real" Competition in Plants

Joseph H. Connell

I. Introduction

The importance of competition in structuring natural communities can be evaluated in various ways (Connell, 1983; Schoener, 1983; Sih *et al.,* 1985). Obviously, the first task is to demonstrate unequivocally its occurrence in nature, yet this has often proved to be difficult. One difficulty is that the evidence often accepted as demonstrating competition can sometimes be produced by other types of interactions. In such cases, the competition supposedly demonstrated by an experiment or set of observations may be more apparent than real.

Competition can be defined most simply as a reciprocal negative interaction between two organisms. The term is traditionally restricted to instances involving only two broad categories of mechanisms: direct interference and indirect exploitation of shared resources (see cases 1 and 2, Fig. 1). However, the reciprocal negative effects could also arise from at least two other types of interactions, both indirect. First, Holt (1977, 1984) pointed out that if two species share one or more common predators, a reciprocal negative interaction could occur between the two prey; he termed this "predator-mediated apparent competition" (Fig. 1, case 3). Holt's model was designed to apply to instances where predators derived benefits from their prey only by consuming them. This chapter extends the concept, since plants also benefit herbivores by providing shelter. Second, I suggest another form of apparent competition, involving mutualism, that could occur within a single trophic level (Fig. 1, case 4). If species P_1 and P_2 interacted positively, and P_2 and P_3 negatively, P_1 and P_3 could show an indirect negative interaction. (The conditions under which cases 3 and 4 could operate will be discussed in detail in Section II,B.) If competition is defined simply as a negative interaction between two organisms on the same trophic level, then all four models in Fig. 1 represent "true" competition. But traditionally, competition has

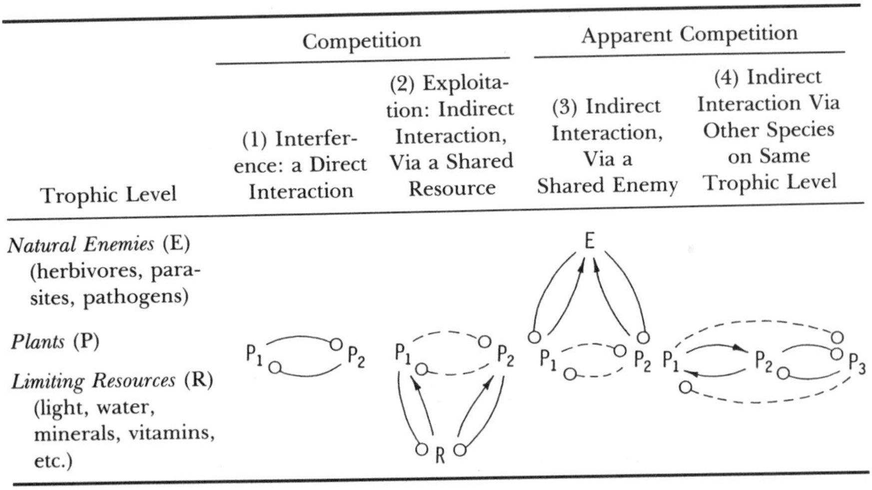

Trophic Level	Competition		Apparent Competition	
	(1) Interference: a Direct Interaction	(2) Exploitation: Indirect Interaction, Via a Shared Resource	(3) Indirect Interaction, Via a Shared Enemy	(4) Indirect Interaction Via Other Species on Same Trophic Level
Natural Enemies (E) (herbivores, parasites, pathogens)				
Plants (P)				
Limiting Resources (R) (light, water, minerals, vitamins, etc.)				

Figure 1   Some possible types of traditional and apparent interspecific competition in plants. Solid lines are direct interactions, dashed lines are indirect ones. An arrowhead indicates a positive effect on that species, a circle indicates a negative effect. In case 4, the apparent competition is between P_1 and P_3. See text for assumptions.

only been applied to cases 1 and 2, direct interference or indirect exploitation. This is the reason why Holt (1977) applied the term "apparent competition" to case 3, and why I include case 4 under the same label. This list of possibilities is probably not exhaustive.

To distinguish the different types of real and apparent competition in nature requires knowledge of the mechanisms involved. Because few studies of competition have unequivocally demonstrated the mechanisms underlying the interaction (Tilman, 1987), it is quite possible that many cases of competition may be apparent, not real. The aim of this chapter is to explore some of the conditions necessary for apparent competition to occur, and then to investigate some of the reported cases of plant competition to see how likely they are to be apparent, not real. Last, I discuss the implications of these findings in understanding the structure of natural communities.

One important semantic point needs to be made here. Burkholder (1952), in classifying the possible ways in which two species could interact, used the term competition for a reciprocal negative interaction $(-, -)$ and the term amensalism for a one-way negative interaction, $(0, -)$. In reviewing a number of studies of field experiments on competition, Lawton and Hassell (1981) and Connell (1983) recorded that in 66 and 61% of the cases, respectively, the interaction found was $(0, -)$. In an interaction between two species sharing the same resources, it is probable that there will be some reciprocal effect, however slight, but with the weaker one being undetectable against the background environmental variation. Therefore, the above authors suggested that, since the term amensalism was never used in these cases, it be replaced by "asymmetrical" competition. In Fig. 1, all interactions are shown as symmetrically reciprocal, for clarity. However, these competitive interactions could also be so asymmetrical as to approach $(0, -)$ in many cases. Some examples of the occurrence of such asymmetrical apparent competition will be discussed in Section III,B.

II. Methods for Demonstrating Competitive Mechanisms

A. Observational Methods and Interpretations

Competition is often inferred from observations of patterns such as character displacement, nonoverlapping "checkerboard" distributions, reciprocal and opposite changes in abundance in space and time, or "natural experiments" in which distributions, abundances, or properties of the niche of a species are observed in sites with and without a putative competitor (see reviews by Diamond, 1975; Arthur, 1982). Such observations may suggest the mechanisms underlying the inferred competition, but will seldom if ever demonstrate them (Connell, 1975).

However, nonexperimental observations of a different kind, i.e., interactions among individuals, may be capable of demonstrating both competition and its underlying mechanisms. For example, direct observations of physical contact among plants on hard substrates, such as crustose lichens, encrusting marine algae, or in sessile aquatic animals, can do so (reviewed by Connell and Keough, 1984). Such observations have been made in two ways. First, if the edge of one individual is seen to be lying directly over that of another, it could be inferred that the first is winning in competition over the second. However, this inference may be incorrect, because the underlying individual may be actively undercutting and displacing the upper one, as has been observed in marine sponges and barnacles. Alternatively, if the change in position of the edges is followed over time, the winner can be identified unequivocally. This latter method also reveals instances in which neither individual wins; such "stand-offs" between corals have been observed to last as long as 8 years (Connell, 1979). The mechanism in such cases is probably direct interference (case 1, Fig. 1), because the overlying individual directly contacts the underlying one, preventing access to resources such as light, nutrients, water, and gases, as well as preventing excretion of waste products. As described below, field experiments will greatly strengthen the conclusions concerning mechanisms.

In contrast, when the growth of short plants is slower in the shade of taller ones than in the open, it is often inferred that the latter is winning in indirect exploitation competition for light. However, this is a less certain demonstration of the mechanism than in the direct overgrowth examples above, since the possibility remains that other aspects such as the abundance or feeding activity of herbivores might be greater in the shade than in the sun (Louda *et al.*, this volume).

B. Experimental Methods and Interpretations

In many species and situations, direct observations are insufficient to demonstrate a competitive mechanism, and field experiments are more effective. However, many previous field experiments were designed only to demonstrate the existence of competition, not the underlying mechanisms. I will now discuss the experimental methods that have been used to test for each of the mechanisms shown in Fig. 1, and some of the problems in their interpretation.

1. Direct Interference (Case 1)   Direct interference involves one individual directly harming a neighbor in various ways, by either releasing toxic substances (Muller *et al.*, 1968); direct contact, e.g., mechanical abrasion when wind or water currents rub branches or algal fronds against each other (Hatton, 1938; Southward, 1953; Jacobs, 1955; Putz *et al.*, 1984; Rebertus, 1988); direct overgrowth (reviewed by Connell and Keough, 1984); mechanical crushing or undercutting (Connell, 1961); or, in animals, by attack with stinging tentacles or digestive filaments (Francis, 1973; Lang, 1973). Controls for unknown mechanisms that might also have caused the deleterious effects involve the removal of neighboring individuals or just the tips of their branches or fronds. To test hypotheses that species interfere by toxic exudates (allelopathy), laboratory experimental bioassays have been used (Muller *et al.*, 1968; McPherson and Muller, 1969; Christensen and Muller, 1975). Experiments in which allelopathic chemicals are administered in the field are possible but more difficult (Williamson, this volume).

2. Exploitative Competition (Case 2)   As shown in Fig. 1, case 2, exploitative competition is an indirect interaction acting through shared resources. Since it is usually impossible to observe this process, field experiments are usually done, e.g., removal of plants in whole or part, or manipulation of particular resources in the field (Tilman, 1984). Such experiments may be difficult to do without also affecting other relevant components of the system such as herbivores or pathogens. If overstory vegetation is removed to increase light levels to the lower story, this may also affect root interactions or grazers. Trenching experiments to reduce root competition may also greatly affect the abundance and activity of microorganisms, fungi, and soil animals. Manipulation of resources may also affect associated species. It is important to keep track of changes in natural enemies when performing field experiments on exploitative competition.

3. Apparent Competition Produced by Interactions with Natural Enemies (Case 3)   Holt (1977, 1984) proposed a model with the following assumptions about conditions under which this type of apparent competition would occur. In relation to plant populations, the conditions seem to be (1) the natural enemies (i.e., herbivores, parasites, or pathogens) are an important source of mortality or lowered fitness (e.g., through loss of plant tissue); (2) The enemies respond to a change in abundance of the plants in the following way: after an increase in P_1, either a consequent rise in the abundance of the enemies (either from increased reproduction or aggregative movements) or a change in their behavior results in an increased per capita rate of attack on P_2; a decrease in P_1 results in a decrease in their per capita rate of attack on P_2. (The opposite response, in which the enemies respond to a decrease in P_1 by switching and increasing their attack on P_2, will, of course, not produce apparent competition.)

If the behavior of the natural enemies satisfies these two assumptions, they could produce apparent competition between plant species. To test the assumptions and to detect this type of apparent competition requires considerable knowledge of the biology of the natural enemies and their effects on the plants, as well as field experiments designed to reveal the interactions. The abundance of each target plant species needs to be altered experimentally (with suitable controls), and any changes recorded in the distribution, abundance, or demographic variables of both the presumed competitors and any shared natural enemies. As we will see, these tasks are seldom accomplished.

4. Apparent Competition Produced by Positive Interactions among Species (Case 4) For this type of apparent competition to occur, there need to be positive interactions between one pair of species (P_1 and P_2 in Fig. 1, case 4), one of which has a negative interaction with a third, P_3. In theory, the latter could be caused by any of the mechanisms in cases 1, 2, or 3. Testing for this type of apparent competition involves ascertaining that the paired interactions are as described above. If so, then this type of apparent competition is possible. If the interactions are strongly asymmetrical, the direction of the stronger effect is crucial. For example, let us assume that P_1 has a strong positive effect on P_2, but the reverse effect is weak, whereas P_2 has only a slight negative effect on P_3, in contrast to a strong negative effect of P_3 on P_2. In this example, the indirect negative effects of P_1 on P_3 shown in Fig. 1, case 4, will be expected to be quite weak, possibly undetectable in field experiments.

III. Evidence for Real versus Apparent Competition in Plants

In this section, I review the evidence for real versus apparent competition using the criteria described in Section II.

A. Direct Interference versus Apparent Competition

Probably the best evidence for the operation of direct interference comes from observations of interactions among individuals or colonies of either plants or sessile aquatic animals on hard substrates. The detailed mechanisms by which these organisms interact are often directly observable. Some examples include observations of interactions among terrestrial lichens on rocks (Hawksworth and Chater, 1979; Pentecost, 1980) and of marine algae, sea grasses, sponges, corals, bryozoans, ascidians, barnacles, and mussels (reviewed by Connell, 1972; Connell and Keough, 1984; Paine, 1984; Buss, 1986; Lang, 1973). The better studies were those with field experiments that controlled for the effects of noncompetitive mechanisms.

It seems unlikely that the direct interference observed in these cases could be apparent competition produced by either of the mechanisms of cases 3 or 4. For example, crustose marine algae competed by direct interference whether grazers were present or experimentally excluded (Paine, 1984). In contrast, some instances of presumed direct interference by toxic exudates among terrestrial plants (i.e., allelopathy) may be more apparent than real. While laboratory bioassays have demonstrated the possibility of allelopathy (Muller *et al.*, 1968), direct field evidence is rare (Williamson, this volume). In one instance of presumed allelopathy, Kaminsky (1981) found strong laboratory evidence that microorganisms in soil around the roots of the shrub *Adenostoma fasciculatum* produce toxins that reduce the germination and growth of herbs in Californian chaparral. This implies that apparent competition via natural enemies, the soil microbes, is a plausible explanation for the negative interactions observed in that system. In terms of Fig. 1, case 3, the direct interactions would be *Adenostoma* $\xrightarrow{+}$ soil microbes $\xrightarrow{-}$ herbs, resulting in the apparent competition between *Adenostoma* and the herbs.

Two other instances of reduced herb abundance near sage bushes (*Salvia, Artemesia*) in Californian chaparral may be due, not to allelopathy, but to herbivores using the bushes as shelter from predators, as suggested by Bartholomew (1970) and by Halligan (1973). Since there is no evidence that the herbivores feed on the sage itself, this, like Kaminsky's (1981) example, could be a case of apparent competition that is asymmetrical. In terms of Fig. 1, case 3, the direct interactions would be sage $\xrightarrow{+}$ herbivores $\xrightarrow{-}$ herbs, resulting in the apparent competition between sage and the herbs. Muller and del Moral (1971) provide some qualitative evidence that opposes Bartholomew's (1970) results. To my knowledge, no published field study has demonstrated direct interference by allelopathy among plants in soil (case 1 in Fig. 1), while excluding the possibility of other indirect interactions with resources, natural enemies, or other competitors (cases 2, 3, and 4).

B. Exploitative versus Apparent Competition

In many instances, patterns suggesting exploitative competition may be produced by mechanisms of apparent competition. For example, spatial segregation between two species in the same area is often attributed to exploitative competition (Diamond, 1975). However, attack by shared predators could produce this habitat segregation (Holt, 1984). For example, Futuyma and Wasserman (1980) found that in a forest in New York, two oak species lived in separate stands with little overlap. In sites dominated by scarlet oaks (*Quercus coccinea*), larvae of a general herbivorous insect, the geometrid moth *Alsophila pometaria*, usually defoliated the rare scattered white oaks (*Quercus alba*), whereas in an adjacent stand dominated by white oaks, the rare scarlet oaks were more heavily attacked than the white oaks by the same insect species. This habitat segregation seems to be due to apparent competition, produced by reciprocal attacks on the rarer plant species by a shared natural enemy.

Another example involves two native composites in relatively undisturbed arid grasslands in New Mexico. Seedlings of the perennial forb *Machaeranthera canescens* occur in the region occupied by the shrub *Gutierrezia sarothae*, but never survive there. Yet, when transplanted among *Gutierrezia* and protected from grazing by the grasshopper *Hesperotettix viridis*, many *Machaeranthera* survived to maturity in both years of the experiment. (*Hesperotettix* also feed on *Gutierrezia*.) No *Machaeranthera* survived when exposed to this herbivory. As in the previous example, apparent competition via shared natural enemies, rather than competition or unfavorable physical conditions, probably maintained the spatial segregation of *Machaeranthera*.

The only experimental demonstration of apparent competition of which I am aware in which the abundance of the natural enemies and both the putative competitors was followed, is that by Schmitt (1987). Although it does not deal with plants, I include it because it provides an excellent example of the steps necessary to demonstrate apparent competition. Two species groups of subtidal marine mollusks show distinct habitat segregation, one living in cobble habitat, the other on rocky reefs. Schmitt performed two field experiments, transplanting each species group into patches of high and low density of its presumed competitor, with appropriate controls. He then assayed the mortality of each group, while also observing changes in abundance of their shared predators. He found that mortality of each presumed competitor increased when the abundance of the other was increased. The increased mortality was associated with increased rates of attack and aggregation of their shared predators on the sites where densities were higher. "Thus, each group of prey was negatively affected by the presence of the other because each alternative prey increased the local density of predators" (Schmitt, 1987).

These examples underline the importance of taking into account the abundances and effects of natural enemies, or positive interactions, when evaluating the role of plant competition in structuring natural communities. To indicate how often this has been done, I reexamined all papers on plant competition contained in two recent surveys of the literature of field experiments on competition (Connell, 1983; Schoener, 1983). There were 54 studies of plant competition, and in 50 of these, the authors claimed to have demonstrated interspecific competition. Of the 54 studies, 46 dealt with terrestrial plants; none investigated whether natural enemies (case 3) or positive interactions (case 4) affected the competitive interaction. The same applies to the two studies of freshwater plants. Of the 6 studies of marine plants, all included some investigation of grazers of the algae. However, none of the authors designed their studies to test the possibility that the experimental results, which they interpreted as evidence of "real" competition, might have been cases of apparent competition, due to the mechanisms modeled as case 3 or 4 in Fig. 1.

Thus, the cases examined are open to an additional interpretation, that the competition supposedly demonstrated may have been more apparent than real. How likely is this alternative? In some instances, the likelihood cannot be estimated, since insufficient information was gathered because the aim of the study was to test a specific hypothesis, e.g., the effects of different ratios of nutrients (Tilman, 1984), rather than to evaluate the role of competition in natural assemblages. In others, herbivores were excluded, or the field conditions had been highly modified by previous cultivating, fertilizing, pesticide use, or heavy grazing, so that application of the results to natural conditions would be doubtful. However, a few studies provide some evidence that is useful in deciding whether the competition demonstrated is apparent or real. Two may represent examples of case 3, apparent competition due to shared natural enemies.

In two studies, the effects of a change in abundance of one competitor on the behavior or abundance of shared natural enemies were discussed. In one of the 46 terrestrial studies, Robertson (1947) found that, where *Artemesia* bushes were removed from a square-shaped acre of land in Arizona, grazing of experimentally placed herbaceous plants by deer, rodents, and insects was less than on herbs placed in narrow strips 3 m wide from which *Artemesia* had been removed. The other species (22 species of herbs) grew better in this large cleared site than in either the narrow cleared strips or the undisturbed sites. Thus, reduction of one plant species (*Artemesia*) over a large area apparently caused a reduction in the per capita rate of attack by natural enemies on another set of species. Since this case appears to satisfy the assumptions of Holt's (1977, 1984) model, apparent competition is probably as likely to be one of the mechanisms causing the difference in herb growth as that suggested by Robertson (1947), exploitative competition for water.

A second example, that of Dayton (1975), is from the marine intertidal zone of Washington state. When the canopy alga *Hedophyllum* was removed, a group of fugitive algae increased. Also, a large grazing mollusk, the chiton *Katherina tunicata*, declined quickly in the five study sites where *Katherina* was originally abundant (Dayton's Figs. 1 and 2), which were also the sites with lesser degrees of wave action. In the more wave-exposed sites, *Katherina* was rare (Dayton's Figs. 2, 3, and 4), so that it could not have influenced the outcome of the experiments. In a later study, Gaines (1985) showed that *Katherina* was capable of reducing one of the fugitive species, *Iridaea cordata*, to very low numbers. Thus the increases in fugitive algae at the sites with less wave action are as consistent with the model of apparent competition (case 3) as with that of interference or exploitative competition; with greater wave action, case 3 seems less likely.

One interesting aspect of both Robertson's (1947) and Dayton's (1975) studies is that the decline in grazer abundance or rate of feeding may have been due, not to a reduction in food supply, but to a decrease in shelter. The *Artemesia* bushes or the *Hedophyllum* canopy may have provided cover or shelter for the grazers rather than, or in addition to, a source of food. In the case 3 model in Fig. 1, the arrow from P_1 to E would then represent a positive effect of shelter rather than of food. However, the result would be the same: apparent competition between P_1 and P_2. This is the same reasoning used by Bartholomew (1970) and by Halligan (1973) as an alternative explanation for allelopathy near chaparral bushes; the bushes offer shelter for vertebrate grazers, which venture out from them only a short distance to feed. Such apparent competition is likely to be asymmetrical.

Eleven of the 46 studies of competition in terrestrial plants involved trenching, i.e., cutting the roots entering study plots to reduce competition for soil nutrients by species outside the plots. By killing some roots, this treatment may also reduce the abundance of natural enemies that had been attacking these roots (soil pathogens, root predators, etc.). If so, it is possible that the increased growth or survival of plants within trenched plots may be a response to a lower rate of attack by natural enemies in the soil, i.e., apparent competition of the case 3 type. While I am unaware of any study of changes in plant pathogens or parasites caused by trenching, at least one study exists of its effect on soil organisms that are not pathogenic. Gadgil and Gadgil (1971) found that the rate of litter decomposition was much higher within than outside trenched plots in a pine plantation in New Zealand. They suggested that saprophytic fungal populations had increased, in abundance or activity or both, as a consequence of trenching, possibly due to a reduction in mycorrhizal fungi on the severed roots. Since trenching apparently had a profound effect on the organisms that decompose litter, it is reasonable to infer that it could also affect organisms pathogenic to plants. This possibility deserves investigation, since positive results of trenching experiments are usually taken to indicate exploitative competition.

The results of Gadgil and Gadgil (1971) also indicate that a more complex version of case 4 could apply to some trenching experiments. Although not on the same trophic level as green plants, the saprophytes and mycorrhizae that make mineral nutrients available to plants could be considered mutualistic associates. Referring to Fig. 1, case 4, P_1 represent the trees whose roots are cut by trenching, P_2 are the mutualistic mycorrhizal fungi, and P_3 are the soil saprophytes that decompose litter. Gadgil and Gadgil (1971) suggest that P_2 negatively affect P_3. The saprophytes, P_3, should have a positive effect (by supplying nutrients through litter decomposition) to the plants that are observed to grow better within the trenched plots. The latter would be an additional group, P_4. The chain of interactions would be $P_1 + P_2 - P_3 + P_4$. Thus, the indirect negative effect of P_1 on P_4 seen in many trenching experiments, and usually inferred to be exploitation competition (case 2), could be apparent competition, case 4. The length of the chain of species in case 4 could thus be longer than shown in Fig. 1, providing that the indirect effect between the species at the ends is negative.

Two of the 54 studies reviewed could also be examples of case 4. Dayton (1975) found that, when the canopy alga *Hedophyllum sessile* was removed, a group of "obligate understory" algae died, while a group of fugitive algae increased. When the fugitive algae were reduced, the understory species had noticeably more branches and *Hedophyllum* increased, as compared to the unmanipulated controls. Thus *Hedophyllum* and the obligate understory algae could correspond to P_1 and P_2, respectively, in case 4, since the former positively affects the latter, and perhaps vice versa, although this possibility was not investigated. The obligate understory and the fugitive algae probably compete for space, judging from the results of experimentally reducing the fugitives. Thus the inferred competition between *Hedophyllum* (P_1) and the fugitive algae (P_3) could also be interpreted as apparent competition in the sense of case 4, Fig. 1. This study is particularly interesting in that the assumptions of both models of apparent competition, cases 3 and 4, appear to be satisfied.

A similar instance was documented in the marine intertidal zone in California by Taylor and Littler (1982). An anemone species had a positive effect on two species of crustose algae; these became bleached and died when the anemone was removed, while a group of fugitive algae increased. Since both the crustose algae and the fugitive algae require rock substrate for attachment, they are likely to be competitors for space. Thus, referring to case 4 in Fig. 1, the anemone represents P_1, the crustose algae P_2, and the fugitive algae P_3. Since anemones P_1 also require space for attachment, they probably compete with fugitive algae; however, this inference is open to an additional explanation, that they are apparent competitors in the sense of case 4.

IV. Discussion

The assumptions of Holt's (1977, 1984) model of apparent competition (case 3 in Fig. 1) seem likely to be satisfied for interactions among terrestrial and aquatic plants and their natural enemies for the following reasons. (1) Herbivores, parasites, and pathogens are important causes of mortality or loss of tissue for plants (Clay and Louda *et al.*, this volume). (2) Natural enemies often interact with several species of plants in a single community, either as sources of food or of shelter (Louda *et al.*, this volume). (3) Natural enemies have been seen to decrease following declines in their prey, and to increase again when the latter increase. Such cycles in both animal predator–prey interactions (Krebs, 1985) and in herbivores and plants (Crawley, 1983) are evidence of such variations. (4) Decreases in the rate of attack by a natural enemy on one plant species as a result of decreases in abundance of another plant species are likely, as indicated in some of the studies discussed above (Robertson, 1947; Dayton, 1975; Futuyama and Wasserman, 1980; Parker and Root, 1981).

The opposite behavior by a natural enemy, in which it switches and increases its per capita rate of attack on P_2 after P_1 decreases, is possible, and has been demonstrated with animals in laboratory experiments (Murdoch, 1969; Murdoch and Bence, 1987). The only example of switching involving plants of which I am aware is that of Murton (1971). Wood pigeons feeding on peas and beans placed in a field at different relative densities ate proportionately more of the commoner one, once it exceeded 80% of the total. Whether these experimental results apply to natural populations is unknown. In two studies of foraging by squirrels on two species of acorns, there was no evidence that they attacked the commoner species at a rate proportionately higher than its relative abundance; thus, they did not show switching behavior (Bakken, cited in Smith and Follmer, 1972; Lewis, 1980). In summary, there is some evidence that herbivores respond to changes in plant abundance in ways that satisfy the assumptions of case 3, while evidence for the opposite behavior is lacking for studies in natural environments. Clearly, more evidence is needed.

Holt's (1977, 1984) model was designed to apply to instances where predators derived benefits from their prey only by consuming them. The present chapter extends the concept, since plants also benefit herbivores by providing shelter. Thus in case 3, Fig. 1, if E feeds only on P_2, but P_1 provides shelter for E, P_1 indirectly harms P_2, but not vice versa, an example of asymmetrical apparent competition. The studies of Robertson (1947), Dayton (1975), and Taylor and Littler (1982) may be examples of this.

The assumptions underlying case 4 are that positive interactions occur between one species pair and negative interactions between one of these and a third species. Mutualisms (+, +) and commensalisms (0, +) between plants have been described many times (see the review by Hunter and Aarssen, 1988). They usually arise when one plant ameliorates the physical or biological conditions in its vicinity, or excretes nutrients, or supports populations of pollinators or seed dispersers or is directly connected to another by root grafts or mycorrhizae. However, for such interactions to apply to case 4, the positive effects must apply only to two of the plant species, not to the third. This is probably most likely when the mutualistic species (P_1 and P_2 in Fig. 1, case 4) have life history characteristics that are quite different from those of P_3. For example, if P_1 and P_2 share insect pollinators, this pattern may support larger pollinator populations and/or increase the regularity and rate of pollinator visitation, which should lead to greater seed production (Waser and Real, 1979; Schemske, 1981). If P_3 were a wind-pollinated grass it would not benefit from this mutualism. If in addition P_3 competes with P_2 but not with P_1 (perhaps because P_2 and P_3 have roots reaching the same depth and so compete for nutrients, whereas P_1 has deeper roots), then case 4 would apply. The same might apply if P_1 and P_2 share animal seed dispersers but P_3 had wind-dispersed seeds. In both instances, the positive interaction between P_1 and P_2 does not extend to P_3. In the two marine examples of case 4 cited above, the positive effects of *Hedophyllum* or the anemone on the crustose algae did not extend to the fugitive algae.

If plants are connected by root grafts or mycorrhizae in species-specific pairs, case 4 may apply. This is more likely with ectomycorrhizae, which tend to be more host-specific than vesicular–arbuscular mycorrhizae (Gerdemann and Trappe, 1974; Trappe and Fogel, 1977). Last, positive effects between plant species have been observed via protection from natural enemies, i.e., plant "defense guilds" (Atsatt and O'Dowd, 1976) and "associational resistance" (Tahvanainen and Root, 1972; Hay, 1986). Whether such positive effects are sufficiently species-specific to qualify for case 4 is unknown at present.

Since the assumptions underlying the models of cases 3 and 4, Fig. 1, seem to be satisfied in several instances, I suggest that it is important to investigate the changes in abundance and behavior of natural enemies or mutualists and commensalists while studying plant competition. In none of the studies of plants surveyed by Connell (1983) and Schoener (1983) were the experiments designed to test for apparent competition of either case 3 or 4. Therefore, in the instances in which competition was inferred to occur in natural conditions, the evidence is insufficient to allow us to decide whether the results point to apparent or "real" competition.

If competition is defined simply as a negative interaction between two organisms on the same trophic level, then all four models in Fig. 1 represent "true" competition. But traditionally, competition has only been applied to cases 1 and 2, direct interference or indirect exploitation. This is the reason why Holt (1977) applied the term "apparent competition" to case 3, and why I include case 4 under the same label. Each of the four cases represents a quite different process, and therefore should be distinguished as far as possible. In so doing, we may find that two or more apply to the same interaction.

For plants on hard substrates, such as crustose and foliose terrestrial lichens and encrusting marine algae, interference competition by overgrowth can be observed directly, so that apparent competition can probably be ruled out in such instances. Natural enemies can affect the degree to which either species wins in competition among such species (Paine, 1984; Buss, 1986), but should not change the process from "real" to apparent competition. However, it seems clear that, for plants that are not encrusting on hard substrates, there are few instances in which apparent versus "real" competition has been distinguished. Clearly, the mechanisms underlying competitive interactions among plants are as yet little understood (Tilman, 1987). In part this is because questions about underlying mechanisms are sometimes not asked, or, if asked, have not included some of the plausible alternatives, such as cases 3 and 4 of Fig. 1.

A common question in ecology is, What is the relative importance of competition, mutualism, predation, herbivory, parasitism, etc. in determining community structure? Given the possibility that, for example, herbivory can produce the effect of competition between plants (i.e., case 3 in Fig. 1), this question can scarcely be answered as stated. Perhaps the question could be more accurately answered if it were posed as How much do direct interactions with resources, direct competitors, herbivores, mutualists, parasites, etc., plus the indirect effects these produce when acting together, affect community structure? If those direct interactions with resources and other species that produce significant effects were measured, the indirect effects could be evaluated as the algebraic sum of the direct ones and a model developed to predict community structure under particular environmental regimes. The second question is a more complex one than the first, but, given what we now know, is probably a more realistic one.

V. Summary

Competition can be defined most simply as a reciprocal negative interaction between organisms. In the traditional view, competition is produced by either of two categories of mechanisms: direct interference or indirect exploitation of shared resources. However, there are other mechanisms that can also indirectly produce reciprocal negative interactions between organisms. These mechanisms produce what has been called "apparent competition" because they differ from the two types of mechanisms usually accepted in the traditional view, but mimic their effects.

I suggest here that the model of apparent competition of Holt (1977), involving shared predators, can be extended in plants to situations in which one plant species provides only shelter to herbivores that feed on another plant. This can result in an indirect negative effect of the first plant on the second, producing apparent competition. I also propose a second type of apparent competition involving positive effects (mutualism or commensalism) between plants.

To distinguish apparent competition from the real (traditional) thing requires that we demonstrate the mechanisms underlying the interactions. This is easier to do in some systems than in others. In this chapter I investigate the problems of accomplishing this for plants in different terrestrial and aquatic systems and discuss the implications of apparent versus real competition for understanding community structure.

Acknowledgments

I would like to thank the following persons who commented on various drafts of the manuscript: L. Aarssen, J. Bence, K. Clay, C. D'Antonio, D. Engle, S. Holbrook, R. Holt, T. Hughes, E. Leigh, D. Lohse, S. Louda, L. Mehrhoff, W. Murdoch, P. Raimondi, D. Reed, P. Ross, O. Sarnelle, R. Schmitt, J. Selwa, A. Stewart-Oaten, S. Swarbrick, R. Turkington, C. Tyler, S. Walde, B. Williamson, and two anonymous reviewers. The National Science Foundation (OCE 86-08829) has supported the research underlying some of these ideas.

References

Arthur, W. (1982). The evolutionary consequences of interspecific competition. *Adv. Ecol. Res.* **12**, 127–187.

Atsatt, P. R., and O'Dowd, D. J. (1976). Plant defense guilds. *Science* **193**, 24–29.

Bartholomew, B. (1970). Bare zone between California shrub and grassland communities: The role of animals. *Science* **170**, 1210–1212.

Burkholder, P. R. (1952). Cooperation and conflict among primitive organisms. *Am. Sci.* **40**, 601–631.

Buss, L. W. (1986). Competition and community organization on hard surfaces in the sea. *In* "Community Ecology" (J. Diamond and T. J. Case, eds.), pp. 517–536. Harper & Row, New York.

Christensen, R., and Muller, C. H. (1975). Effects of fire on factors controlling plant growth in *Adenostoma* chaparral. *Ecol. Monogr.* **45**, 29–55.

Connell, J. H. (1961). The influence of interspecific competition and other factors on the distribution of the barnacle *Chthamalus stellatus*. *Ecology* **42**, 710–723.

Connell, J. H. (1972). Community interactions on marine rocky intertidal shores. *Annu. Rev. Ecol. Syst.* **3**, 169–192.

Connell, J. H. (1975). Some mechanisms producing structure in natural communities; a model and evidence from field experiments. *In* "Ecology and Evolution of Communities" (M. L. Cody and J. M. Diamond, eds.), pp. 460–490. Belknap, Cambridge, Massachusetts.

Connell, J. H. (1979). Tropical rain forests and coral reefs as open nonequilibrium systems. *In* "Population Dynamics" (R. M. Anderson, L. R. Taylor, and B. D. Turner, eds.), Symp. Br. Ecol. Soc., pp. 141–163. Blackwell, Oxford, England.

Connell, J. H. (1983). On the prevalence and relative importance of interspecific competition: Evidence from field experiments. *Am. Nat.* **122**, 661–696.

Connell, J. H., and Keough, M. J. (1984). Disturbance and patch dynamics of subtidal marine animals on hard substrata. *In* "The Ecology of Natural Disturbance and Patch Dynamics" (S. T. A. Pickett and P. S. White, eds.), pp. 125–151. Academic Press, Orlando, Florida.

Crawley, M. J. (1983). "Herbivory: The Dynamics of Animal–Plant Interactions," 437 pp. Univ. of California Press, Berkeley, California.

Dayton, P. K. (1975). Experimental evaluation of ecological dominance in a rocky intertidal algal community. *Ecol. Monogr.* **45**, 137–159.

Diamond, J. M. (1975). Assembly of species communities. *In* "Ecology and Evolution of Communities" (M. L. Cody and J. M. Diamond, eds.), pp. 342–444. Belknap Press, Cambridge, MA.

Francis, L. (1973). Interspecific aggression and its effects on the distribution of *Anthopleura elegantissima* and some related sea anemones. *Biol. Bull. (Woods Hole, Mass.)* **150**, 361–376.

Futuyma, D. J., and Wasserman, S. S. (1980). Resource concentration and herbivory in oak forests. *Science* **210**, 920–922.

Gadgil, R. L., and Gadgil, P. D. (1971). Mycorrhiza and litter decomposition. *Nature (London)* **233**, 133.

Gaines, S. D. (1985). Herbivory and between-habitat diversity: The differential effectiveness of defenses in a marine plant. *Ecology* **66**, 473–485.

Gerdemann, J. W., and Trappe, J. M. (1974). "The Endogonaceae in the Pacific Northwest," Mycol. Mem. 5. Hafner, New York.

Halligan, J. P. (1973). Bare areas associated with shrub stands in grassland: The case of *Artemesia californica*. *BioScience* **23**, 429–432.

Hatton, H. (1938). Essais de bionomie explicative sur quelques especes intercotidales d'algues et d'animaux. *Ann. Inst. Oceanogr. Monaco* **17**, 241–348.

Hawksworth, D. L., and Chater, A. O. (1979). Dynamism and equilibrium in a saxicolous lichen mosaic. *Lichenologist* **11**, 75–80.

Hay, M. E. (1986). Associational plant defenses and the maintenance of species diversity: Turning competitors into accomplices. *Am. Nat.* **128**, 617–641.

Holt, R. D. (1977). Predation, apparent competition, and the structure of prey communities. *Theor. Pop. Biol.* **12**, 197–229.

Holt, R. D. (1984). Spatial heterogeneity, indirect interactions, and the coexistence of prey species. *Am. Nat.* **124**, 377–406.

Hunter, A. F., and Aarssen, L. (1988). Plants helping plants. *BioScience* **38**, 1.

Jacobs, M. R. (1955). "Growth Habits of the Eucalyptus," 262 pp. For. Timber Bur., Dep. Interior, Canberra, Australia.

Kaminsky, R. (1981). The microbial origin of the allelopathic potential of *Adenostoma fasciculatum* H & A. *Ecol. Monogr.* **51**, 365–382.

Krebs, C. J. (1985). "Ecology: The Experimental Analysis of Distribution and Abundance," 800 pp. Harper & Row, New York.

Lang, J. C. (1973). Interspecific aggression by scleractinian corals. II. Why the race is not only to the swift. *Bull. Mar. Sci.* **23**, 260–279.

Lawton, J. H., and Hassell, M. P. (1981). Asymmetrical competition in insects. *Nature (London)* **289**, 793–795.

Lewis, A. R. (1980). Patch use by grey squirrels and optimal foraging. *Ecology* **61**, 1371–1379.

McPherson, J. K., and Muller, C. H. (1969). Allelopathic effects of *Adenostoma fasciculatum* "chamise" in the California chaparral. *Ecol. Monogr.* **32**, 177–198.

Muller, C. H., and del Moral, R. (1971). Role of animals in suppression of herbs by shrubs. *Science* **173**, 462–463.

Muller, C. H., Hanawalt, R. B., and McPherson, J. K. (1968). Allelopathic control of herb growth in the fire cycle of the California chaparral. *Bull. Torrey Bot. Club* **95**, 225–231.

Murdoch, W. W. (1969). Switching in general predators: Experiments on predator specificity and stability of prey populations. *Ecol. Monogr.* **39**, 335–354.

Murdoch, W. W., and Bence, J. (1987). General predators and unstable prey populations. *In* "Predation: Direct and Indirect Impacts on Aquatic Communities" (W. C. Kerfoot and A. Sih, eds.), pp. 17–30. Univ. Press of New England, Hanover, New Hampshire.

Murton, R. K. (1971). The significance of a specific search image in the feeding behaviour of the wood pigeon. *Behaviour* **40**, 10–42.

Paine, R. T. (1984). Ecological determinism in the competition for space. *Ecology* **65**, 1339–1348.

Parker, M. A., and Root, R. B. (1981). Insect herbivores limit habitat distribution of a native composite, *Machaeranthera canescens*. *Ecology* **62**, 1390–1392.

Pentecost, A. (1980). Aspects of competition in saxicolous lichen communities. *Lichenologist* **12**, 135–144.

Putz, F. E., Parker, G. G., and Archibald, R. M. (1984). Mechanical abrasion and intercrown spacing. *Am. Midl. Nat.* **112**, 24–28.

Rebertus, A. J. (1988). Crown shyness in a tropical cloud forest. *Biotropica* **20**, 338–339.

Robertson, J. H. (1947). Responses of range grasses to different intensities of competition with sagebrush (*Artemisia tridentata* Nutt.). *Ecology* **28**, 1–16.

Schemske, D. W. (1981). Floral convergence and pollinator sharing in two bee-pollinated tropical herbs. *Ecology* **62**, 946–954.

Schmitt, R. J. (1987). Indirect interactions between prey: Apparent competition, predator aggregation, and habitat segregation. *Ecology* **68**, 1887–1897.

Schoener, T. W. (1983). Field experiments on interspecific competition. *Am. Nat.* **122**, 240–285.

Sih, A., Crowley, P., McPeek, M., Petranka, J., and Strohmeier, K. (1985). Predation, competition and prey communities. *Annu. Rev. Ecol. Syst.* **16**, 269–311.

Smith, C. C., and Follmer, D. (1972). Food preferences of squirrels. *Ecology* **53**, 82–91.

Southward, A. J. (1953). The ecology of some rocky shores in the south of the Isle of Man. *Proc. Liverpool Biol. Soc.* **59**, 1–50.

Tahvanainen, J. O., and Root, R. B. (1972). The influence of vegetational diversity on the population ecology of a specialized herbivore, *Phyllotreta cruciferae* (Coleoptera : Chrysomelidae). *Oecologia* **10**, 321–346.

Taylor, P. R., and Littler, M. M. (1982). The roles of compensatory mortality, physical disturbance, and substrate retention in the development and organization of a sand-influenced, rocky-intertidal community. *Ecology* **63**, 135–146.

Tilman, D. (1984). Plant dominance along an experimental gradient. *Ecology* **65**, 1445–1453.

Tilman, D. (1987). The importance of the mechanisms of interspecific competition. *Am. Nat.* **129,** 769–774.

Trappe, J. M., and Fogel, R. D. (1977). Ecosystematic functions of mycorrhizae. *Range Sci. Dep. Sci. Ser. (Colo. State Univ.)* **26,** 205–214.

Waser, N. M., and Real, L. A. (1979). Effective mutualism between sequential flowering plant species. *Nature (London)* **281,** 670–672.

3

Components of Resource Competition in Plant Communities

Deborah E. Goldberg

I. Introduction

Over the past two decades, experimental field evidence has accumulated to show that competition between plants in natural communities is a common, although not ubiquitous, phenomenon (see reviews by Connell, 1983; Schoener, 1983; Fowler, 1986). Yet, there is still much debate over what determines which species will be successful in competition under different environmental conditions, and the relative importance

Perspectives on Plant Competition. Copyright © 1990 by Academic Press, Inc. All rights of reproduction in any form reserved.

of competition itself in determining species composition of plant communities (Newman, 1973; Grime, 1977, 1987; del Moral, 1983; Wilson and Keddy, 1986a; Tilman, 1982, 1987a, 1988a; Thompson, 1987; Thompson and Grime, 1988; Keddy, this volume; Grace, this volume). In this chapter, I develop a mechanistic framework (sensu Schoener, 1986; Tilman, 1987b) for studying interactions between plants that I argue clarifies the issues surrounding these questions and suggests new empirical approaches to resolving them.

The framework is based on the observation that most interactions between individual plants actually occur through some intermediary such as resources, pollinators, dispersers, herbivores, or microbial symbionts. Such indirect interactions consist of two distinct processes: one or both plants has an *effect* on abundance of the intermediary and a *response* to changes in abundance of the intermediary (Fig. 1). The type of interaction depends on the identity of the intermediary and the directions of effect and response (Table 1). The focus of this chapter is on competition for resources, which involves negative effects (e.g., light depletion under a plant canopy) and positive responses (e.g., the dependence of growth or survival on available light). However, the framework of effect and response depicted in Fig. 1 could also be useful in analyzing other types of indirect interactions (Table 1) and, indeed, in distinguishing interactions involving resources from other types of interactions (Connell, this volume).

I focus on two important insights that arise from distinguishing between the effect and response components of competition. First, these two components correspond to two ways in which individual plants can be good competitors: by rapidly depleting a resource or by being able to continue growth at depleted resource levels. Because the same traits do not necessarily determine the magnitudes of these two processes (Section II), we need to consider the conditions under which the effect versus the response component of individual competitive ability is more important

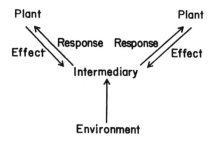

Figure 1   The effect and response components of indirect interactions between plants. The intermediary could be resources, mutualists, natural enemies, or even toxins.

Table 1 Types of Indirect Interactions among Plants[a]

Types of Interaction	Intermediary	Effect	Response	Net
Exploitation competition	Resources	−	+	−
Apparent competition	Natural enemies	+	−	−
Allelopathy	Toxins	+	−	−
Positive facilitation	Resources	+	+	+
Negative facilitation	Resources	−	−	+
Apparent facilitation	Natural enemies	−	−	+

[a] In this classification, resources of plants include mutualists such as pollinators or dispersers as well as abiotic resources such as light, water, mineral nutrients, CO_2. + and − in the Effect, Response, and Net columns indicate the effect of plants on abundance of the intermediary, the response of some "target" plant to abundance of the intermediary, and the net effect of plants on the "target" plant, respectively.

in determining the net outcome of population interactions and therefore the traits of competitively dominant species in a community (Section III).

The second point is equally simple but has often been overlooked: *both* the effect and response components of competition must be significant and of appropriate sign for competition to occur (Table 1). Thus, resolution of controversies about the type of environments under which competition is an important determinant of individual fitness and community structure will depend on trends in *both* the degree of resource limitation and the degree of resource depletion (Section IV). While we know a fair amount about the conditions for limitation by different resources, we know very little about patterns in resource depletion by plants, especially of belowground resources.

II. Traits Related to Effect and Response

A. Effect on Resources

Effect on a particular resource can be operationally defined as the per-individual or per-unit size rate of change in light, water, or nutrient availability, i.e., the slope of a regression of resource availability on plant density or biomass at a given abiotic supply rate (Fig. 2A). The most obvious mechanism through which a plant can affect resource availability is depletion due to uptake, which always results in negative effects. However, for nutrients and water (soil resources), a variety of "nonuptake" mechanisms can also be important and can result in positive as well as negative effects (Table 2; Newman, 1985; Hunter and Aarssen, 1988). Understanding the relative importance of nonuptake versus uptake effects on resources is important because very different plant traits will determine the magnitude of each of these and therefore the relationship

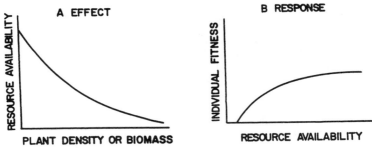

Figure 2   Effect on resources by plants (A) and response to resources by plants (B). Although both relationships are shown as monotonic and of the appropriate sign for competition, they need not be. For example, response to a nutrient could be initially positive but negative at very high levels, or effect on resources could be positive at low biomass of plants but negative at higher biomass.

of particular plant traits to the outcome of competition (see Section III) will differ as well.

Although depletion of light by plants is often measured (Harper, 1977), relatively few studies have reported the magnitude of effects on nutrients or water in the field (Table 3), and none of these has separated uptake from nonuptake effects. However, the many examples in Table 3

Table 2   Examples of Processes and Traits that Determine Magnitudes of Effect on Resources and Response to Resources on a Per-Unit Size Basis

Effect on Resources	Response to Resources
Uptake	Uptake
Physiological activity rates	Physiological activity rates
Allocation to resource-acquiring organs	Allocation to resource-acquiring organs
Architecture of resource-acquiring systems	Architecture of resource-acquiring systems
Nonuptake	Conversion efficiency
Direct addition of available forms	Loss
Association with N-fixing symbionts	Respiration rate
Leaching and throughfall	Transpiration rate
Addition in organic compounds	Tissue longevity
Litter quality and quantity	Leaching
Modification of physical environment	Translocation from senescent tissues
Temperature amelioration	
Reduce evapotranspiration	
Modification of microbial activity	
Temperature and moisture effects, root exudates, and root death	

Table 3 Examples of Effects of Individual Plants or Small Patches of Vegetation on Levels of Soil Resources in the Field

Resource	Effect of Vegetation	Environment	Method	Reference[a]
NO_3/NH_4	−/none	1-year old field	Removal of vegetation in 1 m² plot	Foster et al. (1980)
NH_4, K/PO_4	−/none	Littoral zone	Experimental planting of 2 species of Typha vs. unplanted sediments	Grace (1988)
NO_3, NH_4	−	30-year old field	Removal of all neighbors around individual Schizachyrium scoparium plants	Tilman (1989)
Soil moisture	−	Desert	Removal of individual Hilaria rigida plants	Robberecht et al. (1983)
Soil moisture	−	Semiarid grassland	Removal of tussock grasses	Eissenstat and Caldwell (1988)[*]
CEC, total N	+[b]	Several forests	Under canopies vs. open	Zinke (1962)[*]
NO_3, P_2O_5, total soluble salts	+	Desert	Under canopies vs. open	Turner et al. (1966)[*]
Total inorganic N, exchangeable K	+	Desert	Under canopies vs. open	Nishita and Haug (1973)[*]
Total N	+	Desert	Under canopies vs. open	Tiedemann and Klemmedson (1973)[*], Charley and West (1978)[*], Garcia-Moya and McKell (1970)[*]
Soil moisture	+/−[c]	Semiarid grassland	Comparison of soil depths	Richards and Caldwell (1987)
Soil moisture	−	Desert grassland	Bare vs. vegetated areas	Cable (1969)[*]
Soil moisture, NO_3, NH_4, PO_4	+	Midsuccessional old field	Natural gaps due to moles vs. undisturbed vegetation	Bradshaw and Goldberg (1989)

(continued)

Table 3 (continued)

Resource	Effect of Vegetation	Environment	Method	Reference[a]
Soil moisture	+	Serpentine grassland	Natural gaps due to gophers vs. undisturbed vegetation	Hobbs and Mooney (1985)
NO_3, P, K	+	Coastal prairie	Natural gaps due to gophers vs. undisturbed vegetation	Spencer et al. (1985)
Soil moisture	−	Temperate hard-wood forest	Experimental gaps vs. undisturbed vegetation	Minckler and Woerheide (1965)
Soil moisture, NH_4, NO_3	− or none	Temperate mesic forest	Natural gaps vs. undisturbed vegetation	Mladenoff (1987)*
Soil moisture/ NH_4, NO_3	−/none	Tropical rain forest	Natural gaps vs. undisturbed vegetation	Vitousek and Denslow (1986)

[a] An asterisk indicates that species were compared; in all cases at least some resource levels were different between species.

[b] Effects were negative immediately adjacent to the trunk and positive under the rest of the canopy relative to the area outside canopy influence.

[c] Effects were positive in shallow soil and negative in deeper soil.

of positive effects of the presence of plants on resource availability do suggest that positive nonuptake effects can sometimes overcompensate for negative effects through uptake, although nonuptake mechanisms are generally not included even in models of competition that explicitly include resources (reviewed by Schoener, 1986; for an exception, see Pastor and Post, 1986). It is worth noting from Table 3 that positive effects of the presence of plants on nutrients or water only occurred in nonexperimental studies that compared naturally unvegetated to vegetated areas. This suggests that many of the nonuptake mechanisms of effect on resources are cumulative, so that short-term experiments involving removal of plants may overestimate competitive effects on a longer time scale.

Many of the plant traits that influence the magnitude of nonuptake effects on resource availability are traditionally studied by ecosystem rather than population or community ecologists (Table 2). For example, rates of decomposition and nutrient release from litter depend on leaf $C : N$ ratios and lignin or phenolic concentrations (Schlesinger and Hasey, 1981; Melillo *et al.*, 1982; McClaugherty *et al.*, 1985). The quantity of nutrients and water added to the soil in throughfall and stemflow depends on leaf structure (e.g., cuticle thickness, structural carbon concentration, stomatal density; Tukey, 1970; Gray, 1983; Hollinger, 1986) and bark roughness (Zinke, 1967; Gersper and Holowaychuk, 1971; Brown and Bourn, 1973). Whole-plant architecture also can influence effect on resources; for example, Muller and Muller (1956) found that desert shrubs with multiple stems trapped more litter and, consequently, had higher densities of annual plants beneath their canopies than did single-stem shrubs.

Nonuptake effects on resource availability will also be influenced by environmental conditions. For example, positive effects on soil moisture because of reduced temperatures and evaporation under a plant canopy are more likely to occur in warmer and less humid sites or years (see Table 3). Similarly, positive effects of vegetation on nutrient availability have most often been found in deserts where organic matter accumulates under shrubs ("islands of fertility"; Garcia-Moya and McKell, 1970; Table 3). However, such patterns in nutrient availability have been looked for much more closely in deserts because of the obvious spatial separation of individuals, and may be equally common in more dense vegetation (e.g., Zinke, 1962). Certainly, on the scale of entire watersheds, vegetation in mesic areas can have positive effects on nutrient and water availability (Bormann and Likens, 1979).

The magnitude of uptake effects on resources is determined by total plant size and per-unit size uptake rates, which in turn are determined by physiological activity rates and allocation to and spatial arrangement of resource-acquiring organs (Table 2). Relatively little is known about the relative importance of physiology and morphology in explaining differences between species in per-unit size effects on resources. Caldwell and Richards (1986) have argued that allocation and architecture are of greater importance than physiological activity rates in determining relative effects on resources among similar-sized plants (see also Fitter, 1985, and Chapin *et al.*, 1987). For example, depletion of nutrients or water by plants with similar total root biomass has been shown to be greater for species with thinner roots (Harris, 1967; Caldwell *et al.*, 1985; Caldwell and Richards, 1986; Eissenstat and Caldwell, 1988) or with deeper roots (Gordon *et al.*, 1989).

The magnitudes of both uptake and nonuptake mechanisms of effect will also be strongly influenced by total plant size. Because species and individuals within a species can vary enormously in size, effects of different species on resource availability may sometimes be explained as well or better by differences in the mean and frequency distribution of plant size and growth rate than by differences in traits that determine per-unit size effect on resources. Field studies have found differences among species in total effects (uptake + nonuptake mechanisms) (Table 3). However, none of these studies has simultaneously measured biomass/plant or per plot, so it is impossible to separate species differences (and the traits responsible) from differences in plant size or abundance (Goldberg and Werner, 1983; Goldberg, 1987).

If variations in plant size and abundance often overwhelm species differences in per-unit size effects on resources, an important consequence is that the magnitude of effects on resources should be positively correlated between different resources. Individuals that are large or species that have large mean size in a given habitat can have higher rates of depletion of all resources, even if per-unit size uptake rates of different resources are negatively correlated (Donald, 1963; Grime, 1977; Harper, 1977).

B. Response to Resources

Response to resources can be operationally defined as the relationship of some component of fitness to resource availability, where availability is determined by either or both neighboring plants and the abiotic environment (Fig. 2B). When the resource is limiting (increasing portion of the response curve), response will be determined by the excess of uptake over loss of the resource and the new biomass or seeds that can be produced per unit of internal stores (Vitousek, 1982; Shaver and Melillo, 1984; Berendse *et al.*, 1987; Berendse and Elberse, this volume). Assuming that growth is positively correlated with fitness, this gives three general ways in which a plant could increase its fitness at low resource levels (referred to as high tolerance of low levels): increase resource uptake, decrease resource loss, or increase efficiency of conversion of internal stores to new growth (i.e., reduce requirement) (Table 2; Chapin, 1980; Chapin *et al.,* 1987). There are also mechanisms of tolerance that increase survival but not short-term growth rates at low resource levels, such as desiccation tolerance and luxury consumption and storage during periods of temporary resource abundance.

Efficiency of conversion of internal stores is the most difficult of these to quantify because definitions and therefore empirical data usually combine conversion efficiency and loss rates in a single measure of resource use efficiency (Vitousek, 1982; Chapin *et al.,* 1987). Ideally, one would measure actual growth at a given initial internal concentration and exclude any short-term losses, but such data are not generally available.

Mechanisms that increase uptake rates at low resource availability are known for all three classes of resources (light, water, nutrients). For example, shifts in allocation to roots versus shoots in response to light and soil resource availability will influence per-plant uptake rates of all resources (Chapin *et al.,* 1987). Shade-tolerant plants tend to have higher chlorophyll concentrations, denser grana stacks, and larger, thinner leaves that increase light interception on a per-unit area basis (Boardman, 1977). Drought-tolerant plants can increase water uptake at low soil water potential by adjusting osmotic potential (Larcher, 1980). Low-nutrient-tolerant plants can have high size-specific nutrient uptake rates through high densities of carriers for specific ions (Marschner, 1986).

Reduction of loss rates is also very important for all three resource classes (Chapin, 1980). Shade-tolerant plants typically have low respiration rates and, sometimes, longer-lived leaves (Boardman, 1977). Drought-tolerant plants have many mechanisms that reduce transpiration rates, including thick cuticles, pubescent leaves, small, thick leaves, low stomatal density, small stomata, and rapid stomatal closure (Schulze *et al.,* 1987). Low-nutrient-tolerant plants often have traits that reduce leakage of nutrients from tissues and increase efficiency of translocation of nutrients from senescent tissues and have longer-lived tissues (Chapin, 1980; Chapin *et al.,* 1987; Berendse and Elberse, this volume).

Because having longer-lived tissues could reduce loss rates of all three resource types, it could be argued that response to low levels of different resources should be positively correlated (cf. Grime, 1977). However, a number of other arguments suggest that this will not be generally true. Plants typical of xeric habits often lose leaves and roots in response to drought (Schulze *et al.,* 1987). Traits related to low loss rates of one resource are sometimes negatively correlated with uptake efficiency for another resource. For example, small, thick leaves reduce transpiration rate but are less efficient at gathering light. Efficiency of uptake may often be negatively correlated among resource types. For example, allocation to leaves and stems to intercept more light reduces allocation to roots to obtain nutrients or water (Tilman, 1988a). In some cases, the optimal rooting depth may be different for water versus nutrients (e.g., if nutrients are available mostly near the surface and water is available mostly in deeper soil). So, unlike effect on resources, response to resources may be negatively correlated, at least among broad classes of resources. Remarkably few data exist, however, to test the assertion that species that grow well under low levels of one resource are unlikely also to grow well at low levels of a different resource.

C. Relationship between Effect on and Response to Resources

Whether or not the effect and response components of plant–resource interactions are positively correlated depends on the specific mechanisms of effect and response (Table 2). If uptake is the most important mechanism of both effect and response, they are likely to be positively correlated. However, as I have argued above, uptake may not always or even typically be the most important determinant of the magnitude of either effect or response to resources and therefore the two components of plant–resource interactions are not necessarily correlated. In fact, some evidence suggests that low loss rates (e.g., low respiration, low transpiration, high leaf or root longevity) are generally correlated with low maximum potential growth rates even when resources are high (see reviews by Boardman, 1977; Grime, 1977; Bazzaz, 1979; Chapin, 1980; Chapin *et al.*, 1986, 1987; Shipley and Keddy, 1988). Thus, species that grow relatively well at low resource availability because of low loss rates rather than because of high per-unit size uptake rates are likely to have lower per-plant uptake rates because of lower growth rates. This should lead to a negative correlation between ability of individuals to deplete resources when they are abundant and ability to tolerate low resource levels.

III. Resource Effect/Response and Competitive Ability

Before discussing how the effect and response components of plant–resource interactions relate to the outcome of competition between individuals, we must first define competitive ability in terms of plant–plant interactions. Individuals of different species can be ranked in competitive ability either by how strongly they suppress other individuals (net competitive effect) or by how little they respond to the presence of competitors (net competitive response) (cf. Jacquard, 1968; Goldberg and Werner, 1983). On first examination, it would seem that these two types of individual competitive ability directly map onto the two components of the process of competition: plants that strongly suppress other plants must be good at depleting resources and making them unavailable to others, while plants that are indifferent to the presence of competitors are good at tolerating depleted resource levels. This will be true for the effect component of competition: ranking of *net competitive effect* of a group of species on individuals of a single "target" species will always be determined by their rankings of effect on resource availability. However, I argue below that ranking of *net competitive response* among a group of target species to a single species of neighboring plants can be determined by either or both effect on resources and response to resources, depending on the size of the target plant relative to the neighboring plants and the potential for resource preemption.

If the target individuals are very small relative to the neighbor individuals, they are unlikely to cause any significant depletion of resources to either themselves or their neighbors. Therefore, they are unlikely to have any net effect on growth of their neighbors which could later feed back on their neighbors' effects on them. In this case, the rankings of net response of different species of targets to a single neighbor species should be determined by the rankings of their response to resources at the level to which that particular neighbor species can deplete them. In contrast, when competition is between individuals of similar size, effect on resources and response to resources of both species must all be taken into account to determine their net interaction. In this case, a species could have a relatively strong net response (be relatively indifferent to the presence of neighbors) either by rapid uptake and hence preemption of resources (large effect on resources) and/or by growing well despite depleted resources (little response to resources). If there is a tradeoff between tolerance of low resources and rapid depletion of resources (Section II,C; Grime, 1977; Chapin, 1980), because ability to deplete resources conveys both strong net competitive effect and response under size-symmetric competition, traits associated with large effects on resources should be selected for in plants that typically occur in conditions with size-symmetric competition.

This discussion suggests that which traits are related to individual success in a competitive environment depends on the size structure of the competing populations. Because seedling establishment is often the critical life history stage for population persistence (Harper, 1977; Goldberg, 1982a; Gross and Werner, 1982; Peart, 1989b), the important consideration is whether seedlings are primarily competing with other seedlings (size-symmetric competition, resource preemption possible) or with mature vegetation (size-asymmetric competition, resource preemption not possible). Seedling–seedling interactions are more likely to be important when all plants in the community arise from seed at approximately the same time, as would happen early in succession, in gaps within non-successional communities, or in annual communities. Therefore, traits related to strong effect on resource availability should predominate in species characteristic of early succession or that require gaps for regeneration. Traits related to tolerance of low resource levels should predominate in species characteristic of later successional or equilibrium communities (which resource would depend on the abiotic environment). Persistence of species without such tolerance but with ability to deplete resources rapidly in equilibrium communities would depend on the frequency of gap formation.

This scenario is consistent with those described both by Grime (1977, 1987) and Tilman (1982, 1985, 1988a), although their differences in definitions of competitive ability have led to much confusion about what plant traits determine competitive ability (see Grace, this volume, for further discussion). Tilman (1982) defines a superior competitor for a resource as one whose *population* can deplete the resource to a lower level at equilibrium. Although this superficially sounds like the effect component of competition, for individuals in size-structured populations it can be interpreted as equivalent to tolerance of low resource levels because, at equilibrium, individual seedlings of a superior competitor must be able to establish at the low levels to which the adult population has depleted the resource. Clearly, there can be species whose adults can deplete a resource to levels lower than their seedlings can tolerate—this is the basis of the tolerance model of succession (Connell and Slatyer, 1977). However, when the community is not at equilibrium, traits other than ability of individuals to tolerate low resource levels become important. In particular, high growth rate is predicted to lead to dominance during the early stages of succession (Tilman, 1985, 1988a).

In contrast, Grime (1977) associates competitive ability with traits that maximize resource capture by *individuals,* i.e., effect on resources. Species with traits that convey tolerance of low resources are called stress tolerators, even if resources are low because of depletion by other plants. This leads to the conclusion that dominant species in equilibrium (non-successional) communities are called good stress tolerators rather than good competitors (Fig. 4 in Grime, 1977). Nevertheless, the sequence of traits of plants over succession predicted by Grime is similar to that predicted by Tilman and in this chapter—from fast-growing species with rapid uptake rates to slower-growing species that are tolerant of low resource levels. Bazzaz (1979) has reviewed the empirical evidence for such patterns in growth rate and tolerance of low light during succession.

The differences in definition of competitive ability between Grime and Tilman also are consistent with their differing assumptions about tradeoffs in competitive ability for different resources. Grime (1977, 1987) argues that competitive abilities for different resources are positively correlated because faster growth or larger size results in greater uptake rates of all resources. This is consistent with the arguments made above that Grime's definition is close to that for effects on resources (at least through uptake) and that abilities to deplete different resources are positively correlated. In contrast, Tilman (1982, 1988a) argues that competitive abilities for different resources must be negatively correlated. This assumption is consistent with my previous interpretation that Tilman's definition of competitive ability of populations at equilibrium corresponds to individual tolerance of low resources and the arguments made earlier that response is likely to be negatively correlated among resources.

The difference in traits related to competitive ability between strongly size-asymmetric and size-symmetric competition also has important implications for the design and interpretation of experiments about the role of interspecific competition in determining community structure. Most greenhouse and field experiments with pairs of species use initially similar-sized individuals and follow a single generation. Hence the result that rapid growth rate or large size conveys both strong net competitive effect and strong net competitive response is not surprising (e.g., Wilson and Keddy, 1986b; Goldberg and Fleetwood, 1987; Miller and Werner, 1987). However, experiments allowed to run more than one generation or that start with seeds added to mature vegetation would be more likely to show a lack of correlation or even a negative correlation between net competitive effect and net competitive response (e.g., Peart, 1989a).

The discussion so far has assumed that resource supplies are constant. However, in nature, soil resources are often supplied in pulses due to sporadic rainfall and temperature and moisture effects on microbial activity. When resources are pulsed, species with rapid uptake are more likely to be able to take advantage of the pulse than are slower-growing species with low uptake rates but tolerance of low levels (e.g., Bunce *et al.*, 1977; Sommer, 1985; see Grime *et al.*, 1986, for a discussion of consequences of different types of resource pulses). Thus, even under size-asymmetric conditions, it is possible that strong effect but weak response competitors can persist if resources are sufficiently pulsed.

IV. Importance of Competition over Environmental Gradients

The distinction between the effect and response components of competition points out the almost trivial observation that both effect on resources and response to resources must be significantly different from zero and of appropriate sign for competition to occur (Table 1). Thus, resource limitation as indicated by positive response to addition of a resource does not necessarily mean that plants compete for that resource if they are unable to deplete the resource significantly for some reason. For example, in a first-year old field in Michigan dominated by *Ambrosia artemisiifolia*, I found no significant effect of *Ambrosia* biomass on soil moisture in 0.5 m diameter plots, but a significant positive response to soil moisture by transplanted *Plantago lanceolata* seedlings in the same plots (Fig. 3A,C). The reverse situation was found for light: increasing *Ambrosia* biomass was associated with decreasing irradiance at seedling height but *Plantago* seedlings showed no significant response to light (Fig. 3B,D). Thus, despite significant responses to water by *Plantago* and significant effects on light by *Ambrosia*, *Ambrosia* had no significant net competitive effect on *Plantago* (Fig. 3E). The most likely explanation for the lack of effect of *Ambrosia* on soil moisture is that increasing water use by increasing *Ambrosia* biomass was roughly balanced by decreasing evaporation from the soil surface because of cooler temperatures under larger *Ambrosia* canopies. Under circumstances where a positive nonuptake mechanism of effect balances a negative effect through uptake, actual availability of the resource will be determined by abiotic spatial variation. In this case, the important abiotic control on soil moisture appeared to be soil microtopography (old plow lines) and consequent spatial variation in drainage.

The fact that effect on and response to resources can be decoupled has important implications for questions about the conditions under which competition is an important determinant of individual fitness and community structure. Two general viewpoints pervade the ecological literature: competition is most intense for a given resource where that resource is most limiting (Wiens, 1977; Chapin and Shaver, 1985; see also most ecology textbooks) and competition for all resources is most intense where density or biomass is greatest (Kruckeberg, 1954, 1969; Gankin and Major, 1964; Grime, 1977; Goldberg, 1982b; Wilson and Keddy, 1986a). These correspond to where response to a given resource is greatest (greatest potential for an increase in growth with an increase in resource availability) and to where effect on all resources is greatest (greatest absolute magnitude of resource depletion), respectively. Because both processes must occur for competition to occur, the question can be rephrased as Under what conditions are *both* effect on and response to a particular resource likely to be large?

This question can be most directly applied to productivity gradients due to underlying abiotic gradients in nutrient or water availability. Along such a gradient, as productivity increases, the magnitude of light depletion will increase in the presence of vegetation because of increas-

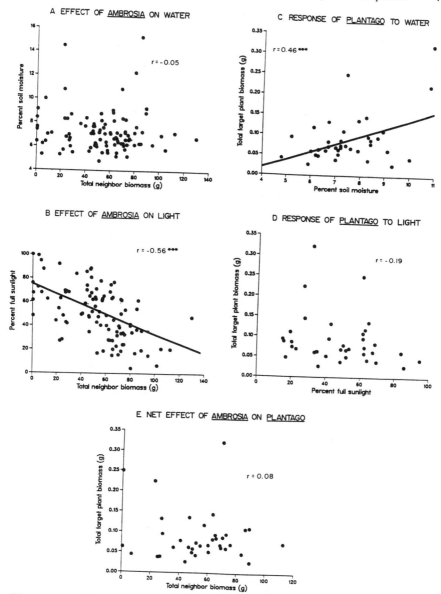

Figure 3   Effect of *Ambrosia artemisiifolia* biomass in 0.5 m diameter plots on percent soil moisture (A) and percent of full sunlight (B), response of transplants of *Plantago lanceolata* seedlings to soil moisture (C) and sunlight (D), and the net interaction between *Plantago* seedlings (target) and *Ambrosia* plants (neighbors) in the same plots (E). Only significant regression lines are drawn.

ing standing crop (Fig. 4A; e.g., Tilman, 1984; Werner, 1990). Because the greatest depletion of light and the lowest absolute light levels both occur at high-productivity sites, it is unambiguous that competition for light will be strongest under high-productivity conditions.

The situation for soil resources (nutrients and water) is more complicated. In the absence of vegetation, soil resource availability will increase along an abiotic productivity gradient (Fig. 4B). Presumably, availability will also increase along this gradient in the presence of vegetation, but not as much because some or even all of the increase will be taken up by the increased standing crop (Fig. 4B). This means that the magnitude of depletion will increase as standing crop increases. Thus, effect on soil resources is greatest at high productivity, but the potential for response to soil resources will be greatest where resource availability is lowest, at the low-productivity end of the gradient. This negative correlation between the magnitudes of effect and of response for soil resources is the

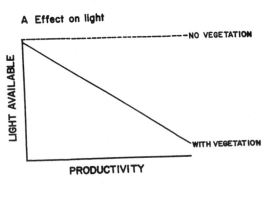

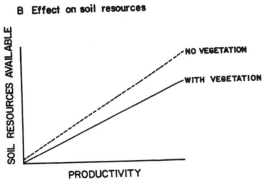

Figure 4 Hypothetical effects of the presence versus absence of vegetation on (A) light and (B) soil resource availability along a productivity gradient, where total productivity and standing crop of vegetation increase because of an underlying abiotic gradient in nutrient or water availability.

source of what Grime (1977) refers to as "the vexed question of competition in unproductive environments." Studies addressing this question have examined the magnitude of net competitive effects over productivity gradients and typically find smaller (and sometimes even positive) net effects on less-productive sites (Grace and Wetzel, 1981; del Moral, 1983; Goldberg, 1985; Gurevitch, 1986; Wilson and Keddy, 1986a; Werner, 1990). Nevertheless, numerous competition experiments within unproductive environments have shown that significant net competitive effects often occur (see Fowler, 1986, for a review).

An alternative approach that could prove more general is to address directly the question of total vegetation effects on resources over productivity gradients. No complete data set exists but one hypothesis is suggested from consideration of the nonuptake mechanisms of effect on resources discussed earlier. Competition will be less important in environments or plants in which nonuptake positive effects on a limiting resource are large. Because there are no mechanisms by which plants can have positive effects on light, this is most likely in environments in which nutrients or water, but not light is limiting, i.e., in unproductive environments with low standing crop. For example, in xeric environments, both soil moisture and nutrients are often higher under a plant canopy than between canopies or in vegetation gaps because of reduced evapotranspiration and accumulation of organic matter under canopies (Table 3; Whitford, 1986). This would lead to facilitation rather than competition for soil resources, and indeed a number of studies in deserts have found higher biomass of annual plants under plant canopies (Muller, 1953; Halvorson and Patten, 1975; Parker *et al.*, 1982) or higher survival of woody seedlings (Turner *et al.*, 1966). Nevertheless, significant depletion of at least soil moisture has been found in arid and semiarid areas (Robberecht *et al.*, 1983; Eissenstat and Caldwell, 1988). Insufficient data are available to generalize at this point how common positive versus negative effects on soil resources are. How does the balance between facilitation and competition depend on the type of species (e.g., shallow versus deep rooted) or the type of environment (e.g., frequent but low rainfall versus infrequent rainfall, low water versus low nutrient, different limiting nutrients)?

V. Conclusions

Recognizing the distinction between the effect and response components of competition identifies a number of major gaps in our understanding of the connection between resource use by plants and the outcome of competition across environmental resource gradients. Filling in these gaps is essential for developing mechanistic models of the role of competition in structuring plant communities.

First, separation of effect and response in theoretical and empirical analyses of resource competition is critical if the magnitudes of each of these are determined by different plant traits. I have argued that effect on and response to a given resource will be positively correlated only to the extent that both are a function of uptake rates, and have speculated that uptake will often be relatively unimportant in determining the magnitude of effect on and response to resources. However, we know very little about the conditions under which the total effect of a plant on resource availability is determined mostly by nonuptake versus uptake mechanisms or ability of plants to tolerate low resource levels is determined mostly by low loss rates versus high uptake rates of resources. How do the relative importance of these different mechanisms of effect and response vary among resources, among environments, and among plant species?

Second, we need to explore the relationships between effect on and response to resources by individuals and measurements of net competitive ability. I have speculated that species with traits that result in large negative effects on resources will be dominant in communities where persistence is determined by size-symmetric competition, as would occur early in succession or in vegetation gaps. In contrast, species with traits that result in tolerance of low resource levels will be dominant when persistence is determined by size-asymmetric competition, as would occur for seedlings germinating in mature vegetation. To test these hypotheses, we need to conduct competition experiments that vary the size structure of competing populations and that use species of known effect and response in terms of resources.

Third, we need to document patterns in the magnitude of total effect of plants on resources in different types of environments. Response to resources of species from different environments has been analyzed extensively in greenhouse experiments and in resource addition experiments in the field. However, measurements of effect on resources in the field are much rarer. How common are environments or species in which depletion of resources by plants is negligible even when the resource is severely limiting to plant growth? What kinds of environments or plants are these? Such information is critical to developing accurate generalizations about the kinds of environments in which competition is important—both as a selective agent on individuals and as a determinant of community structure.

Although I have speculated on the answers to some of these questions, the data to test these speculations are severely limited. If I have stimulated field workers to address these issues, the major aim of this chapter will have been satisfied.

Acknowledgments

I am grateful to the many people who read drafts of this chapter and improved it in various ways: Lonnie Aarssen, Drew Barton, Norma Fowler, Dan Friedus, Doria Gordon, Jim Grace, Phil Grime, Kay Gross, Paul Keddy, Richard Kiesling, Betsy Kirkpatrick, Mathew Liebold, Tom Miller, Gary Mittelbach, Dave Tilman, Roy Turkington, John Vandermeer, and Earl Werner. I also thank the participants in the Cedar Creek Symposium for stimulating discussion and challenging questions. Research reported in this chapter was supported by grants from the Rackham Graduate School of the University of Michigan and the National Science Foundation.

References

Bazzaz, F. A. (1979). The physiological ecology of plant succession. *Annu. Rev. Ecol. Syst.* **10,** 351–371.

Berendse, F., Oudhof, H., and Bol, J. (1987). A comparative study of nutrient cycling in wet heathland ecosystems. I. Litter production and nutrient losses from the plant. *Oecologia* **74,** 174–184.

Boardman, N. K. (1977). Comparative photosynthesis of sun and shade plants. *Annu. Rev. Plant Physiol.* **28,** 355–377.

Bormann, F. H., and Likens, G. E. (1979). "Pattern and Process in a Forested Ecosystem." Springer-Verlag, New York.

Bradshaw, L., and Goldberg, D. E. (1989). Resource levels in undisturbed vegetation and mole mounts in old fields. *Am. Midl. Nat.* **121,** 176–183.

Brown, J. H., and Bourne, T. G. (1973). Patterns of soil moisture depletion in a mixed oak stand. *For. Sci.* **19,** 23–30.

Bunce, J. A., Miller, L. N., and Chabot, B. F. (1977). Competitive exploitation of soil water by five eastern North American tree species. *Bot. Gaz.* **138,** 168–173.

Cable, D. R. (1969). Competition in the semi-desert grassland shrub type as influenced by root systems, growth habits, and soil moisture extraction. *Ecology* **50,** 27–38.

Caldwell, M. M., and Richards, J. H. (1986). Competing root systems, morphology and models of absorption. *In* "On the Economy of Plant Form and Function" (T. J. Givnish, ed.), pp. 251–273. Cambridge Univ. Press, Cambridge, Massachusetts.

Caldwell, M. M., Eissenstat, D. M., Richards, J. H., and Allen, M. F. (1985). Competition for phosphorus, differential uptake from dual-isotope-labelled soil interspaces between shrub and grass. *Science* **229,** 384–386.

Chapin, F. S., III (1980). The mineral nutrition of wild plants. *Annu. Rev. Ecol. Syst.* **11,** 233–260.

Chapin, F. S., III, and Shaver, G. (1985). Individualistic growth response of tundra plant species to environmental manipulations in the field. *Ecology* **66,** 564–576.

Chapin, F. S., Vitousek, P. M., and Van Cleve, K. (1986). The nature of nutrient limitation in plant communities. *Am. Nat.* **127,** 48–58.

Chapin, F. S., Bloom, A. J., Field, C. B., and Waring, R. H. (1987). Plant responses to multiple environmental factors. *BioScience* **37,** 49–57.

Charley, J. L., and West, N. E. (1978). Plant-induced soil chemical patterns in some shrub-dominated semi-desert ecosystems in Utah. *J. Ecol.* **63,** 945–963.

Connell, J. H. (1983). On the prevalence and relative importance of interspecific competition: Evidence from field experiments. *Am. Nat.* **122,** 661–696.

Connell, J. H., and Slatyer, R. O. (1977). Mechanisms of succession in natural communities and their role in community stability and organization. *Am. Nat.* **111,** 1119–1144.

del Moral, R. (1983). Competition as a control mechanism in subalpine meadows. *Am. J. Bot.* **70**, 232–245.

Donald, C. M. (1963). Competition among crop and pasture plants. *Adv. Agron.* **15**, 1–118.

Eissenstat, D. M., and Caldwell, M. M. (1988). Competitive ability is linked to rates of water extraction: A field study of two aridland tussock grasses. *Oecologia* **75**, 1–7.

Fitter, A. H. (1985). Functional significance of root morphology and root system architecture. *In* "Ecological Interactions in Soil" (A. H. Fitter, ed.), pp. 87–106. Blackwell, Oxford, England.

Foster, M. M., Vitousek, P. M., and Randolph, P. A. (1980). The effects of ragweed (*Ambrosia artemisiifolia* L.) on nutrient cycling in a 1st year old field. *Am. Midl. Nat.* **103**, 106–112.

Fowler, N. (1986). The role of competition in plant communities in arid and semiarid regions. *Annu. Rev. Ecol. Syst.* **17**, 89–110.

Gankin, R., and Major, J. (1964). *Arctostaphylos myrtifolia,* its biology and relationship to the problem of endemism. *Ecology* **45**, 792–808.

Garcia-Moya, E., and McKell, C. M. (1970). Contribution of shrubs to the nitrogen economy of a desert-wash plant community. *Ecology* **51**, 81–88.

Gersper, P. L., and Holowaychuk, N. (1971). Some effects of stem flow from forest canopy trees on chemical properties of soils. *Ecology* **52**, 691–702.

Goldberg, D. E. (1982a). Comparison of factors determining growth rates of deciduous vs. broad-leaf evergreen trees. *Am. Midl. Nat.* **108**, 133–143.

Goldberg, D. E. (1982b). The distribution of evergreen and deciduous trees relative to soil type, an example from the Sierra Madre, Mexico, and a general model. *Ecology* **63**, 942–951.

Goldberg, D. E. (1985). Effects of soil pH, competition, and seed predation on the distribution of two tree species. *Ecology* **66**, 503–511.

Goldberg, D. E. (1987). Neighborhood competition in an old-field plant community. *Ecology* **68**, 1211–1223.

Goldberg, D. E., and Fleetwood, L. (1987). Comparison of competitive effects and responses among annual plants. *J. Ecol.* **75**, 1131–1144.

Goldberg, D. E., and Werner, P. A. (1983). Equivalence of competitors in plant communities, a null hypothesis and a field experimental approach. *Am. J. Bot.* **70**, 1098–1104.

Gordon, D. R., Welker, J. M., Menke, J. W., and Rice, K. J. (1989). Neighborhood competition between annual plants and blue oak (*Quercus douglasii*) seedlings. *Oecologia* **79**, 533–541.

Grace, J. B. (1988). The effects of nutrient additions on mixtures of *Typha latifolia* L. and *Typha domingensis* Pers. along a water-depth gradient. *Aquat. Bot.* **31**, 83–92.

Grace, J. B., and Wetzel, R. G. (1981). Habitat partitioning and competitive displacement in cattails (*Typha*): Experimental field studies. *Am. Nat.* **118**, 463–474.

Gray, J. T. (1983). Nutrient use by evergreen and deciduous shrubs in southern California. I. Community nutrient cycling and nutrient-use efficiency. *J. Ecol.* **71**, 21–41.

Grime, J. P. (1977). Evidence for the existence of three primary strategies in plants and its relevance to ecological and evolutionary theory. *Am. Nat.* **111**, 1169–1194.

Grime, J. P. (1987). Dominant and subordinate components of plant communities: Implications for succession, stability and diversity. *In* "Colonization, Succession and Stability" (A. J. Gray, M. J. Crawley, and P. J. Edwards, eds.), pp. 413–428. Blackwell, Oxford, England.

Grime, J. P., Crick, J. C., and Rincon, J. E. (1986). The ecological significance of plasticity. *Soc. Exp. Biol. Symp.* **40**, 5–29.

Gross, K. L., and Werner, P. A. (1982). Colonizing abilities of "biennial" plant species in relation to ground cover: Implications for their distributions in a successional sere. *Ecology* **63**, 921–931.

Gurevitch, J. (1986). Competition and the local distribution of the grass *Stipa neomexicana*. *Ecology* **67**, 46–57.

Halvorson, W. L., and Patten, D. T. (1975). Productivity and flowering of winter ephemerals in relation to Sonoran desert shrubs. *Am. Midl. Nat.* **93**, 311–319.

Harper, J. L. (1977). "Population Biology of Plants." Academic Press, New York.

Harris, G. A. (1967). Some competitive relationships between *Agropyron spicatum* and *Bromus tectorum*. *Ecol. Monogr.* **37**, 89–111.

Hobbs, R. J., and Mooney, H. A. (1985). Community and population dynamics of serpentine grassland annuals in relation to gopher disturbance. *Oecologia* **67**, 342–351.

Hollinger, D. Y. (1986). Herbivory and the cycling of nitrogen and phosphorus in isolated California trees. *Oecologia* **70**, 291–297.

Hunter, A. F., and Aarssen, L. W. (1988). Plants helping plants. *BioScience* **38**, 34–40.

Jacquard, P. (1968). Manifestation et nature des relations sociales chez les vegetaux superieurs. *Oecol. Plant.* **3**, 137–168.

Kruckeberg, A. R. (1954). The ecology of serpentine soils. III. Plant species in relation to serpentine soils. *Ecology* **35**, 267–274.

Kruckeberg, A. R. (1969). Soil diversity and the distribution of plants, with examples from western North America. *Madrono* **20**, 129–154.

Larcher, W. (1980). "Physiological Plant Ecology," 2nd ed. Springer-Verlag, Berlin.

Marschner, H. (1986). "Mineral Nutrition of Higher Plants." Academic Press, London.

McClaugherty, C. A., Pastor, J., and Aber, J. D. (1985). Forest litter decomposition in relation to soil nitrogen dynamics and litter quality. *Ecology* **66**, 266–275.

Melillo, J. M., Aber, J. D., and Muratore, J. F. (1982). Nitrogen and lignin control of hardwood leaf litter decomposition dynamics. *Ecology* **63**, 621–626.

Miller, T. E., and Werner, P. A. (1987). Competitive effects and responses between plant species in a first-year old-field community. *Ecology* **68**, 1201–1210.

Minckler, L. S., and Woerheide, J. D. (1965). Reproduction of hardwoods 10 years after cutting as affected by site and opening size. *J. For.* **63**, 103–107.

Mladenoff, D. J. (1987). Dynamics of nitrogen mineralization and nitrification in hemlock and hardwood treefall gaps. *Ecology* **68**, 1171–1180.

Muller, C. H. (1953). The association of desert annuals with shrubs. *Am. J. Bot.* **40**, 53–60.

Muller, W. H., and Muller, C. H. (1956). Association patterns involving desert plants that contain toxic products. *Am. J. Bot.* **43**, 354–361.

Newman, E. I. (1973). Competition and diversity in herbaceous vegation. *Nature (London)* **244**, 310–311.

Newman, E. I. (1985). The rhizosphere: Carbon sources and microbial populations. *In* "Ecological Interactions in Soil" (A. H. Fitter, ed.), pp. 107–121. Blackwell, Oxford, England.

Nishita, N., and Haug, R. M. (1973). Distribution of different forms of nitrogen in some desert soils. *Soil Sci.* **116**, 51–58.

Parker, L. W., Fowler, H. G., Ettershank, G., and Whitford, W. G. (1982). The effects of subterranean termite removal on desert soil nitrogen and ephemeral flora. *J. Arid Environ.* **5**, 53–59.

Pastor, J., and Post, W. M. (1986). Influence of climate, soil moisture, and succession on forest carbon and nitrogen cycles. *Biogeochemistry* **2**, 3–27.

Peart, D. R. (1989a). Species interactions in a successional grassland. II. Colonization of vegetated sites. *J. Ecol.* **77**, 252–266.

Peart, D. R. (1989b). Species interactions in a successional grassland. III. The effects of canopy gaps, gopher mounds, and grazing on colonization. *J. Ecol.* **77**, 267–289.

Richards, J. H., and Caldwell, M. M. (1987). Hydraulic lift: Substantial nocturnal water transport between soil layers by *Artemesia tridentata* roots. *Oecologia* **73**, 486–489.

Robberecht, R., Mahall, B. E., and Nobel, P. S. (1983). Experimental removal of intraspecific competitors—Effects on water relations and productivity of a desert bunchgrass, *Hilaria rigida. Oecologia* **60**, 21–24.

Schlesinger, W. H., and Hasey, M. M. (1981). Decomposition of chaparral shrub foliage, losses of organic and inorganic constituents from deciduous and evergreen leaves. *Ecology* **62**, 762–774.

Schoener, T. W. (1983). Field experiments on interspecific competition. *Am. Nat.* **122**, 240–285.

Schoener, T. W. (1986). Mechanistic approaches to community ecology: A new reductionism? *Am. Zool.* **26**, 81–106.

Schulze, E.-D., Robichaux, R. H., Grace, J., Rundel, P. W., and Ehleringer, J. R. (1987). Plant water balance. *BioScience* **37**, 30–37.

Shaver, G. R., and Melillo, J. M. (1984). Nutrient budgets of marsh plants, efficiency concepts and relation to availability. *Ecology* **65**, 1491–1510.

Shipley, B. and Keddy, P. A. (1988). The relationship between relative growth rate and sensitivity to nutrient stress in twenty-eight species of emergent macrophytes. *J. Ecol.* **76**, 1101–1110.

Sommer, U. (1985). Comparison between steady state and non-steady state competition: Experiments with natural zooplankton. *Limnol. Oceanogr.* **30**, 335–346.

Spencer, S. R., Cameron, G. N., Eshelman, B. D., Cooper, L. C., and Williams, L. R. (1985). Influence of pocket gopher mounds on a Texas coastal prairie. *Oecologia* **66**, 111–115.

Thompson, K. (1987). The resource ratio hypothesis and the meaning of competition. *Funct. Ecol.* **1**, 297–303.

Thompson, K., and Grime, J. P. (1988). Competition reconsidered—A reply to Tilman. *Funct. Ecol.* **2**, 114–116.

Tiedemann, A. R., and Klemmedson, J. O. (1973). Nutrient availability in desert grassland soils under mesquite (*Prosopis juliflora*) trees and adjacent open areas. *Soil Sci. Soc. Am. Proc.* **37**, 107–111.

Tilman, D. (1982). "Resource Competition and Community Structure," Monogr. Pop. Biol. Princeton Univ. Press, Princeton, New Jersey.

Tilman, D. (1984). Plant dominance along an experimental nutrient gradient. *Ecology* **65**, 1445–1453.

Tilman, D. (1985). The resource-ratio hypothesis of plant succession. *Am. Nat.* **125**, 827–852.

Tilman, D. (1987a). On the meaning of competition and the mechanisms of competitive superiority. *Funct. Ecol.* **1**, 304–315.

Tilman, D. (1987b). The importance of the mechanisms of interspecific competition. *Am. Nat.* **129**, 769–774.

Tilman, D. (1988a). "Plant Strategies and the Dynamics and Structure of Plant Communities," Monogr. Pop. Biol. Princeton Univ. Press, Princeton, New Jersey.

Tilman, D. (1989). Competition, nutrient reduction and the competitive neighborhood of a bunchgrass. *Funct. Ecol.* **3**, 215–220.

Tukey, H. B., Jr. (1970). The leaching of substances from plants. *Annu. Rev. Plant Physiol.* **21**, 305–324.

Turner, R. M., Alcorn, S. M., Olin, G., and Booth, J. A. (1966). The influence of shade, soil and water on saguaro seedling establishment. *Bot. Gaz.* **127**, 95–102.

Vitousek, P. M. (1982). Nutrient cycling and nutrient use efficiency. *Am. Nat.* **119**, 553–572.

Vitousek, P. M., and Denslow, J. S. (1986). Nitrogen and phosphorus availability in treefall gaps of a lowland tropical rainforest. *J. Ecol.* **74**, 1167–1178.

Werner, P. A. (1990). The effects of vegetation and edaphic gradients on goldenrods

(*Solidago* spp.) in virgin prairie and old-field habitats: A field experiment using clonal reciprocal transplants. *Ecol. Monogr.*, in press.

Whitford, W. G. (1986). Decomposition and nutrient cycling in deserts. *In* "Pattern and Process in Desert Ecosystems" (W. G. Whitford, ed.), pp. 93–118. Univ. of New Mexico Press, Albuquerque, New Mexico.

Wiens, J. A. (1977). On competition and variable environments. *Am. Sci.* **65,** 590–597.

Wilson, S. D., and Keddy, P. A. (1986a). Measuring diffuse competition along an environmental gradient: Results from a shoreline plant community. *Am. Nat.* **127,** 862–869.

Wilson, S. D., and Keddy, P. A. (1986b). Species competitive ability and position along a natural stress/disturbance gradient. *Ecology* **67,** 1236–1242.

Zinke, P. J. (1962). The pattern of influence of individual forest trees on soil properties. *Ecology* **43,** 130–133.

Zinke, P. J. (1967). Forest interception studies in the United States. *In* "Forest Hydrology" (W. E. Sopper and H. W. Lull, eds.), pp. 137–161. Pergamon, Oxford, England.

4

On the Relationship between Plant Traits and Competitive Ability

James B. Grace

I. Introduction

It is good thus to try in imagination to give to any one species an advantage over another. Probably in no single instance should we know what to do (Darwin, 1859, Chapter III, p. 85–86).

Dating back to Darwin, a long-standing goal of ecologists has been to be able to predict the outcome of competition from an analysis of the characteristics of species in isolation. In the years since Darwin, our understanding of what features can give one species an advantage over another has improved considerably. Nonetheless, it is still true that our ability to predict competitive outcomes is quite limited for higher plants. A variety of approaches have been used to predict competitive outcome. At the most general level, life history theories have typically described syndromes of characteristics that are correlated with high or low competitive ability (Baker, 1965; Gadgil and Solbrig, 1972; Grime, 1977). More sophisticated models such as those of Tilman (1982, 1988) have predicted competitive success using specific assumptions about the mechanisms of resource use. In some cases, statistical approaches have been used to predict competitive ability (Austin, 1982; Grace, 1988a). At the most detailed level, complex simulation models have, at times, been used to predict the relationships between plant traits and competitive ability (Baldwin, 1976). At present, the two theories that are most widely discussed are those of Grime (1979) and Tilman (1982, 1988), in part because of the apparently conflicting views they offer on the relationships between plant traits and competitive ability. Because of the fundamental importance of the issues on which these theories differ, this chapter presents an analysis of these two theories and the bases for their differences.

Expanding on the theory of r- and K-selection (MacArthur and Wilson, 1967), Grime (1977) proposed a more refined interpretation of life histories developed specifically for higher plants. In this scheme and its further elaborations (Grime, 1979, 1981; Grime and Hodgson, 1987), he proposed that plants differed dramatically in the life history characteristics of their established phases depending on the degrees of "stress" ("phenomena which restrict photosynthetic production," Grime, 1979, p. 7) and "disturbance" ("partial or total destruction of plant biomass," Grime, 1979, p. 7) to which they were adapted. According to this system, those plants adapted to low levels of both disturbance and stress are referred to as "competitive," those adapted to low disturbance and high stress are "stress-tolerant," and those adapted to high disturbance and low stress are "ruderal."

Although not stated in quantitative terms, Grime's theory is based on a set of mechanistic assumptions about how plants interact. He defines competition as the tendency for neighboring plants to utilize the same resources and argues that success in competition is largely a reflection of the capacity for resource capture. According to Grime, one of the key characteristics of plants that is positively correlated with competitive ability is the maximum relative growth rate (RGR_{max}). Coupled with low sexual reproductive effort, the rapid growth of good competitors translates into a rapid development of absorptive surface area which leads to a preemption of both above- and below-ground resources.

Of equal importance to his assumptions about how plants compete are Grime's assumptions about evolutionary tradeoffs among traits. According to Grime, there are strong tradeoffs between the ability to tolerate low resource supplies and the ability to grow rapidly and to exploit resources. It is this basic tradeoff and its attending physiological constraints that result in a division between "competitive" and "stress-tolerant" species. This tradeoff has been discussed in some detail for the case of adaptation to nutrient limitation by Chapin (1980). Because of the tradeoff assumed between tolerance to low resource supply (high stress sensu Grime) and RGR_{max} (maximum relative growth rate), species are constrained from being both tolerant to resource shortages and also highly effective at exploiting resources. A central assumption in this relationship is that the ability to compete is determined by the ability to exploit resources rapidly rather than by the ability to tolerate resource depletion.

Tilman (1982) has proposed a resource-based theory of competition for plants that is based on a quantitative, mechanistic model (Table 1). In its simplest form this model consists of a pair of equations that describe the changes that occur in population size and resource concentration as

Table 1 Terms in Tilman's 1982 Model, Including Some of the Individual Plant Traits That Are Subsumed within the Population Parameters

The Model:

$$dN/Ndt = rR/(R + k) - m$$

and

$$dR/dt = a(S - R) - (dN/dt + mN/Y)$$

where

N = population density
R = concentration of limiting resource
r = maximum growth rate for population, includes
    maximum growth capacity
    maximum rate of seed production
    dispersal rate
    regeneration requirements
k = half-saturation constant, includes resource uptake and use for
    established plants
    seeds
    seedlings
m = mortality rate, includes density-independent mortality rate for
    established plants
    seeds
    seedlings
    plant parts such as leaves, roots, and stems
Y = resource requirement per individual
S = amount of resource supplied to system
t = time
a = resource supply rate

species compete. A critical feature of Tilman's model is the assumption that, when resources are used, the concentration is drawn down to a level $R*$, which is defined as the equilibrium resource concentration or the level below which the population is unable to maintain itself. Because of the structure of the equations, the species with the lowest $R*$ will competitively displace all other species at equilibrium.

Tilman and co-workers have validated this model and some of its extensions for algal species in a number of cases (Tilman, 1977; Tilman *et al.*, 1981). Although the generalizations from this model have been extended to higher plants (Tilman, 1982), to date there has not been a complete assessment of the assumptions of this model using higher plants.

Several features of Tilman's original model are unrealistic for higher plants and he has now developed a model for size-structured populations that describes plants in terms of their allocation to roots, stems, leaves, and seeds (Tilman, 1988). This model (referred to as ALLOCATE) is substantially more complex than Tilman's original model and will not be recapitulated here. However, there are several major features of this model that determine its basic behavior: First, plants grow in size to a maximum value and then allocate all further photosynthate to seeds. Second, the population is divided into cohorts based on the sizes of individuals. Third, reproduction is continuous throughout the growing season. Fourth, plants compete for light through shading one another (i.e., light available to a plant is determined by the density of leaves belonging to plants of greater stem height). And fifth, plants compete for nutrients by Michaelis–Menten type kinetics.

The behavior of ALLOCATE is substantially more sophisticated than Tilman's original model and is used primarily to consider the changes in plant form that are to be expected during autogenic succession (i.e., based on the assumption that competitive interactions drive the successional process). In terms of the competition mechanism, however, ALLOCATE behaves in very similar ways to the earlier model. Importantly, the key feature that is unchanged is that the species with the lowest minimum resource requirement, $R*$, is still the species predicted to be the superior competitor at equilibrium.

II. The Conflict between Grime's and Tilman's Theories

Both Grime's and Tilman's theories have received widespread attention, although not universal support (Solbrig, 1979; Harper, 1982; Grubb, 1985; Huston and Smith, 1987; Loehle, 1988). It is important to recognize that these theories were developed with somewhat different objectives in mind and, not surprisingly, are in some ways difficult to compare directly. However, examination of the behavior of Grime's and Tilman's theories reveals apparent conflicts in their predictions about what traits contribute to competitive superiority and the nature of evolutionary tradeoffs associated with competitive ability (Thompson, 1987; Tilman, 1987a; Thompson and Grime, 1988). In brief, Grime's theory predicts that the species with the highest maximal growth rate of vegetative tissues (maximum capacity for resource capture) will be the superior competitor while Tilman's theory predicts that the species with the minimum resource requirement ($R*$) will be the superior competitor.

In a recent exchange between Thompson (1987) and Tilman (1987a), the issue of semantics was discussed to some degree. According to Thompson, a major cause for at least some of the disagreements between Grime's and Tilman's theories is the different definitions of competition being used. He argued that the primary difference in definitions is that Grime defines competition in terms of resource capture while Tilman defines competition in terms of tolerance to low resource levels (for a further discussion of this important point, see the chapter in this volume by Goldberg). Tilman, however, argued that the real reason for the dispute stems not from the differences in the definition of competition but instead, "from the different traits that we believe allow plants to be competitively superior in particular habitats" (Tilman, 1987a). In this chapter, I present an analysis of both the semantic and mechanistic issues that contribute to this conflict.

III. The Meaning of Competitive Success

Any attempt to define competitive success must begin with a definition of competition. The variety of possible definitions of competition have been discussed numerous times and it is safe to say that there is no universally accepted definition. Nonetheless, it can be argued that a conventional definition does exist, based on the methologies used to study competition. Practically speaking, there exists a body of experimental data that constitutes our observational basis for discussing competition. In nearly all cases, these data were collected by allowing plants to grow either with or without neighbors of another species and, in many cases, by demonstrating that the plants were limited by some common set of resources. As a result, it can be argued that there exists a "conventional" definition of interspecific competition that is exemplified by the definition offered by Begon *et al.* (1986): "an interaction between individuals, brought about by a shared requirement for a resource in limited supply, and leading to a reduction in the survivorship, growth and/or reproduction of the competing individuals concerned." Within this general definition it can be recognized that there are various types of competition (resource competition, interference competition, scramble competition, contest competition, etc.) and by using appropriate modifiers it is possible to restrict discussion to specific mechanisms of interaction. In the subsequent discussion, I will compare how the definitions used by Tilman and Grime compare with this conventional usage.

In principle, Tilman defines interspecific competition as the utilization of shared resources in short supply by two or more species. His ultimate criterion for competitive success is ability of one species to drive another to extinction. It is important to note that this definition includes all phases of the life cycle and focuses on the interaction between competing populations. In practice, the comparison of model predictions to patterns of community structure in nature leads to an operational definition of competition that is a bit more general than the theoretical definition. When used in this way, competitive success is defined based on the dominance of the species in the community, and entire successional sequences are described in terms of changes in competitive outcome resulting from changes in the ratios of resources. This is seen by Tilman (1977) as the simple, logical extension of the Lotka–Volterra description of the phenomenon of competition.

Within this theory, forces that are sometimes considered by others to work in opposition to competition, such as disturbance and herbivory, simplify influence the resource levels at which the species compete. Thus, early-successional annuals and perennials are seen as being competitively superior in habitats for which high disturbance rates cause a high ratio of light to soil resources. Later successional species such as hardwood trees are, in contrast, competitively superior where low disturbance rates allow light to become scarce and the ratio of light to soil resources to decline.

While internally consistent, Tilman's theory uses a very broad definition of competitive success in which the plant traits and environmental conditions that lead to dominance by a species are seen to do so through the mechanism of competition. That this is so is seen by the fact that there is no set of conditions within the context of the model under which the species can survive but not compete (no minimum density required for competition). That this definition of competition is not always accepted by others can be seen by the arguments of Huston and Smith (1987) and Thompson and Grime (1988), who seem not to disagree with Tilman in their predictions of which species should dominate a site but, instead, dispute the role of competition in that dominance. Further, the traditional debate about the relative roles of competition, disturbance, and herbivory seems incompatible with a model where disturbance and herbivory (both of which contribute to the "loss rate" in Tilman's theory) only act to determine the resource level at which plants compete [contrast, for example, with the discussion by Connell (1975) and the model of Shmida and Ellner (1984)]. To his credit, however, Tilman does make explicitly clear how he operationally defines competition.

Grime, in contrast to Tilman, offers a much more restricted definition of competition, "the tendency of neighbouring plants to utilize the same quantum of light, ion of mineral nutrient, molecule of water, or volume of space." He clarifies this definition somewhat by pointing out that, "competition refers exclusively to the capture of resources and is only part of the mechanism whereby a plant may suppress the fitness of a neighbour by modifying its environment." However, an added complexity of Grime's definition of competition is that he classifies plants that possess a particular suite of traits as "competitors." As a result, his operational definition of "competition" is "what 'competitors' do best."

A related issue linking Grime's definitions of competition with his description of a "competitor" is that the concept of stress tolerance includes tolerance to both biotic and abiotic stress. Thus, a plant able to overcome the low resource levels imposed by another species is classified as a stress-tolerant species rather than as a good competitor. Because of this operational definition, Grime considers a late successional species that replaces other species not to be a good "competitor" and, therefore, not to owe its dominance to "competition" despite the fact that it has replaced the earlier species by denying it resources. This definition is both nonoperational (not based on the outcome of interactions) and inconsistent with the conventional usage of the term competition. Thus, it would seem that some other term such as resource exploitation would more appropriate for what Grime has referred to as "competition." Further, perhaps a term such as "exploiter" would be more accurate in describing the syndrome of traits that has been labeled as "competitor."

IV. The Semantics of Populations versus Individuals

Although the definition of competition lies at the heart of the semantic confusion surrounding the debate between Grime and Tilman, an additional semantic issue of importance is the distinction between populations and individuals. Even though Tilman's theory has been based around mathematical and graphical models, it is my perception that a substantial amount of confusion has arisen about the meaning of Tilman's predictions (e.g., Huston and Smith, 1987; Thompson, 1987). This confusion may stem, in part, from the abstract nature of the population terms used in Tilman's original model. Table 1 presents the terms used in Tilman's original model (Tilman, 1982) and some of the individual traits that are subsumed within those population parameters. There are important differences between the "minimum resource requirement" or "maximum growth rate" for populations versus individuals. For example, in Tilman's model, a high rate of density-independent mortality results in the species with the highest population growth rate (and in the model ALLOCATE the highest individual growth rate) having the lowest minimum resource requirement for the population (though not necessarily the minimum resource requirement for adult individuals). Only at low rates of density-independent mortality will the species with the lowest resource requirements for individuals also have the lowest minimum resource requirements for the population.

A related issue is the effect of timespan on competitive success. Tilman's model is explicitly an across-generation model that requires population turnover through the death of adults and recruitment of new individuals into the population. As a result, Tilman (1988) has shown that, at moderate to low mortality rates, transient dominance is predicted whereby species may initially dominate due to their superior growth rate but will eventually be replaced by slower growing species with lower resource requirements. Short-term competition experiments that do not allow for population turnover will not be able to test for this kind of competitive interaction. Tilman's (1987b) field results are a caution against the indiscriminate extrapolation of short-term pot experiments to long-term field processes (see also Berendse and Elberse, this volume).

V. Evolutionary Tradeoffs and Competitive Ability

In addition to the above semantic differences between Grime's and Tilman's theories, there exist other differences in their views on evolutionary tradeoffs. Tilman's theory operates within the context where competition in unproductive habitats is primarily for soil resources because the plant biomass is insufficient to result in light limitation. Conversely, competition in productive habitats becomes primarily light competition once the vegetation develops a dense canopy. In this theory, nonresource factors (such as temperature) can act to affect habitat productivity as well as the species' rate variables (e.g., R^*) but are not explicit variables in the model (Tilman *et al.*, 1981). Not surprisingly, Tilman considers evolutionary tradeoffs in terms of the relative ability of a species to compete for different ratios of resources. Tradeoffs in biomass allocation to roots, stems, and leaves, for example, result in a straightforward tradeoff in the abilities to compete for different resources. Further, changes in resource ratios that occur during secondary succession are viewed to drive the succession of species based on compromises in their abilities to compete at various ratios of light to soil resource. In effect, all environmental factors affecting productivity, be they resource supply rates or nonresource stresses such as soil toxins, are modeled by how they influence $R*$ and the shape and position of resource-dependent growth isoclines. Thus, Tilman has focused on tradeoffs that are compromises between abilities to compete for different resources.

Grime, in contrast, considers a wide range of factors including drought, infertile soil, shade from a higher canopy, or low temperature as "stresses." In his approach to tradeoffs, Grime considers that adaptation to any suboptimal set of environmental conditions reduces the ability of a species to compete for all resources by diverting energy away from adaptations contributing to resource capture. As an example, adaptation to saline soils by salt-tolerant plants involves the tolerant species expending energy to either maintain osmotic balance or exclude salt in order to survive. A critical component to this view of adaptation to nonresource factors (nonresource stress sensu Grime) is that the adapted species has a lower growth rate under nonstressful conditions than does the nonadapted species, regardless of what resource is limiting. Grime does not appear to distinguish fully the consequences of adaptation to low resource supply from adaptation to unfavorable nonresource conditions (resource stress versus nonresource stress sensu Grime).

One result of the above differences between Grime's and Tilman's views of evolutionary tradeoffs is a difference in their predictions about the correlations among a species' ability to compete for different resources. Grime's emphasis on resource capture is consistent with the idea that plants with rapid growth rates will be simultaneously good at trapping all resources, at least initially. Tilman's emphasis on minimum resource requirements is likewise consistent with a negative correlation among competitive abilities. However, Grime's and Tilman's assumptions lead to contrasting predictions about evolutionary tradeoffs. Because Tilman's theory focuses on resources, it predicts that there will be tradeoffs among the abilities to compete for different resources. Grime's theory, in contrast, considers that adaptation to unproductive conditions reduces a species' ability to capture and efficiently use all resources and thus, that there should be a positive correlation among competitive abilities for different resources.

Relatively few studies of competition have been conducted in such a way as to be useful in comparing Grime's and Tilman's theories. Studies of two species of cattails (*Typha*) that segregate along a gradient in water depth (Fig. 1) yield some insight into the nature of evolutionary tradeoffs involving unfavorable nonresource conditions. In this study, the availability of nitrogen (the limiting soil resource) and light were found to be

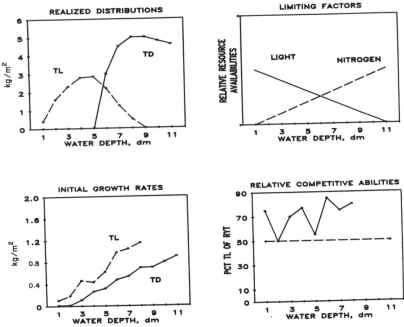

Figure 1 Competition between two *Typha* species, *T. latifolia* (TL) and *T. domingensis* (TD), along a water depth gradient (modified from Grace, 1987, 1988b). Realized distributions are based on both experimental pond and field populations. Limiting factors are based on measurements of sediment nitrogen and incident light. Initial growth rates of monocultures were obtained in experimental pond studies 10 months after planting. Relative competitive abilities were obtained by comparing monocultures and mixtures. (The dashed line represents the values expected if competitive abilities were equal.)

inversely correlated along the water depth gradient (Grace, 1988b) and a resource ratio interpretation of the observed segregation would lead to the expectation that the deep-water species is the superior competitor in deeper water, where its greater height would be an advantage in acquiring light. Actual measurements of competitive performance found that the shallow-water species was the superior competitor at all depths where it could survive and that the deep-water species was restricted to a refuge from competition. Parallel field studies have demonstrated that this result was not a short-term phenomenon but reflects the long-term competitive outcome. As such, these results are consistent with the tradeoffs posited by Grime. The deep-water species is adapted to a nonresource stress and as a result has a lower growth rate (Grace, 1987), a higher requirement for sediment ammonia (Grace, 1988b), and a higher light requirement (unpublished observations). Thus, based on what is known for these two species, they appear to represent a case of a tradeoff between tolerance to a nonresource stress and the overall ability to compete. It should be noted here that this example does not dispute the ability of Tilman's model to predict competitive success. Rather, it represents a pattern of adaptation not predicted by his usual assumptions about evolutionary tradeoffs (but see Tilman *et al., 1981*).

When dealing with the tradeoffs associated with infertile conditions (resource stress), distinguishing Grime's and Tilman's predictions appears to be quite difficult. The chapter by Berendse and Elberse in this volume considers in detail the plant traits contributing to competitive success in nutrient-rich or nutrient-poor sites. Interestingly, their analysis of the components of plant nutrient budgets and competition appears to be largely consistent with Grime's theory without actually refuting Tilman's theory. The primary reason that it appears to be so difficult to distinguish these two theories is that it is seldom known if competition at elevated nutrient levels is actually competition for nutrients or competition for light. In their work with *Molinia caerulea* and *Erica tetralix,* there is clearly a tradeoff between the ability to compete at high and low nutrient levels but it is unclear if this is a tradeoff between the abilities to compete for nutrients and light (as predicted by Tilman). Further, the traits involved in the observed tradeoff are precisely those proposed to distinguish plants adapted to low nutrient levels and those that are associated with competing in productive environments. Thus, any apparent discrepancy between these findings and Grime's theory results only from the semantic differences in definitions of competition. At present there exists some empirical evidence to support the assumptions about evolutionary tradeoffs of both theories (Mahmoud and Grime, 1976; Tilman, 1984, 1987b; see also Keddy, this volume). To resolve this matter further, tests of the correlations among competitive abilities will be necessary where species are forced to compete either exclusively for light or exclusively for nutrients at high and low levels of resource supply. Also, additional tests of the intensity of competition along gradients of fertility are needed to determine if competition among species at low levels of fertility plays an important role in controlling species dominance.

VI. Conclusions

Both Grime's and Tilman's theories provide insight into how species interact for limiting resources. Because of the different emphasis each has on plant traits (Tilman on population traits, Grime on established plant traits), their perspectives on competition likewise differ. Once the differences in their definitions of competition are taken into account, the two theories can be seen to be largely compatible and the remaining differences are comparatively subtle (though not unimportant). At this point, several things could contribute to the utility of these theories. Tilman has shown that his theory is amenable to modification through the incorporation of more specific plant traits. The further inclusion of nonresource variables, environmental fluctuations, and a greater variety of plant traits could only act to make his theory more applicable to natural communities. Grime's theory, on the other hand, would benefit from a less rigid labeling system of plant syndromes and, in particular, the substitution of titles such as "exploiters" in place of "competitors." In addition, a greater emphasis on the distinctions among different types of limiting factors (particularly resource versus nonresource factors and biotic versus abiotic factors) would allow for a greater variety of syndromes to be recognized. To obtain the maximum benefit from Grime's theory it would be best if its main propositions could be quantified into a mathematical framework which would permit more explicit evaluation of the link between assumptions and their implications.

VII. Summary

The current controversy between the theories of Grime and Tilman about how plants compete is based on a variety of apparent conflicts about the traits that determine competitive ability. Grime's theory predicts that the species with the greatest capacity for resource capture will be the superior competitor. Further, his theory predicts a positive correlation between the ability to compete for different resources. Tilman's theory predicts that the species with the lowest minimum resource requirement will be the superior competitor and that there should be a negative correlation among the abilities to compete for different resources.

Analysis of both the theoretical and operational definitions of competition used by Grime and Tilman suggests that many of the apparent contradictions are actually semantic differences. Grime defines competition as the capacity to capture resources while Tilman defines it as a net negative relationship between the abundances of competing species that involves both resource capture and tolerance to low resource levels. It is argued that Grime's definition of competition is not operational and not consistent with conventional usage. Tilman's theoretical definition of competition is consistent with conventional usage but his operational definition (based on his mathematical model) is such that competition is the only factor leading to dominance (regardless of disturbance rate or nonresource conditions).

Behind the differences in definitions are differences in the authors' assumptions about evolutionary tradeoffs. Tilman's theory seeks primarily to explain adaptation to temporal and spatial gradients in resources, and focuses on tradeoffs among abilities to compete for different resources. In his theory, the ratios of resources are the primary selective factor, both in space and in time. Grime's theory, in contrast, seeks to explain adaptation to gradients in productivity, regardless of the cause of unproductive conditions (either resource levels or nonresource conditions). In his theory, the degree to which conditions are unproductive ("stressful") is the primary selective factor (note that both theories also consider disturbance or loss rates).

A limitation of both theories is the failure to distinguish between adaptation to resource levels and adaptation to nonresource conditions. It is argued here that adaptation to gradients in fertility (per se) is expected to result in tradeoffs between the abilities to compete for nutrients versus light (where competition is defined as a negative interaction between species of the same trophic level). However, it is also argued that adaptation to gradients in nonresource conditions may result in tradeoffs between the ability to tolerate extreme conditions and the ability to compete for either nutrients or light. An example of the latter case is presented, showing for two species of *Typha* a tradeoff between tolerance to deep water and the ability to compete for either nutrients or light.

Overall, both theories contribute to our understanding of plant traits and competitive ability and, semantic differences notwithstanding, make generally similar predictions about the types of plants that will dominate under various environmental conditions. The primary differences between the theories lie in the role of various forces that lead to dominance. Further refinement and modification of these theories is needed in order to reduce confusion and extend their utility.

Acknowledgments

I wish to thank Glenn Guntenspergen, Deborah Goldberg, Phil Grime, Paul Keddy, Janet Keough, Betsy Kirkpatrick, Peter Jordan, Jim McGraw, Steve Pacala, Bill Platt, and Dave Tilman for reviews of versions of the manuscript. Supported in part by a grant from the National Science Foundation (BSR-8604556).

References

Austin, M. P. (1982). Use of a relative physiological performance value in the prediction of performance in multispecies mixtures from monoculture performance. *J. Ecol.* **70,** 559–570.

Baker, H. G. (1965). Characteristics and modes of origin of weeds. *In* "The Genetics of Colonizing Species" (H. G. Baker and G. L. Stebbins, eds.), pp. 147–172. Academic Press, New York.

Baldwin, J. P. (1976). Competition for plant nutrients in soil: A theoretical approach. *J. Agric. Sci.* **87,** 341–356.

Begon, M., Harper, J. L., and Townsend, C. R. (1986). "Ecology." Sinauer, Sunderland, Massachusetts.

Chapin, F. S., III (1980). The mineral nutrition of wild plants. *Annu. Rev. Ecol. Syst.* **11,** 233–260.

Connell, J. H. (1975). Some mechanisms producing structure in natural communities: A model and evidence from field experiments. *In* "Ecology and Evolution of Communities" (M. L. Cody and J. M. Diamond, eds.), pp. 460–490. Harvard Univ. Press, Cambridge,

Darwin, C. (1859). "The Origin of Species," reprinted in 1958. New Am. Libr. World Lit., New York.

Gadgil, M., and Solbrig, O. T. (1972). The concept of r- and K- selection: Evidence from wild flowers and some theoretical considerations. *Am. Nat.* **106,** 14–31.

Grace, J. B. (1987). The impact of preemption on the zonation of two *Typha* species along lakeshores. *Ecol. Monogr.* **57,** 283–303.

Grace, J. B. (1988a). The effects of plant age on the ability to predict mixture performance from monoculture growth. *J. Ecol.* **76,** 152–156.

Grace, J. B. (1988b). The effects of nutrient additions on mixtures of *Typha latifolia* L. and *Typha domingensis* Pers. along a water-depth gradient. *Aquat. Bot.* **31,** 83–92.

Grime, J. P. (1977). Evidence for the existence of three primary strategies in plants and its relevance to ecological and evolutionary theory. *Am. Nat.* **111,** 1169–1194.

Grime, J. P. (1979). "Plant Strategies and Vegetation Processes." Wiley, London.

Grime, J. P. (1981). Plant strategies in shade. *In* "Plants and the Daylight Spectrum" (H. Smith, ed.). Academic Press, New York.

Grime, J. P., and Hodgson, J. G. (1987). Botanical contributions to contemporary ecological theory. *New Phytol.* **106,** 283–295.

Grubb, P. J. (1985). Plant populations and vegetation in relation to habitat, disturbance and competition: Problems of generalization. *In* "The Population Structure of Vegetation" (J. White, ed.), pp. 595–621. Junk, Dordrecht, The Netherlands.

Harper, J. L. (1982). After description. *In* "The Plant Community as a Working Mechanism" (E. I. Newman, ed.), pp. 11–25. Blackwell, Oxford, England.

Huston, M., and Smith, T. (1987). Plant succession: Life history and competition. *Am. Nat.* **130,** 168–198.

Loehle, C. (1988). Problems with the triangular model for representing plant strategies. *Ecology* **69,** 284–286.

MacArthur, R. H., and Wilson, E. O. (1967). "The Theory of Island Biogeography." Princeton Univ. Press, Princeton, New Jersey.

Mahmoud, A., and Grime, J. P. (1976). An analysis of competitive ability in three perennial grasses. *New Phytol.* **77,** 431–435.

Shmida, A., and Ellner, S. (1984). Coexistence of plant species with similar niches. *Vegetatio* **58,** 29–55.

Solbrig, O. T. (1979). Ecological classification [book review of Grime, 1979]. *Science* **206,** 1176–1177.

Thompson, K. (1987). The resource ratio hypothesis and the meaning of competition. *Funct. Ecol.* **1,** 297–303.

Thompson, K., and Grime, J. P. (1988). Competition reconsidered—A reply to Tilman. *Funct. Ecol.* **2,** 114–116.

Tilman, G. D. (1977). Resource competition between planktonic algae: An experimental and theoretical approach. *Ecology* **58,** 338–348.

Tilman, G. D. (1982). "Resource Competition and Community Structure." Princeton Univ. Press, Princeton, New Jersey.

Tilman, G. D. (1984). Plant dominance along an experimental nutrient gradient. *Ecology* **65,** 1445–1453.

Tilman, G. D. (1987a). On the meaning of competition and the mechanisms of competitive superiority. *Funct. Ecol.* **1,** 304–315.

Tilman, G. D. (1987b). Secondary succession and the pattern of plant dominance along experimental nitrogen gradients. *Ecol. Monogr.* **57,** 189–214.

Tilman, G. D. (1988). "Plant Strategies and the Dynamics and Structure of Plant Communities." Princeton Monographs, Princeton, New Jersey.

Tilman, D., Mattson, M., and Langer, S. (1981). Competition and nutrient kinetics along a temperature gradient: an experimental test of a mechanistic approach to niche theory. *Limnol. Oceanogr.* **26,** 1020–1033.

Werner, P. A. (1975). Predictions of fate from rosette size in Teasel (*Dipsacus fullonum* L.). *Oecologia* **20,** 197–201.

5

The Application of Plant Population Dynamic Models to Understanding Plant Competition

John A. Silander, Jr.

Stephen W. Pacala

I. Introduction

The broad aim of population or community ecology is to understand the way different kinds of interactions affect the dynamics and structure of a particular system or systems. For example, one needs to understand the contribution of density- and frequency-dependent interactions (i.e., competition, predation, and compensatory interactions) to community structure and dynamics. Does the population or community reach an equilibrium? Is it stable or unstable? What is the nature of the equilibrium? Is the dynamical behavior oscillatory or nonoscillatory? What are the conditions necessary for species coexistence? Likewise, one may need to evaluate the contribution of temporal or spatial heterogeneity in the environment. What is the role of disturbance or the contribution of environmental versus demographic stochasticity?

For plant systems the most common approach to addressing problems and issues in population and community ecology has been almost exclusively an empirical one. However, empirical approaches are inadequate by themselves for addressing many of the questions posed above. What has been missing is the development of an appropriate theory for plant population and community dynamics on which to base empirical studies. It has only been relatively recently that significant attempts have been made to develop theoretical models which can be used to understand and predict the abundance and distribution of plants through time. This contrasts sharply with the studies of animal systems, where there has been a long and rich history of population and community theory underpinning empirical studies.

A theoretical model of plant population dynamics is an explicit description of the processes governing plant population size change. The model may either be a mathematically tractable, analytical one that will at least provide a good qualitative prediction of the behavior for the system (i.e., conditions for equilibrium, stability, persistence, etc.), or a computer simulation that provides a more quantitatively exact prediction for a particular case. Analytical models are most useful for discovering general principles, often sacrificing precision and detail for simplicity and generality. On the other hand, simulation models can provide exact, quantitative predictions for the outcome of a system case by case. There are obvious advantages to the development of a mathematical model that combines the attributes of both approaches. In practice, however, this is often difficult to achieve. Both analytical and simulation models provide predictions of future events. If the models can be empirically calibrated in the field, one obtains an explicit description of the processes governing the community under study, together with a prediction of the dynamics and change in structure of the community. Moreover, the predictions can be tested by simple observations of natural populations over time. The description of the population dynamic process may be dissected to assess the contribution of intra- and interspecific competitive interactions (magnitude, symmetry, etc.) to population fluctuation, or more specifically assess the importance of such interactions at different stages in the life cycle of an individual. Competition is thus placed in the context of population or community dynamics, rather than studied as a static phenomenon in isolation.

One simple way to develop a model of plant population dynamics is to link together separate phenomenological (sub)models, each of which describes an individual phase in the life cycle of an individual. For simple annuals this may include separate mathematical functions that describe and predict germination and dormancy, survival from seedling to adult, fecundity, and dispersal. When considering more complex systems such as perennials with overlapping generations or clonal plants, additional components may be needed. By constructing a model that closes the life cycle of the plant, it is straightforward to project the dynamics of the population. In a simulation, one simply iterates the model. If the model is analytically tractable, it can be relatively straightforward to characterize the dynamical behavior of the population (i.e., stability and equilibrium conditions). Despite the inherent simplicity of such an approach, there has been little attempt to develop Lotka–Volterra analogs for plant populations. In part this reflects certain characteristics of plants which set them apart from animal systems, and which at first look seem to make plant models more intractable.

The most obvious, unique features of plants include the following:

1. *Sedentariness.* A plant is at the mercy of its local environment such that plant performance and hence population dynamics tend to be affected by spatial heterogeneity at many scales: from that of the seed to that of the whole population.

2. *Circumscribed interactions.* Interactions among plants are spatially local (among neighbors) and thus population dynamics will be affected by the spatial distribution of individuals in the population which may change with population growth. An important consequence of this is that one may need to specify the spatial location of individuals in any model in order to describe the population dynamics.

3. *Plasticity.* Individual plants within a population may vary by several orders of magnitude in growth and fecundity. This is an inevitable consequence of the above two features. One is left with the impression that population numbers mean little unless scaled by size.

4. *Abiotic niche resources.* The resources all plants use are essentially the same—water, light, and mineral nutrients. One might easily assume that there is little latitude for potential niche differentiation and hence coexistence. However, Tilman (1986, 1988) has clearly demonstrated that this is not necessarily the case.

Plant population dynamic models have been developed, particularly within the past few years, that include one or more of the above traits. However, none include all of these attributes and yet remain analytically tractable. In Section II we review briefly some of the major plant population dynamic models that have been developed to date, listing some of the advantages and disadvantages of the various approaches. We then focus most of our attention on the neighborhood models, showing how these are developed and calibrated, and how the predictions are generated and tested in the field. Specific empirical examples are drawn from a simple two-species community of annuals (velvet leaf and pigweed) that we have studied in some detail. The generalities of our empirical and theoretical findings are discussed and, finally, we offer a prospectus for future directions.

II. Plant Population Dynamic Models

Yield models were developed as phenomenological (nonspatial) models to describe plant growth in agricultural plots over the course of a single season. The Japanese (Kira *et al.*, 1953) were among the first to examine the dynamics of within-season population growth. These and subsequent studies (see Firbank and Watkinson, this volume, for details) have led to a general understanding of intraspecific competition in the context of within-season population dynamics, and an understanding of the source of individual variation in size. Watkinson (1980) has extended these models to produce a discrete generation (nonspatial) model of population dynamics. The dynamics of a population and intraspecific interference can thus be characterized from empirically calibrated relationships of total plant yield with sowing density, and seed number with plant weight.

Alternative yield models were developed by de Wit (1960). These are static replacement series models designed to predict the competitive outcome of agronomic components raised over a single season. The relative yields of two components raised at a constant density (but varied initial proportions) are compared with the respective pure stand yields at equivalent densities. The results are static descriptions relating seeds sown to the total weight or numbers of seed harvested at the end of the season. This was of course the original agronomic objective of the approach. Ratio diagrams were developed as a way to translate the static predictions of replacement series analysis to population dynamical predictions over time. This and related approaches have come under heavy criticism (Inouye and Schaffer, 1981; Connolly, 1986; Law and Watkinson, 1987). The equilibria predicted are constrained by fixed densities in a system in which the qualitative and quantitative outcome is density and frequency dependent. Thus, the dynamics predicted by ratio diagrams are at best misleading and at worst invalid.

Holsinger and Roughgarden (1985) have developed a model analogous to that of Watkinson (1980) that allows one to examine within- and between-season dynamics for single-species or multispecies populations.

The model provides theoretical insight into the dynamical consequences of variation in plant performance and environmental heterogeneity; however, the model has not been specifically calibrated or tested in the field. Firbank and Watkinson (1986) have also developed a phenomenological, nonspatial model of the dynamics of two-species populations as an extension of Watkinson (1980). This model can be calibrated in the field from observed survivorship and reproduction of populations raised in monoculture and in mixtures at a range of densities, together with observed germination success. Unfortunately, one test of the model failed to predict the dynamics of a system of one weed species growing with wheat (reported by Firbank and Watkinson, this volume; see also Law and Watkinson, 1987).

Demographic models of the type developed by Leslie (1945) can of course be used to project the dynamics of single-species plant populations over time using the matrix method. Law (1983) has elegantly shown how this can be done for a plant population classified by size and age. Such models are phenomenological and nonspatial. One drawback is that these are linear models which lack density dependence.

Mechanistic models represent an alternative modeling approach to examining plant population dynamics. Tilman (1986, 1988) has developed models that include the dynamics of abiotic resources together with the dynamics of plants competing for these resources (light, water, nutrients) in a heterogeneous environment. The models are extensions of standard predator–prey models, and seem to explain plant species coexistence on limited resource types. Tilman has shown that the models can account for observed patterns of species abundances and distributions (see Tilman, this volume). The models have not been fully validated in the field or extended to include space.

The forest simulation models of Shugart (1984) are mechanistic and include spatial dynamics of multispecies systems with plastic growth but not spatial heterogeneity. These models can be calibrated despite the large number of parameters that need to be estimated, and they do have remarkable predictive powers. Nevertheless, the models remain analytically intractable and there has been little attempt to examine them theoretically.

Several models have been developed that could be characterized as spatial or quasi-spatial. Some of these are extensions of the spatial cell models of Skellam (1951), and most are phenomenological. The Markovian models (Usher, 1966; Waggoner and Stephens, 1970; Horn, 1975) developed to predict community dynamics, particularly of forests, can be considered at least quasi-spatial. These have yielded some theoretical insight into the dynamics of complex, multispecies communities. In addition, they are easy to calibrate and have proven predictive power for at least some systems, particularly those near equilibrium (cf. Stephens and Waggoner, 1980). However, these models have been criticized because they specify constant replacement probabilities, and because the Markovian assumptions do not hold under some conditions (Usher, 1979).

Shmida and Ellner (1984) have developed a phenomenological, spatial model of community dynamics that includes spatially local interactions, nonuniform dispersal, lottery competition for juvenile occupancy of microsites, and spatiotemporal environmental heterogeneity. The probabilities of coexistence are strongly affected by dispersal characteristics and environmental heterogeneity. The model has not been specifically calibrated or tested in the field, but it has been evaluated in the light of observed patterns predicted in natural communities.

We have developed neighborhood models of single and multispecies population dynamics (Pacala and Silander, 1985, 1989; Pacala, 1986a) that include spatially local interactions, plastic growth, and environmental heterogeneity. The models are analytically tractable and have been calibrated and tested in the field. They remain at this point phenomenological models. We discuss these neighborhood models in detail below.

III. Neighborhood Models of Annual Plant Population Dynamics

We developed neighborhood population models as a way to translate empirical observations on plant performance and interference into a precise description of population and community dynamics. Most of our efforts to date have focused on simple two-species populations of annuals. Our neighborhood models are constructed from a set of four submodels, which we have variously called predictors or descriptors, for each species in the population. The submodels do in fact predict independently the fate of individuals in the population at successive stages in their life cycle. The germination submodel gives the probability that a seed produced in one year will survive to and germinate in each of k subsequent years. The survivorship submodel predicts the probability of survival from seedling to adult, given the local density about each seedling. We define the local density (or neighborhood) of species-j about a species-i individual as the number of species-j individuals that occur within a circle of radius r_{ij} from the focal species-i individual. Similarly, there is an equivalent intraspecific neighborhood about the focal species-i individual, defined as the number of species-i individuals within another circle of radius r_{ii} from the focal species-i individual. For an N-species model, there will be a series of concentric circles defined about all individuals in the population of radii $r_{ij}, j = 1, 2, \ldots, N$. Seed production of surviving adults is governed by a fecundity submodel which is similar to the survivorship submodel except that it predicts seed set as a function of the local (neighborhood) density about each individual. The concentric circles of $r_{ij}, j = 1, 2, \ldots , N$, about species-$i$ individuals in the population are used in constructing the fecundity submodels. The survivorship and fecundity radii need not be the same. In practice, our fecundity submodels comprise two functions: a biomass predictor that gives adult biomass as a function of neighborhood crowding and a function that relates biomass to seed production. At maturity the seeds are dispersed. The dispersal submodel predicts the spatial location of seeds given the location of the mothers.

A. Field Calibration of Neighborhood Models

Germination submodels (including dormancy) can be calibrated on field plots lacking a seed pool of the species of interest. One simply initiates the plots with a given number of seed and censuses the plots periodically during the first and subsequent seasons while preventing the recruitment of additional seed. The fraction of seeds of species-i produced in year $t - k - 1$ alive in year t (k = the age of the seed) is the germination predictor g_{ik}. See Pacala (1986b) and Pacala and Silander (1989) for further details.

Survivorship and fecundity submodels can be calibrated together on another set of field plots. Seeds of the species are scattered together on the plots in a heterogeneous manner to span a broad range of densities and frequencies. Each emerging seedling may then be tagged and numbered. At the end of the growing season following seed maturity, the aerial portion of each plant is harvested and the dry weight determined. Spatial maps (x, y coordinates) for all plants in the plot are determined from digitized overhead photographs or direct digitization of the numbered tags. These maps allow a characterization of each plant's survivorship and fecundity neighborhood. Biomass–seed set conversion can be obtained from seed counts taken from a subset of the harvested plants that spanned a range of plant sizes.

The construction of survivorship and fecundity predictors from field data is a bit more complex than germination or dispersal predictors. For either survivorship or fecundity predictors, neighborhood radii are set to some series of initial values $[r_{ij}, (j = 1, 2, \ldots , N)]$. In practice we often set these approximately equal to mean plant size. One tallies the number of species-j neighbors for each neighborhood in the population and regresses the number of individuals per neighborhood (n_j) against focal plant size, selecting an appropriate functional form $b(n_i, n_j, \ldots n_n)$. Tallying the number of neighbors for each individual in the population is facilitated by a fast-sorting computer algorithm that we have developed for this purpose (see Pacala and Silander, 1985, for details). Neighborhood attributes, in addition to the number of neighbors may be included if the added complexity is warranted. Generally, biomass versus local crowding relationships tend to be concave, and we have found that simple hyperbolic functions work well for fecundity predictors. The form for the two-species case is

$$w_i = M_i/(1 + c_{ii}n_i + c_{ij}n_j) \tag{1}$$

where w_i is the biomass for a species-i plant, M_i is the biomass of a species-i plant with no neighbors, and c_{ij} is an interference coefficient giving the effect of a species-j neighbor on a species-i focal plant fecundity. These simple hyperbolic biomass predictors are easily interpreted biologically and provide as good as or better fits than various linear or exponential functions for many plant species (Weiner, 1982; Law and Watkinson, 1987; Pacala and Silander, 1987).

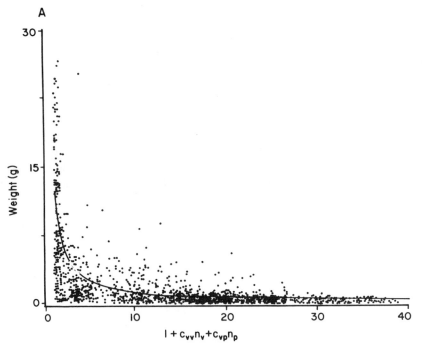

Figure 1 Biomass predictors for (A) velvet leaf and (B) pigweed. The horizontal axes give, as indices of local crowding, the denominator of the biomass predictor Eq. (4) using the following parameter values from one of the fecundity–survivorship field calibration plots (designated West 1984): $M_v = 43.0$, $c_{vv} = 0.46$, $c_{vp} = 0.0$, and $M_p = 20.0$, $c_{pp} = 3.3$, $c_{pv} = 3.5$ for, respectively, $r_{vv} = 20$ cm, $r_{vp} = 10$ cm and $r_{pp} = 5$ cm, $r_{pv} = 20$ cm. Each curve was estimated using the gamma regression method described in the text.

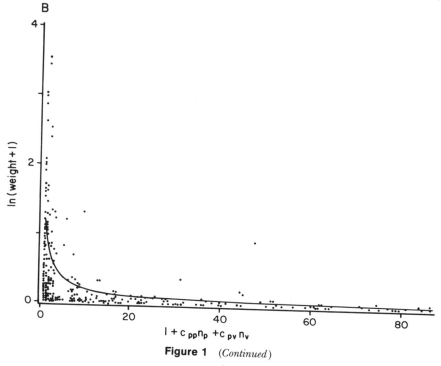

Figure 1 (*Continued*)

An inspection of plots of biomass versus neighborhood crowding (Fig. 1) typically reveals a skewed and heteroskedastic distribution of residuals which is not easily dealt with by normal-based statistics. Others have pointed out this difficulty and concluded that neighborhood models may be inappropriate or inadequate for understanding population dynamic processes (Firbank and Watkinson, 1987). To deal with this problem, we have developed a maximum likelihood estimator for nonlinear regressions where the residuals follow a gamma rather than a normal distribution. Details are given in Pacala and Silander (1989). When we have followed this procedure, we have obtained excellent statistical fits to experimental data with well-balanced residuals and negligible autocorrelations (Pacala and Silander, 1989). In so doing we obviate the statistical problem that dominance and suppression generate when one tries to use normal-based statistical analyses.

One evaluates the fit for a series of different neighborhood radii and identifies the "best" neighborhood radii for a given species as those which yield the greatest likelihood. Biomass predictors are converted to fecundity submodels by simply regressing seed set versus biomass. Survivorship submodels are generated in an analogous manner, except one uses a multiple nonlinear binomial maximum likelihood regression estimator for density-dependent survivorship models.

Dispersal submodels are calibrated from maps of seedlings that recruit from isolated mothers of varying sizes. Functions can be fitted to plots of seedling numbers versus distance to mother plants. Or, alternatively, probability density functions can be fitted to plots of seedling positions.

With spatially and temporally replicated calibration plots, one can assess the contribution of spatial and temporal variation in parameter values to population dynamic processes.

B. Predicting and Testing Population Dynamics

The four submodels (germination, survivorship, fecundity, and dispersal) together can be used to forecast the fate of an individual species-i plant throughout its life cycle. There are two routes to predicting the population dynamics with this information: simulation or analytical models. A neighborhood simulation of population dynamics includes modeled plants as points on a model plot. Given some initial seedling population size and spatial distribution, the seedlings survive to reproduce with some probability that is dependent on local neighborhood crowding given by the survivorship model. The number of seeds produced by each surviving individual is similarly dependent on local crowding defined by the fecundity submodel. The seeds are dispersed as specified by the dispersal submodel and germinate with a probability specified by the germination submodel. By determining the fate of every plant in each modeled generation, the population sizes and spatial distributions of individuals are predicted.

The analytically tractable neighborhood population dynamic models are identical to the simulation models except that probability density functions (giving the probability that a randomly chosen species-i plant has n neighbors) are used to specify the spatial distribution of individuals in the population at each generation. These may be derived explicitly or estimated experimentally.

Both simulation and analytical models provide specific predictions of the dynamical behavior of the population. If these predictions are based on field-calibrated submodels the predictions can easily be tested. One can simply set up independent test plots in the field, allow the establishment of the appropriate assemblage of species from seed, and census the plots periodically, comparing predicted with observed values.

C. Population Dynamics of Pigweed and Velvet Leaf

We have followed the above protocol for a simple two-species population of annuals by developing simulation and analytical models from field-calibrated submodels and testing these in the field (for details, see Pacala and Silander, 1989). The system comprised velvet leaf (*Abutilon theophrasti*) and red root pigweed (*Amaranthus retroflexus*), two common annual weeds that occur together on or near agricultural fields throughout eastern North America. Spatially replicated sets of calibration plots were set up in each of three years (1984–1986) following the protocol outlined above. This yielded three sets of calibration data. In addition, replicated field test plots (spatially paired with the calibration plots) were initiated in 1984 and censused yearly through 1987. This replication allowed us to examine the contribution of spatial and temporal heterogeneity to population fluctuation.

An important result of this study is that it is indeed possible to calibrate fully and test a density-dependent plant population dynamic model in the field. The general form of the full neighborhood model is

$$S_{it} = \sum_{k=0}^{\infty} g_{ikt} S_{i,t-k-l} \sum_{n_{if}=0}^{\infty} \sum_{n_{jf}=0}^{\infty} \sum_{n_{is}=0}^{\infty} \sum_{n_{js}=0}^{\infty} D_{it}(\underline{n}) U_i(\underline{n}) \tag{2}$$

$$(i, j) = (v, p) \text{ and } (p, v) \qquad \underline{n} = (n_{if}, n_{jf}, n_{is}, n_{js})$$

where S_{it} is the population density of species-i seeds produced in year t, n_{if} and n_{jf} are the numbers of species-i and species-j neighbors in the fecundity neighborhood submodels, n_{is} and n_{js} are the corresponding quantities for the survivorship neighborhood submodels, $U_i(\underline{n})$ is the product of species-i's survivorship and fecundity submodels, and g_{ikt} is the germination submodel for k-year-old species-i seeds in year t. The $D_{it}(\underline{n})$ function gives the fraction of species-i plants with n_{if}, n_{jf}, n_{is}, and n_{js} neighbors in year t.

The explicit formulation of the survivorship submodel for pigweed is a hyperbolic function:

$$Q_p = P_p/(1 + c_{pp(s)}n_p + c_{pv(s)}n_v) \tag{3}$$

where P_p is the survivorship of a pigweed plant with no neighbors and $c_{ij(s)}$ are interference coefficients giving the effect of a species-j neighbor on a species-i focal plant (Fig. 2). [Watkinson *et al.* (1983) use an identical formulation.] Survivorship of velvet leaf was density independent and thus estimated as a simple constant, P_v.

The fecundity submodels for pigweed and velvet leaf were formulated as follows. The biomass predictors were hyperbolic functions analogous to Eq. (1):

$$W_p = M_p/(1 + c_{pp(f)}n_p + c_{pv(f)}n_v) \tag{4a}$$

$$W_v = M_v/(1 + c_{vv(f)}n_v + c_{vp(f)}n_p) \tag{4b}$$

where w_i is the adult above-ground biomass of a species-i plant, M_i is the

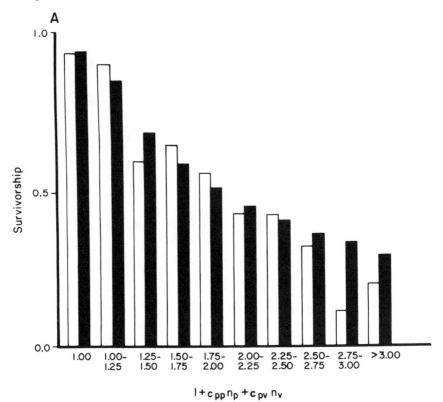

Figure 2 Survivorship predictors for pigweed. The horizontal axes give, as indices of local crowding, the denominator of the pigweed survivorship predictor Eq. (3) using the following parameter values: (A) $P_p = 0.94$, $c_{pp} = 0.079$, $c_{pv} = 0.037$ for $r_{pp} = 5$ cm, $r_{pv} = 10$ cm (from calibration plot West 1984) and (B) $P_p = 0.48$, $c_{pp} = 0.20$, $c_{pv} = 0.018$ for $r_{pp} = 5$ cm, $r_{pv} = 10$ cm (from calibration plot West 1985). The open bars give the observed survivorships and the solid bars give values predicted by the relevant survivorship predictors. Note the hyperbolic dependence of survivorship on local crowding and the close correspondence between observed and predicted values.

biomass of a species-i plant with no neighbors, and the $c_{ij(f)}$ are interference coefficients. Nonlinear gamma regressions were used to fit the data as described above (Fig. 1).

Linear regressions were used to set biomass–seed set relations:

$$S_p = a_p w_p \tag{5a}$$

$$S_v = a_v w_v - b_v \quad \text{for } w_v > b_v/a_v; \text{ otherwise } S_v = 0 \tag{5b}$$

where a_i and b_i are, respectively, the slope and intercept of the relation for species-i. Since even the smallest pigweed plant produced at least one

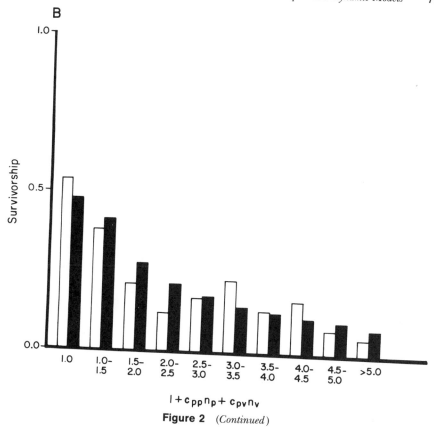

Figure 2 *(Continued)*

seed, the *y*-intercept (seed set) was not significantly different from zero. Velvet leaf plants must reach a minimum threshold size before they set seed and there was thus a significant, negative (seed set) intercept.

To assess the predictive power of the neighborhood population dynamic model (2), we used the estimated parameter values obtained from the calibration plots. The predictions from these neighborhood models were then compared with the values observed independently from censuses of the test plots. We found remarkable agreement among the predicted population densities in all replicates, and between predicted and observed densities. Two representative cases are given in Fig. 3A,B. Note that the observed population sizes in each year fall within or close to the 95% error bars of the model predictions.

There are small deviations of observed from predicted values in several instances. We believe that underlying spatial and temporal variations in demographic parameters are partially responsible for this dis-

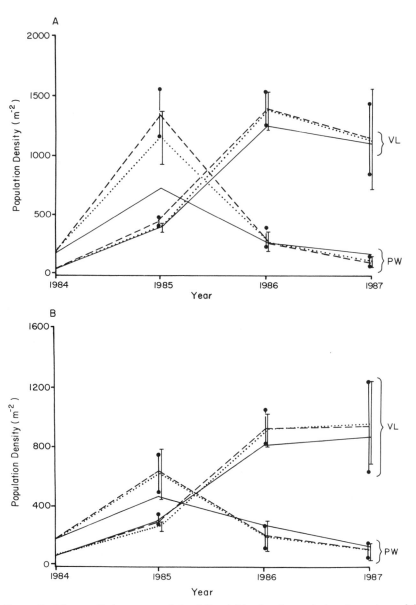

Figure 3 The predictive power of the full neighborhood population dynamic model (2) and the mean model (6) compared with the observed population densities in the field test plots. Dashed lines are the densities predicted by the full neighborhood model, dotted lines are for the mean model, and the solid lines are the observed densities for each of the years 1985–1987. Vertical bars are 95% prediction error limits that translate the statistical uncertainty about the values of the estimated parameters into statistical uncertainty about predicted population sizes. Bars that end in solid circles are for the neighborhood model and bars that end in lines are for the mean model. (A) The population dynamics test plot designated SW, (B) the population dynamics plot ME.

crepancy: We have shown elsewhere (Pacala and Silander, 1989) that fecundity and survivorship do vary spatially and temporally while germination primarily varies temporally. Nevertheless, Fig. 3 effectively demonstrates that one can predict the dynamics of plant populations from neighborhood models. Indeed, it is important to note that, even if we use any combination of calibration site and year values, we get good, qualitative dynamical predictions.

Details of the spatial interactions can be obtained from a closer look at results obtained from the individual submodels. We observed that neighborhood interactions occurred over small distances (best neighborhood radii: 20 cm $\geq r_{ij} \geq$ 5 cm for both survivorship and fecundity predictors) with little or no variation across plots or years. Each species showed significant levels of intraspecific interference ($c_{ii} > 0$). However, interspecific interference was asymmetrical, with velvet leaf affecting pigweed more than vice versa ($c_{pv} > c_{vp} \geq 0$). In most cases $c_{vp} \sim 0$, indicating that the interaction between velvet leaf and pigweed was essentially amensalistic. The differences in mean survivorship and fecundity neighborhood radii ($r_{pp} < r_{vv}$ and $r_{vp} \leq r_{pv}$) probably reflected morphological differences between the two species. Velvet leaf plants tended to be larger on average than pigweed plants. The leaves of velvet leaf tended to be restricted to the canopy level, while those of pigweed were continuously distributed along the stem. With these attributes one might expect velvet leaf plants to shade larger areas than pigweed plants of a similar biomass and have larger mean neighborhoods (cf. Pacala and Silander, 1987).

In projecting the population dynamics of pigweed and velvet leaf, we initially assumed that the spatial distribution of seedlings was random. We were able to check this by examining the spatial distribution of individuals in the test plots in the years after they were initiated. Partial mappings of these plots allowed us to estimate an observed mean number of neighbors in each replicate plot in each of 2 years. We developed a dispersion statistic as simply the observed mean number of neighbors divided by the expected mean number of neighbors for a random distribution. Values greater than one indicate underdispersion and values less than one indicate overdispersion. In virtually all cases the dispersion statistic was slightly greater than one, indicating a weak spatial aggregation at least at the scale of the neighborhood. Spatial autocorrelation analyses within these plots also indicated weak aggregation at a small scale. Using the spatial autocorrelation statistic (I) (see Cliff and Ord, 1981), significant autocorrelations always occurred at scales less than 25 cm and in most cases less than 10 cm. Since this scale of aggregation is small relative to mean dispersal distances (approximately 50 cm in both species), this probably represents seed responses to soil heterogeneities rather than clumping of seed about parents (see Harper, 1977, Ch. 2).

The aggregation is sufficiently low and at such a small scale that the distribution of individuals can be considered nearly random.

In any study that entails analytical modeling one seeks simplifying assumptions to make the models analytically tractable as well as general. As it stands, the full population dynamic neighborhood model (2) is analytically intractable. However, this complex expression is closely approximated by a simpler expression:

$$S_{it} = \sum_{k=0}^{\infty} g_{ikt} S_{i,t-k-1} U_i(\bar{n}) \tag{6}$$

where $(\bar{n})$ is a vector of mean numbers of neighbors. This expression is valid if, over the range of variation in number of neighbors within a population, the $U_i(n)$ function from Eq. (2) is linear and if the spatial distributions of individuals are approximately random. Pacala and Silander (1989) show that the $U_i(n)$ functions are approximately linear because the survivorship and biomass predictors are hyperbolic, and because neighborhood densities in the field tend to be high and coincide with the flat portion of the predictor curves (e.g., $x > 20$ in Fig. 1a,b). Moreover, since the spatial distributions are close to random (see above), expression (6) is approximately valid.

The explicit expanded expressions of Eq. (6) for pigweed and velvet leaf are

$$S_{vt} = S_{vt}^* \{ [M_v'/(1 + C_{vv(f)} S_{vt}^* + C_{vp(f)} S_{pt}^*)] - b_v' \} \tag{7a}$$

$$S_{pt} = S_{pt}^* [M_p'/(1 + C_{pp(f)} S_{pt}^* + C_{pv(f)} S_{vt}^*)(1 + C_{pp(s)} S_{pt}^* + C_{pv(f)} S_{vt}^*)] \tag{7b}$$

where $S_{it}^* = \sum_{k=0}^{\infty} g_{ikt} S_{i,t-k-1}$, $M_v' = M_v a_v P_v$, $M_p' = M_p a_p P_p$, $b_v' = b_v P_v$, $C_{ij(s)} = c_{ij(s)} A_{ij(s)}$, and $C_{ij(f)} = c_{ij(f)} A_{ij(f)}$; $A_{ij(s)}$ and $A_{ij(f)}$ are, respectively, survivorship and fecundity neighborhood areas. We call Eqs. (7a) and (7b) "mean" population dynamic models. Note that it is not completely independent of spatial information on individuals in the population. The spatial scale of interference is set by the neighborhood areas defining the interference coefficient values.

It is important to ask if there is any loss of predictive power as a consequence of using this simpler formulation. In Fig. 3 the population projections of the mean model with error bars are plotted together with the full neighborhood model projections and the observed values from the test plots. Note that there is very little difference between the predictions of the full neighborhood model and the mean model. It is clear from this that information on the spatial distribution of individuals, or factors that effect this, bring relatively little to predicting the dynamics of a pigweed–velvet leaf system. Simplifying rules such as this are essential if one is to have any hope of developing general, analytically tractable models in population and community theory.

D. General and Theoretical Findings with Extensions to Other Systems

Nonspatial models similar to our mean neighborhood model have also been developed by Firbank and Watkinson (1986) to describe plant population dynamics. Their models are based on density-dependent population level performance rather than individual performance as in our neighborhood models. Our results imply that the approach used by Firbank and Watkinson may indeed be appropriate for systems such as the pigweed–velvet leaf system.

We have concluded that, in our annual system, information on spatial distributions has relatively little effect on the dynamics of this system. Why is this the case? Recall that the spatial information in the $D_{it}(\underline{n})$ functions of Eq. (2) drops out when neighborhood densities are sufficiently high and span a limited range, and when spatial distributions are random, overdispersed, or weakly aggregated. When this holds, the full spatial population dynamics model reduces to the mean model. This does *not* mean that density dependence or neighborhood interactions are unimportant. Recall that intense neighborhood interactions are an integral part of the mean model (i.e., c_{ij} values may be large). The complex neighborhood model (2) that includes information on spatial distributions of individuals reduces to essentially a nonspatial model that includes only information on spatial scale.

How generally applicable are nonspatial population dynamic models for a range of population or community types? It is evident from the previous paragraph that nonspatial models will likely fail to describe adequately the dynamics of populations that are sufficiently clumped, or that have low or variable numbers of neighbors. What kinds of populations, if any, have such attributes? A number of possible examples come to mind: Short propagule dispersal distances (seed or ramet) can produce populations with aggregated spatial distributions (cf. Pacala, 1987). Harper (1977, Ch. 2) gives seed dispersal profiles for a number of species, and for many of these the vast majority of seeds land below the mother plant. Mean dispersal distances for these are undoubtedly less than the neighborhood radii. A likely consequence of this is an aggregated spatial distribution of individuals. In many clonal plant populations, ramets, genets, and species tend to be spatially aggregated (Austin, 1968, 1980; Kershaw, 1959; Silander, 1979; Symonides, 1985). These aggregated patterns may be a consequence of the clonal morphology and short ramet dispersal distances, or the response to heterogeneous physical or biotic environment (Kershaw, 1963; Symonides, 1985). However, ramet distributions for at least a few clonal species may tend to be uniform (Bell, 1984). In forest tree populations, the number of neighbors per neighborhood often appears to be low (Opie, 1968; Weiner, 1984). However, there have been few attempts to estimate neighborhood size or

the functional form of density-dependent performance in tree populations. Seed dispersal distances for a number of tree species (including
hemlock and other conifers, *Liriodendron*, basswood, and others) are
known to be low (Harper, 1977; Woods, 1984). This likely contributes to
aggregated spatial distributions of individuals. The size, and spatial and
temporal distribution of disturbances or gaps, and physical heterogeneities in the environment will also affect the spatial distributions of individuals in a population (cf. Barclay-Esterup and Gimmingham, 1969;
Grubb, 1977). Gap phase species such as *Liriodendron* or yellow birch may
tend to be aggregated to the extent that large gaps tend to be spatially
and temporally segregated. Both hemlock and basswood may show aggregated spatial distributions as a consequence of short dispersal distances and high self-replacement probabilities (Woods, 1984). The cyclical gap phase dynamics found in *Calluna* communities appears to
promote species aggregations (Barclay-Esterup and Gimmingham,
1969). During the process of succession in some communities the spatial
distribution of individuals and species may actually become more aggregated (Symonides, 1985). Although none of the factors discussed above
were important in the pigweed–velvet leaf system, it is obvious that these
factors may be important in many other plant populations and communities. We therefore caution against the indiscriminate application of
nonspatial population dynamic models (such as our mean model) to
plant communities in general. We need to ask first where along a continuum in the relative importance of spatial processes a particular population or community falls. With this information one can then develop the
appropriate modeling approach.

A close look at Fig. 1 reveals large residual variances. This implies
considerable statistical uncertainty in predicting fecundities with neighborhood models alone. In addition, survivorship and germination predictors do not allow us to identify who will germinate or survive. Firbank
and Watkinson (this volume) point this out as a critical fault with neighborhood models of competitive interaction. How is it that we can predict
community structure and dynamics with deterministic models despite
the great stochasticity in germination, survival, and reproduction? There
are two sources of stochastic variation in population sizes according to
May (1973). Demographic stochasticity occurs because of the stochastic
nature of births and deaths even in a constant environment. Environmental stochasticity occurs because of temporal stochastic variation in
the environment. We have shown elsewhere (Pacala and Silander, 1989)
that demographic stochasticity has little effect in all but the smallest
populations of velvet leaf or pigweed, and that environmentally driven
population fluctuations will generally overwhelm demographic stochasticity. As evidenced by our ability to predict population fluctuations us-

ing deterministic neighborhood models, the unpredictability of individual plants does not imply unpredictable population dynamics.

We have seen that both the full, spatial neighborhood model and the mean model provide good predictions of the short-term dynamics of pigweed and velvet leaf. What can we learn of the long-term dynamics of these two species populations? A local stability analysis (details in Pacala and Silander, 1989) predicted that velvet leaf competitively excludes pigweed with no internal equilibrium (coexistence) for any combination of the field-calibrated submodel parameter values. The competitive dominance of velvet leaf is a consequence primarily of small $c_{vp(f)}$ and large $c_{pv(f)}$ values. These results are shown graphically in Fig. 4, which plots a numerical solution of the mean model using parameter values from one of the predictor plots in one of the years. Notice that the approach of velvet leaf to a stable equilibrium population size shows damped oscillations. We have shown elsewhere (Pacala, 1986b; Pacala and Silander, 1989; Thrall *et al.*, 1989) that populations, such as those of velvet leaf, that are characterized by one or more of the following attributes are likely to show oscillatory dynamics: low seed dormancy, high germination success, high soil fertility, minimum plant size thresholds

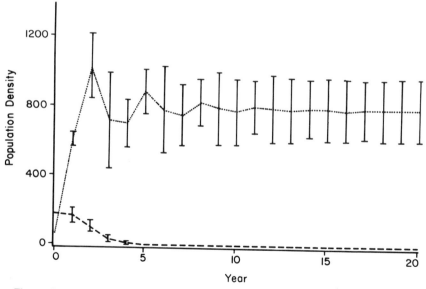

Figure 4 Long-term dynamics predicted by the mean model (6). The dotted line is for velvet leaf and the dashed line is for pigweed. Note the rapid exclusion of pigweed and the oscillatory approach of velvet leaf to equilibrium.

for seed production, and large fruited individuals with many large seeds. Contrary to Watkinson (1980) and Crawley and May (1987), oscillatory dynamics may be a common feature of some annual plant populations.

Our population dynamic models predict that velvet leaf will eventually drive pigweed extinct in the field plots. Indeed, this prediction is upheld by the census data collected to date from the test plots. One can ask by what processes or under what conditions is coexistence likely to be promoted for these species or any set of annuals in general. Two factors that may qualitatively affect species coexistence are dispersal distances and spatial heterogeneity in the environment. We have explored theoretically the effect that these factors have on the structure of annual communities (Pacala, 1987). Our results show that dispersal abilities of the species present have a striking effect on community structure, even in a homogeneous environment. In some cases, dispersal is as important as the relative magnitudes of intra- and interspecific interference in determining the outcome of competition. Theoretical expectations are that, given asymmetrical (compensatory) interactions between species-i and species-j (i.e., $c_{ij} > c_{ji}$ and $c_{ij} > c_{ii}$), species-i should exclude j unless j has a higher equilibrium population size in monoculture (i.e., as in the Montgomery effect). If the latter case holds, coexistence is possible. Similarly, in the case of negative interference (mutual antagonism) where interspecific competitive interactions are greater than intraspecific interactions (i.e., $c_{ii} < c_{ij}$ and $c_{jj} < c_{ji}$), coexistence should not occur. However, if dispersal distances are sufficiently small (relative to neighborhood size), most individuals will occur in clumps of conspecifics. If at the same time intraspecific interferences (c_{ii}'s) are sufficiently large, population densities of the clumped species will decline since most individuals in the crowded clumps will produce few seeds. In the case of a single species with short dispersal distances, a second species may successfully invade between the clumps and coexist. If both species have short dispersal distances, interspecific segregation occurs and interspecific interference declines since most of the interference is intraclump and hence intraspecific. Coexistence may thus occur in spite of predictions from asymmetrical or negative interference considered alone.

Furthermore, in a heterogeneous environment, the community structure is in part determined by the relationship between dispersal distance or neighborhood radius and patch size. Given an environment with two patch types in which species-i wins in one and species-j wins in the other, the two species may coexist only if the dispersal distance is sufficiently small relative to patch size (Pacala, 1987). Coexistence may also require that the patch size be sufficiently larger or sufficiently smaller than the neighborhood diameter. If patch sizes are large relative to neighborhood size, most of the interplant interactions will be intrapatch (i.e., among plants in the same patch). If the patches are small relative to neighborhood size, most neighborhoods will average over several patches. In many cases, coexistence is possible only when one or the other of the above conditions hold (Pacala, 1987). If plants alter the local resource quality or quantity, then they may contribute to spatial heterogeneity (Kershaw, 1963; Symonides, 1985). It may therefore be important to include the dynamics of spatially distributed resources in theoretical or empirical studies if we are to understand the structure and diversity of plant communities. In addition, because coexistence may be mediated by sufficiently small patch sizes, it may be important to assess the effects of fine-scale environmental heterogeneity.

IV. General Discussion and Conclusions

We have shown that it is possible to develop spatial models of plant population and community dynamics that include plastic growth, sedentariness, and local interactions. These neighborhood models are analytically tractable and can be calibrated and tested in the field. For a simple, two-species population of annuals we have been able to predict accurately the dynamical behavior and the equilibrium conditions. It is indeed ironic that the very features of plants which were thought to make population dynamic models intractable turn out to be most advantageous in modeling and understanding plant systems.

By constructing population dynamic models from submodels of performance at each stage in the life cycle of the individual, we have been able to evaluate the contribution of any change in performance at each of these stages to the dynamics of the system. For example, a change in mean dispersal distances can affect probabilities of coexistence. The morphology of one of the species contributes in an important fashion to the dynamical behavior of the system. The minimum threshold size for reproduction in velvet leaf results in a humped density versus yield relationship, and this tends to produce oscillatory dynamics in the system. In addition, we have shown how temporal and spatial variation in the environment can affect the dynamical predictions and likelihood of coexistence. A consequence of the fact that our neighborhood model is analytically tractable and has been field calibrated and tested is that a variety of simplifying assumptions can be made and their consequences evaluated. For example, the analysis is greatly simplified if we consider just mean density effects instead of the full spatial neighborhood effects. The predictions of both models turn out to be virtually identical and both match well the observed dynamics of the community.

The next challenge is (1) to understand the dynamics of more complex systems—systems characterized by species with more complex life cycles such as perennial or clonal species, as well as more diverse communities, and (2) to obtain a more mechanistic understanding of the structure and dynamics of plant communities. Some of our current efforts are focused on the development of neighborhood models of perennial, clonal plant communities. We anticipate little difficulty in accomplishing this objective for simple systems. However, it becomes prohibitively difficult to calibrate neighborhood models for diverse communities. There are just too many parameters to fit. What is needed is some way to find simplifying rules that will reduce the number of parameters that are included in any model. In principle, this is possible using spatial models. After all, we were able to show that space dropped out of our two-species model of annuals.

An alternative approach with considerable appeal is one that explores, more fully, mechanistic models of plant population dynamics or combines these with analytically tractable, spatial models, such as ours, that can be field calibrated and tested. Mechanistic models that have been developed to date are either analytically intractable, with large numbers of parameters, and have not been field calibrated, or they are nonspatial. Tilman's mechanistic models (1986, 1988) are attractive because they include competition for specific resources in a spatially and temporally heterogeneous environment, but they are nonspatial. Similarly, the forest models of Shugart (1984) are attractive because they are mechanistic and can predict the dynamics and structure of diverse communities. However, Shugart's models remain analytically intractable and they are also nonspatial. Thus, we are left in a position of being unable to understand the general features that are controlling the structure and dynamics of the system. Spatial processes are evidently important in controlling the structure and dynamics of many community types and may thus warrant inclusion in mechanistic models. It is therefore apparent that we need to come up with models that are spatial and mechanistic, but remain analytically tractable, and can be calibrated and tested in the field. To this end, there is a need to focus some attention on where various communities fall along a continuum vis-à-vis the importance of spatial processes in controlling community structure and dynamics. Indeed, this may be of prime importance in understanding how communities are organized. An approach that combines empirically based spatial and mechanistic models will perhaps provide the best hope for understanding the structure and dynamics of diverse plant communities.

V. Summary

A significant impediment to advances in the field of plant population and community dynamics has been the slow development of a general theory that will enable one to understand and predict abundances and distributions of plants through time. We review some of the major modeling approaches developed to date, which can be characterized as mechanistic or phenomenological, and spatial or nonspatial. We then focus primarily on neighborhood (phenomenological, spatial) models, outlining their general development, calibration, prediction generation, and empirical field testing. Specifics are drawn from a case study of a two-species community of annual weeds, velvet leaf and pigweed. We found a remarkable agreement between the model predictions for this system and the population dynamics observed in the field. We discuss the generality and the implications of the empirical and theoretical findings. Finally, we offer a prospectus for future directions, pointing out the need for an approach that combines the development of spatial, mechanistic models of population dynamics with empirical studies in the field.

References

Austin, M. P. (1968). Pattern in a *Zerna erecta* dominated community. *J. Ecol.* **56,** 734–759.

Austin, M. P. (1980). An exploratory analysis of grassland dynamics: An example from a lawn succession. *Vegetatio* **43,** 87–94.

Barclay-Esterup, P., and Gimmingham, L. H. (1969). The description and interpretation of cyclical processes in a heath community. I. Vegetational changes in relation to the *Calluna* cycle. *J. Ecol.* **57,** 737–758.

Bell, A. D. (1984). Dynamic morphology: A contribution to plant population ecology. *In* "Perspectives on Plant Population Ecology" (R. Dirzo and J. Sarukhan, eds.), pp. 48–65. Sinauer, Sunderland, Massachusetts.

Cliff, A. D., and Ord, J. K. (1981). "Spatial Processes, Models and Applications." Pion, London.

Connolly, J. (1986). On difficulties with replacement-series methodology in mixture experiments. *J. Appl. Ecol.* **23,** 125–137.

Crawley, M. J., and May, R. M. (1987). Population dynamics and plant community structure: Competition between annuals and perennials. *J. Theor. Biol.* **125,** 475–489.

de Wit, C. T. (1960). On competition. *Versl. Landbouwkd. Onderz.* **66,** 1–82.

Firbank, L. G., and Watkinson, A. R. (1986). Modelling the population dynamics of an arable weed and its effect upon crop yield. *J. Appl. Ecol.* **23,** 147–159.

Firbank, L. G., and Watkinson, A. R. (1987). On the analysis of competition at the level of the individual plant. *Oecologia* **71,** 308–317.

Grubb, P. J. (1977). The maintenance of species richness in plant communities: The importance of the regeneration niche. *Biol. Rev.* **52,** 107–145.

Harper, J. L. (1977). "Population Biology of Plants." Academic Press, New York.

Holsinger, K. E., and Roughgarden, J. (1985). A model for the dynamics of an annual plant population. *Theor. Pop. Biol.* **28,** 288–313.

Horn, H. S. (1975). Markovian properties of forest succession. *In* "Ecology and Evolution of Communities" (M. L. Cody and J. M. Diamond, eds.), pp. 196–211. Belknap, Cambridge, Massachusetts.

Inouye, R. S., and Schaffer, W. M. (1981). On the ecological meaning of ratio (de Wit) diagrams in plant ecology. *Ecology* **62,** 1679–1681.

Kershaw, K. A. (1959). An investigation of the structure of a grassland community. III. Discussion and conclusions. *J. Ecol.* **47**, 31–53.

Kershaw, K. A. (1963). Pattern in vegetation and its causality. *Ecology* **44**, 377–388.

Kira, T., Ogawa, H., and Sakazaki, N. (1953). Intraspecific competition among higher plants. I. Competition–yield–density relationships in regularly dispersed populations. *J. Inst. Polytech., Osaka City Univ., Ser. D* **4**, 1–16.

Law, R. (1983). A model for the dynamics of a plant population containing individuals classified by age and size. *Ecology* **64**, 224–230.

Law, R., and Watkinson, A. R. (1987). Response-surface analysis of two-species competition: An experiment on *Phleum arenarium* and *Vulpia fasciculata. J. Ecol.* **75**, 871–886.

Leslie, P. H. (1945). On the use of matrices in certain population mathematics. *Biometrika* **33**, 182–212.

May, R. M. (1973). "Stability and Complexity in Model Ecosystems." Princeton Univ. Press, Princeton, New Jersey.

Opie, J. E. (1968). Predictability of individual tree growth using various definitions of competing basal area. *For. Sci.* **14**, 314–323.

Pacala, S. W. (1986a). Neighborhood models of plant population dynamics. II. Multi-species models of annuals. *Theor. Pop. Biol.* **29**, 262–292.

Pacala, S. W. (1986b). Neighborhood models of plant population dynamics. IV. Single and multi-species models of annuals with dormant seed. *Am. Nat.* **128**, 859–878.

Pacala, S. W. (1987). Neighborhood models of plant population dynamics. III. Models with spatial heterogeneity in the physical environment. *Theor. Pop. Biol.* **31**, 359–392.

Pacala, S. W., and Silander, J. A., Jr. (1985). Neighborhood models of plant population dynamics. I. Single-species models of annuals. *Am. Nat.* **125**, 385–411.

Pacala, S. W., and Silander, J. A., Jr. (1987). Neighborhood interference among velvet leaf, *Abutilon theophrasti,* and pigweed, *Amaranthus retroflexus. Oikos* **48**, 217–224.

Pacala, S. W., and Silander, J. A., Jr. (1989). Tests of neighborhood population dynamic models in field communities of two annual weed species. *Ecol. Monogr.,* in press.

Shmida, A., and Ellner, S. (1984). Coexistence of plant species with similar niches. *Vegetatio* **58**, 29–55.

Shugart, H. H. (1984). "The Theory of Forest Dynamics." Springer-Verlag, New York.

Silander, J. A., Jr. (1979). Microevolution and clonal structure in *Spartina patens. Science* **203**, 658–660.

Skellam, J. G. (1951). Random dispersal in theoretical populations. *Biometrika* **38**, 196–218.

Stephens, G. R., and Waggoner, P. E. (1980). A half century of natural transitions in mixed hardwood forests. *Bull. Conn. Ag. Exp. Sta., New Haven* **783**.

Symonides, E. (1985). Population structure of psamophyte vegetation. In "The Population Structure of Vegetation" (J. White, ed.). Junk, Dordrecht, The Netherlands.

Thrall, P., Pacala, S. W., and Silander, J. A., Jr. (1989). Oscillatory dynamics in populations of an annual weed species (*Abutilon theophrasti*). *J. Ecol.,* in press.

Tilman, D. (1986). Evolution and differentiation in terrestrial plant communities: The importance of the soil resource : light gradient. In "Community Ecology" (J. M. Diamond and T. J. Case, eds.), pp. 359–380. Harper & Row, New York.

Tilman, D. (1988). "Plant Strategies and the Dynamics and Structure of Plant Communities." Princeton Univ. Press, Princeton, New Jersey.

Usher, M. B. (1966). A matrix approach to the management of renewable resources with special reference to selection forests. *J. Appl. Ecol.* **3**, 355–367.

Usher, M. B. (1979). Markovian approaches to ecological succession. *J. Anim. Ecol.* **48**, 413–426.

Waggoner, P. E., and Stephens, G. R. (1970). Transition probabilities for a forest. *Nature (London)* **255,** 1160–1161.

Watkinson, A. R. (1980). Density-dependence in single-species populations of plants. *J. Theor. Biol.* **83,** 345–357.

Watkinson, A. R., Lonsdale, W. M., and Firbank, L. G. (1983). A neighbourhood approach to self-thinning. *Oecologia* **56,** 381–384.

Weiner, J. (1982). A neighborhood model of plant interference. *Ecology* **63,** 1237–1241.

Weiner, J. (1984). Neighbourhood interference amongst *Pinus rigida* individuals. *J. Ecol.* **72,** 183–195.

Woods, K. D. (1984). Patterns of tree replacement: Canopy effects on understory pattern in hemlock–northern hardwood forests. *Vegetatio* **56,** 87–107.

6

Competition and Nutrient Availability in Heathland and Grassland Ecosystems

Frank Berendse   **Wim Th. Elberse**

I. Introduction

In many natural environments, nutrient availability is a major factor affecting species composition and dynamics of plant communities. Field studies frequently reveal significant correlations between available soil nutrients and species composition (e.g., Kruijne *et al.*, 1967; Vermeer and Berendse, 1983; Pastor *et al.*, 1984). Fertilization experiments in natural ecosystems often show dramatic shifts in dominance after the application of nutrients such as nitrogen or phosphorus (e.g., Tilman, 1984; Elberse *et al.*, 1983; Vermeer, 1985). In large parts of Western Europe the inputs of nitrogen by precipitation have strongly increased during the past 20–30 years because of volatilization of ammonia from farm manure and the production of nitrogenous oxides by the combustion of fossil fuels. The increased levels of available nitrogen brought about by these developments are reflected in dramatic shifts in the species composition of many ecosystems on formerly nutrient-poor soils and have led to the local extinction of numerous plant species (Mennema *et al.*, 1980). Basic understanding of the effects of these raised levels of nitrogen availability on the dynamics of plant communities is urgently needed, for such knowledge might allow manipulation of our last remaining nutrient-poor ecosystems to reduce further losses of biotic diversity.

This chapter focuses on the long-term effects of the rate of nutrient supply on competition between plant populations and community composition. A central question in this context is, Which physiological or morphological plant features cause a population to succeed in nutrient-poor or in nutrient-rich habitats? In the analysis that is presented here, the long-term effects of increased levels of nutrient availability on plant growth are derived from a description of the short-term effects, by taking into account the loss of nutrients from the plant. The question put forward above is addressed step by step, by considering, in turn, the growth of single plants, the nutrient balance of the plant, and its competitive ability as affected by the supply of nutrients. Finally, we analyze the trade-off between plant properties that serve as adaptations to nutrient-poor environments and those that serve as adaptations to relatively nutrient-rich circumstances.

II. Vegetation Dynamics and the Growth of Single Plants

In recent years we have studied two different ecosystems in which the effects of increased nutrient supply were pronounced: wet heathlands on sandy soils and hayfields on basin clay. In the first half of this century most wet heathlands in the Netherlands were dominated by the evergreen dwarf shrub *Erica tetralix* L. Other plant species that occurred at low frequencies in these communities were *Scirpus caespitosus* L., *Gentiana pneumonanthe* L., *Drosera intermedia* Hayne, and *Lycopodium inundatum* L. During the past 20–30 years the perennial grass *Molinia caerulea* (L.) Moench has increased strongly in these heathlands. Now most *Erica*-dominated communities have been replaced by monocultures of *Molinia*. In order to test the hypothesis that an increase in availability of nutrients caused this shift in dominance, a fertilization experiment in the field was carried out. In vegetation dominated by *Erica*, but containing a small proportion of *Molinia*, an experiment was laid out with five replicated blocks of plots that received different nutrient treatments. One treatment was not fertilized; two other treatments of plots received nitrogen fertilizer or phosphate fertilizer. Over 3 years we measured the cover of the different plant species in each plot, using a photographic method. A two-factor analysis of variance showed a significant effect of time and fertilizer treatment, the interaction between these two factors being not significant (Aerts and Berendse, 1988). In the plots receiving the two fertilized treatments a significant decrease in the cover of *Erica* was measured. The increase in the cover of *Molinia* was significant in the plots that received phosphate (Fig. 1). We therefore concluded that increased rates of nitrogen or phosphorus supply could have been important factors causing *Erica*-dominated communities to be replaced by *Molinia* stands.

Boot (unpublished observations) compared the growth of single plants of the dwarf shrubs *Erica tetralix* L. and *Calluna vulgaris* (L.) Hull and the

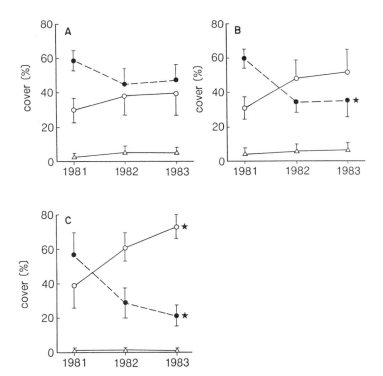

Figure 1 Percent cover of *Erica tetralix* (solid circles), *Molinia caerulea* (open circles), and *Scirpus caespitosus* (triangles) in plots that were unfertilized (A), or fertilized with 20 g N m^{-2} year^{-1} ammonium nitrate (B) or with 4 g P m^{-2} year^{-1} sodium biphosphate (C). Bars represent standard errors of the mean. Differences between cover in 1981 and 1983 were tested using an a-posteriori test after analysis of variance. Stars indicate significant differences ($P < 0.05$). After Aerts and Berendse (1988).

grass *Molinia* during 2 years. The plants were grown in pots that were either unfertilized or fertilized with NPK. In the plots receiving the unfertilized treatment there were no large differences in the biomass production of the three species. *Molinia* showed a slightly higher production than the two other species that are characteristic of nutrient-poor sites. *Molinia* showed a much stronger growth response to the application of nutrients than the two ericaceous dwarf shrubs did. During the growing season of the second year of the experiment, Boot measured relative growth rates based on total plant weights. In the unfertilized treatment plots there were no clear differences in relative growth rate between the three species (*Erica:* 0.36 month^{-1}; *Calluna:* 0.31 month^{-1}; *Molinia:* 0.32 month^{-1}). However, in the fertilized treatment plots *Molinia* had a higher relative growth rate (0.85 month^{-1}) than the two dwarf shrubs (0.37 month^{-1} for *Erica* and 0.57 month^{-1} for *Calluna*).

Other ecosystems where increased levels of available nutrients have altered species composition are the hayfields and meadows. At the beginning of this century, most grasslands in the Netherlands harbored a large number of plant species, including both grasses and many dicotyledonous species. In 1958 an experiment was started in a hayfield on basin clay (Van den Bergh, 1979; Elberse *et al.*, 1983). Two unfertilized plots were compared with two plots that received a NPK-fertilization each year. In a period of about 30 years, the number of species (observed in 50 samples, each with an area of 25 cm^2) in the fertilized plots has dropped to less than 25% (9 species) of the initial species number (38 species). In the unfertilized plots there has also been a slight decrease in the number of species, but in these plots 28 species are still present after 30 years as compared with 36 species at the beginning of the experiment. Figure 2 reflects part of the changes in species composition after fertilization. The grasses *Festuca rubra* L. and *Anthoxanthum odoratum* L. had almost disappeared 10 years after fertilization had started, but remain at high frequencies in the unfertilized plots. Two other grass species, *Arrhenatherum elatius* (L.) Beauv. ex J. & C. Presl and *Alopecurus pratensis* L., show the opposite behavior. *Alopecurus* responds rapidly to fertilization and occurs after 4 years in almost all samples. *Arrhenatherum* is absent at the start of the experiment. After 8 years it has established itself permanently in the plots and after 14 years the frequency of this species starts to increase to rather high values. The long lag in response to fertilization of this species can be attributed to the time period that it needed to establish itself in these hayfields. In the unfertilized plots *Arrhenatherum* was almost absent, whereas *Alopecurus* fluctuates here at intermediate frequencies. Both the reduction in species number and the increasing dominance of a few grass species are common phenomena in grasslands after being fertilized and have also been very well documented in the Rothamsted grassland experiment (Tilman, 1982).

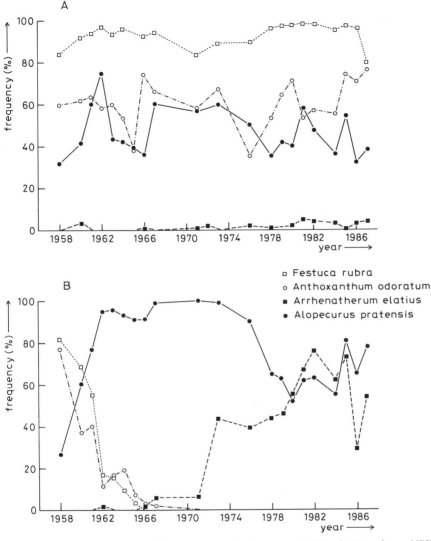

Figure 2 The frequency of four grass species in two unfertilized (A) and two NPK fertilized plots (B) during 30 years in a hayfield on basin clay. The fertilized plots received 160 kg N, 52 kg P, and 332 kg K per hectare annually. Frequency has been measured as percentage of 2 × 50 samples with an area of 25 cm². After Berendse and Elberse (1989).

An important question is, Which properties enable *Festuca* and *Anthoxanthum* to maintain themselves in the unfertilized treatment and which properties allow *Arrhenatherum* and *Alopecurus* to increase in the fertilized plots. The two groups of grass species were grown in pots under extremely nutrient-poor conditions and with nutrients supplied at a rate enabling maximum relative growth rates to be attained. The rate of dry matter production was clearly higher in *Alopecurus* and *Arrhenatherum* than in *Festuca* and *Anthoxanthum*, both in the nutrient-poor and in the nutrient-rich series (Fig. 3). The difference was, however, larger in the nutrient-rich series. There are no clear differences in relative growth rate between the two groups of species, neither in the nutrient-poor treatment nor in the nutrient-rich treatment. Maximum relative growth rates were, for *Anthoxanthum* and *Festuca*, 0.21 and 0.19 day^{-1}, respectively, and, for *Alopecurus* and *Arrhenatherum*, 0.21 and 0.18 day^{-1}, respectively. *Anthoxanthum* and *Alopecurus*, which start vegetative growth and flowering under natural conditions much earlier than the two other species, show slightly higher relative growth rates. Growth analysis showed that the differences between the two species of nutrient-poor sites and those of nutrient-rich sites, as shown in Fig. 3, were mainly caused by differences in embryo and endosperm weight. The production of relatively heavy seeds and the resulting higher initial growth rates after germination seem to be important adaptive features of plant species characteristic of nutrient-rich, productive hayfield communities.

Ingestad (1979) used another approach in order to analyze the effects of the rate of nitrogen supply on plant growth. He introduced the concept of nitrogen productivity, defining it as the dry weight production per unit nitrogen in the plant (g DW g^{-1} N day^{-1}). In his experiments, he was able to induce stable nitrogen concentrations in the plant by increasing the rate of nutrient supply exponentially. He found a close linear

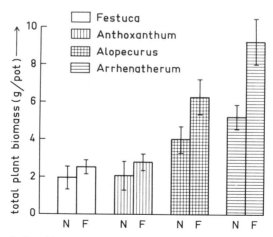

Figure 3 Total plant biomass of 16 plants of four grass species grown in pots under unfertilized (N) and fertilized conditions (F). Plants were harvested after 4 weeks. Bars represent standard deviations.

relationship between relative growth rate (dW/Wdt) and nitrogen concentration in the total plant (n):

$$(1/W)(dW/dt) = An \qquad (1)$$

Rearranging this equation yields

$$dW/dt = AN \qquad (2)$$

where N is the total amount of nitrogen in the plant and A the nitrogen productivity. In single growing plants the nitrogen productivity seems to be an appropriate parameter to measure the efficiency with which the nitrogen present in the plant is used for carbon assimilation or dry matter production. We combined data from three different articles (Ingestad, 1979; Hui-jun and Ingestad, 1984; Ingestad and Kähr, 1985) that reported on experiments measuring the effect of nitrogen supply on the growth of seedlings of three different tree species. These experiments were carried out in climate chambers under similar conditions. The studied species were *Pinus sylvestris* L. (which occurs on sandy, nutrient-poor sites), *Paulownia tomentosa* (Thunb.) Steud. (which is characteristic of nutrient-rich habitats), and *Betula verrucosa* Ehrh. (which occurs under intermediate conditions). The slopes of the regression lines of the relative growth rate versus nitrogen concentration in the plant are different, indicating different values of nitrogen productivity (Fig. 4). *Pinus sylvestris* is characterized by a low value of nitrogen productivity, whereas *Paulownia* (and also other tree species that are characteristic of nutrient-rich environments such as *Populus* spp.; cf. Ingestad and Kähr, 1985) show much higher values of nitrogen productivity.

Summarizing, we conclude that, in nutrient-poor treatments of short-term experiments, species that are successful in nutrient-rich environments show growth rates equal or even higher than those of plant species of nutrient-poor habitats. When fertilized, the growth of plant species of nutrient-rich environments responds much more strongly to the application of nutrients. Their potential growth rates are clearly higher than those of plant species that are found in nutrient-poor environments. This important difference between species of nutrient-poor and those of nutrient-rich environments has already been described (Grime, 1979; Chapin, 1980). Differences in maximum growth rate are one of the important distinctions between the competitive and stress-tolerant strategies that are distinguished within Grime's conceptual framework. However, variation in different plant features seems to be responsible. The differences in absolute growth rate between the two groups of grass species characteristic of nutrient-poor and nutrient-rich hayfield communities were found to be caused by variation in seed weight. In the two other examples, clear differences in maximum relative growth rate and

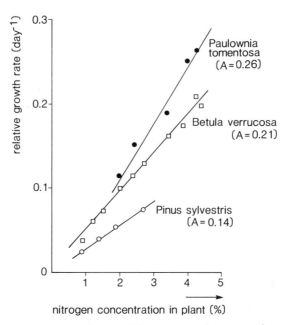

Figure 4   The relative growth rate of three tree species versus nitrogen concentration in the total plant. The values of the nitrogen productivity A (g DW g^{-1} N h^{-1}) are given by the regression coefficients of the presented lines. After Ingestad (1979), Hui-jun and Ingestad (1984), and Ingestad and Kähr (1985).

in nitrogen productivity were measured. Possibly, such differences may be explained by differences in the rate of net photosynthesis per unit of nitrogen in the leaf caused by differences in, for instance, stomatal resistance or respiration. Moreover, plant species may vary widely in the allocation of nitrogen to different plant organs (e.g., leaves or roots) or organelles (e.g., chloroplasts), which may also explain a large part of the variation in relative growth rate and nitrogen productivity that was observed.

III. The Nutrient Balance of the Plant

The advantage of a high potential growth rate for a plant growing in a nutrient-rich environment is evident. However, it may be questioned as to what the adaptive significance is of a low potential growth rate in nutrient-poor habitats. In examining this, it is important to realize that the success of a perennial plant population is determined not only by the amount of nutrients absorbed, but also by the amount of nutrients lost from the population. In environments where nutrients limit plant growth, the dynamics of populations of perennial plants are largely determined by the balance between the uptake and the loss of nutrients. Plants living in a natural environment are continuously losing nutrients. These losses may occur through a variety of different pathways: mortality of plant parts (e.g., abscission of leaves and flowers, mortality by disturbance, turnover of roots), herbivory (large herbivores, phytophagous insects, root nematodes, parasitic fungi), leaching from leaves etc. (especially of cations), production of seeds and pollen, and root exudation (amino acids, amino sugars). Most nutrients that return to the soil by one of these pathways enter the general soil nutrient pool and cannot simply be taken up again by the same plant. The plant has to compete with other plants and with microorganisms for the nutrients that remineralize from the litter. Moreover, a portion of the nutrients that return to the soil may be lost by leaching from the soil profile, denitrification, or volatilization of ammonia or be bound in humus compounds, which can have a turnover time of more than 500 years (Campbell *et al.*, 1967).

A plant that annually loses a large part of the nutrients in its biomass must absorb more nutrients to maintain its biomass than a plant that is more economical of the nutrients that it has acquired. In order to measure and compare the nutrient uptake needed by different plant species in natural environments, we introduced the concept of the relative nutrient requirement (Berendse, 1985; Berendse *et al.*, 1987b). The nutrient requirement (NL) is defined as being the loss of nutrients from the individual or population during a given time interval (e.g., mg N m^{-2} yr^{-1}). This amount must be absorbed by the individual or the population just to maintain or replace its biomass. If the population absorbs more, its biomass will increase; if it absorbs less, its biomass will decline. The relative nutrient requirement (L) is defined as the nutrient requirement per unit biomass, i.e., the amount of nutrients that is needed to maintain or replace each unit biomass during a given time period (e.g., mg N g^{-1} dry wt yr^{-1}). The relative nutrient requirement measures the costs of biomass maintenance in terms of the nutrients that are required. In some cases it may be useful to calculate the nutrient requirement per unit of nutrients present in the plant (L_n; e.g., mg N mg^{-1} N yr^{-1}). Under steady-state conditions the inverse of this parameter is equal to the mean residence time of the nutrient in the plant.

During 1982 and 1983, we carried out a detailed study on nutrient cycling and litter production in wet heathland communities. Stands of *Erica* and stands of *Molinia* occurring under similar conditions were compared (Berendse *et al.*, 1987a,b). Since direct measurement of litter production was not possible, we studied the demography of the different plant organs and calculated mortality rates of each plant part on the basis of the data collected. The withdrawal of nutrients preceding abscission of leaves, culms, and flowers was measured as well. Unfortunately, it was not possible to measure the retranslocation of nutrients from dying roots. In Table 1 the percent nitrogen losses and the relative nitrogen requirements as calculated from these data are summarized. Minimum and maximum estimates for the nitrogen losses from the total plant were calculated. Minimum losses were estimated assuming that 50% of the nitrogen in dying roots was withdrawn before abscission, and maximum estimates were calculated assuming that no retranslocation occurred before root death.

The annual losses of nitrogen that we measured are quantitatively significant and should have an important impact on the dynamics of perennial plant populations in environments where nutrients limit plant growth. *Erica* and *Molinia* have different strategies to restrict their nutrient losses. *Molinia* is a perennial grass that dies off above ground at the end of the growing season. The main mechanism by which *Molinia* retains nutrients is withdrawal of nutrients from dying leaves and culms. At the end of the growing period about 60% of the nitrogen in leaves is withdrawn, stored in basal internodes and roots, and reused at the beginning of the new growing season. *Erica* is a woody evergreen and retains its nutrients by the inherently low mortality of its stems and the relatively long life span of its leaves. About 40% of the nitrogen in the leaves is

Table 1   Percentage Losses of Nitrogen by Litter Production from the Above-Ground Biomass and from the Total Plant and the Relative Nitrogen Requirements of *Erica tetralix* and *Molinia caerulea* as Measured During 1982 and 1983

	Erica		*Molinia*	
	1982	1983	1982	1983
Percentage losses[a]				
Above-ground	19	27	46	63
Total	22–32	—	64–100	—
Relative nitrogen requirement[b]				
Above-ground	2.0	2.6	6.0	7.5
Total	2.3–3.4	—	7.4–11.7	—

After Berendse and Elberse (1989).
[a] Percentage losses are expressed as percentage of the amount of nitrogen (yr^{-1}) in above-ground biomass or total biomass at the end of the growing season.
[b] Expressed as mg N g^{-1} dry wt yr^{-1}

withdrawn before abscission, which is significantly less than in *Molinia*. In spite of these different strategies, the percentage of nitrogen lost from *Molinia* is two to three times higher than the losses from *Erica* (Table 1). The percentage losses were higher during 1983 than in 1982. [Note that in 1983 we used much more detailed and accurate techniques to measure actual mortality and nutrient withdrawal (Berendse *et al.*, 1987b).] However, the relative difference between *Erica* and *Molinia* was observed to be the same during 1983. The relative nitrogen requirement of *Molinia* is about three times as high as that of *Erica*.

IV. Competition between Perennial Plant Populations

The next question that needs to be answered is how these differences in relative nitrogen requirement affect the outcome of the competition between the two species. In order to answer this question, we carried out a theoretical analysis of the competition between two perennial populations. Let us first consider the partitioning of a limiting resource between two identical plant populations. Plant densities are assumed to be sufficiently high for the total flow of the resource that becomes available during one time interval (e.g., the mineralization of nutrients) to be absorbed. The absorption of the limiting resource by each of the two populations (U_1 or U_2) is then proportional to the fraction each species composes of total plant biomass:

$$U_1 = \frac{B_1}{B_1 + B_2} \text{NF} \tag{3a}$$

$$U_2 = \frac{B_2}{B_1 + B_2} \text{NF} \tag{3b}$$

where NF is the flow of the limiting resource that becomes available for plant uptake during one time interval and B_1 and B_2 represent the biomasses of species 1 and species 2, respectively. If the two populations are not identical, the biomass of each of the two populations should be multiplied by a weighting coefficient that takes into account the different competitive abilities of the two species with respect to the uptake of the limiting resource. Under conditions where plant growth is mainly light-limited and the two competing plant species are identical except, for instance, in the ratio between leaf area and biomass (LAR), the biomass of each species in Eqs. (3) should be multiplied by its LAR, so that B_1 and B_2 are replaced by the leaf area of species 1 and species 2, respectively. If competition is predominantly for, e.g., nitrate, the biomass of each species should be weighted by a coefficient that converts plant weight into total root length or total active root surface. Multiplying the biomass of each population by a weighting coefficient b_1 or b_2 and substituting the relative competition coefficient k_{12} for b_1/b_2, we obtain the following equations for the partitioning of the limiting resource between two non-identical species:

$$U_1 = \frac{k_{12}B_1}{k_{12}B_1 + B_2} \, \text{NF} \tag{4a}$$

$$U_2 = \frac{B_2}{k_{12}B_1 + B_2} \, \text{NF} \tag{4b}$$

These equations are the competition hyperbolas that were introduced by De Wit (1960, 1961) by analogy with Raoult's law for the relationship between the composition of liquor and that of vapour. In the competition during one short time interval between two ideal plant species, i.e., species that differ in only one feature, the relative competition coefficient has a simple physiological or morphological meaning (e.g., the ratio between the LAR of species 1 and that of species 2). In most cases, in the competition during a longer time period between plant species that are different in many respects, k_{12} is a phenomenological coefficient that is measured in replacement experiments. The relative competition coefficient measured in such a way is determined to a large part by the competitive ability of the two species with respect to each other. If k_{12} is higher than unity, species 1 is the superior competitor; if k_{12} is below unity, species 2 is superior. De Wit (1961) pointed out that the relative competition coefficient as measured in replacement experiments depends not only on the competitive ability of the two species, but also on total plant density and on the duration of the experiment (cf. Spitters, 1979; Berendse, 1981).

We assume that the dynamics of two competing perennial plant populations in a nutrient-poor environment are determined by the balance between the uptake of nutrients and the loss of nutrients. Furthermore, we assume that only nutrients limit plant growth, so that biomass is correlated linearly with the amount of nutrients in the plant. The absorption of nutrients (U) by each of the two species is described by means of expressions (4). The loss of nutrients from the population during one time interval is given by the product of its biomass (B) and its relative nutrient requirement (L). The dynamics of each of the two populations are given by:

$$\Delta B_1/\Delta t = \left(\frac{k_{12}B_1}{k_{12}B_1 + B_2} \, \text{NF} - L_1 B_1 \right)/n_1 \tag{5a}$$

$$\Delta B_2/\Delta t = \left(\frac{B_2}{k_{12}B_1 + B_2} \text{ NF} - L_2B_2\right)/n_2 \qquad (5b)$$

where L_1 and L_2 are the relative nutrient requirements of the two species and n_1 and n_2 are the concentrations of the growth-limiting nutrient in the plant (e.g., mg N/g dry wt). NF is the flux of nutrients that become available for plant uptake (e.g., mg N m^{-2} yr^{-1}). Here, k_{12} is defined with respect to one year or growing season and, moreover, with respect to the uptake of nutrients only.

We analyzed Eqs. (5) for steady-state conditions, where the uptake of nutrients equals the loss of nutrients. In the competition between species 1 with a high relative nutrient (e.g., *Molinia*) and species 2 with a lower relative nutrient requirement (e.g., *Erica*), the first species will ultimately be dominant if

$$k_{12} > L_1/L_2 \qquad (6a)$$

and species 2, with the lower relative nutrient requirement, will become dominant if

$$k_{12} < L_1/L_2 \qquad (6b)$$

The validity of these conditions was verified by a large number of numerical simulations in situations where uptake did not equal nutrient loss. In these simulations an equilibrium between the loss and uptake of nutrients was always established, indicating that the conditions presented predict the final outcome of the competition between the two species.

Our equations may be seen as an extension of the classical competition theory of De Wit (1960). If the relative nutrient requirements of the two competing species are equal to each other, conditions (6a) and (6b) are reduced to the analogous condition in classical theory. The relative nutrient requirement of *Molinia* is higher than that of *Erica* ($L_m/L_e = {\sim}3$). In this case, it is possible that *Molinia* wins with respect to the uptake of nutrients, but ultimately loses in the competition with *Erica* if k_{me} is still below the critical limit ($1 < k_{me} < 3$). When the relative competition coefficient exceeds the critical ratio between the relative nutrient requirements of the two species, *Molinia* will be able to replace *Erica* as the dominant species. It can therefore be concluded that, in order to predict the outcome of the competition between perennial plant populations, we not only need to know their competitive abilities with respect to each other, but must also take into account possible differences between the nutrient economy of the two species. It appears that the relative nutrient requirement is a biologically meaningful parameter for achieving this purpose.

V. Competitive Ability and Nutrient Supply

Before further conclusions can be drawn, the effects of the nutrient supply on the relative competition coefficient must be examined. In order to study the effects of an increased nutrient availability on the competition between *Erica* and *Molinia,* we carried out a competition experiment during one growing season in which we measured the relative competition coefficient with respect to the uptake of nutrients at different supplies of nitrogen and phosphorus (Berendse and Aerts, 1984). The relative competition coefficient of *Molinia* with respect to *Erica,* k_{me}, is close to unity under unfertilized conditions, whereas it clearly increases with increasing nitrogen or phosphorus supply (Fig. 5). From comparison with conditions (6), we can conclude that, under nutrient-poor conditions, the relative competition coefficient is still below the critical ratio of the relative nutrient requirements of the two species ($L_m/L_e = \sim 3$), so that *Erica* will be able to maintain itself as the dominant species. After an increase in the nutrient supply the relative competition coefficient exceeds this critical limit and *Molinia* will replace *Erica* as the dominant species. This mechanism has probably underlain the replacement of the former *Erica*-dominated communities by the grass *Molinia* that has taken place in most wet heathlands in The Netherlands during the past 30 years.

These results suggest that *Erica* is more successful in nutrient-poor environments because it is more economical of the nutrients that it has acquired. After an increase in the nutrient supply other plant properties become more favorable for the plant. *Molinia* is able to respond much more rapidly than *Erica* to an increase in the nutrient availability by investing more carbohydrates and nutrients in photosynthetic tissues.

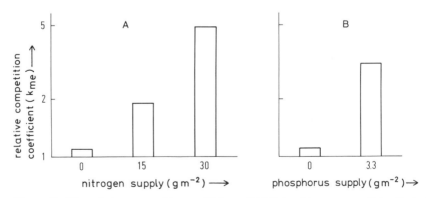

Figure 5   The relative competition coefficient of *Molinia* with respect to *Erica* (k_{me}) at different (A) nitrogen and (B) phosphorus supplies. See text for further explanation. After Berendse *et al.* (1987b).

This difference between the two species is crucial, as a plant needs a sufficiently large photosynthetic apparatus to convert rapidly an increased nutrient uptake into a higher biomass production. Another important difference between *Molinia* and *Erica* concerns the vertical distribution of leaf area in the two species. In the habitats where we carried out our study, *Erica* has about 80% of its total leaf area in the lower 10 cm and the remaining 20% in the layer between 10 and 20 cm. *Molinia* plants are much higher and have 35% of their total leaf area in the lower 10 cm, but 45% in the layer between 10 and 20 cm. This difference enables *Molinia* to overtop *Erica*, especially at raised nutrient levels. Naturally, competition for soil nutrients might be expected to be important in these environments. However, it must be remembered that root features, such as specific root length, root hair density, and degree of mycorrhizal infection, are not the only determinants of the competitive ability with respect to the absorption of nutrients: the flow of carbohydrates to the roots seems to be one of the major factors determining root growth (Brouwer, 1962a,b) and hence the plant's competitive ability to take up nutrients. Thus, even in nutrient-poor ecosystems, the investment in photosynthetic tissues and the vertical distribution of leaf area might be important factors determining a plant's competitive ability.

We also tried to analyze the effects of increased nutrient supply on the competitive ability of the grasses occurring in the hayfields that were studied. A pot experiment was carried out to study the effect of the nutrient supply on the competition between *Arrhenatherum elatius* and *Festuca rubra*. This experiment consisted of a factorial design including two factors: (1) a series on an extremely nutrient-poor sandy soil versus a series on the same soil receiving a nutrient solution at regular time intervals; (2) a treatment where root competition was excluded versus a treatment where both below- and above-ground competition were possible. The plants were clipped after 33, 54, 82, and 110 days. At each harvest the plants were clipped 5 cm above soil surface. In Table 2 the relative competition coefficients of *Arrhenatherum* with respect to *Festuca* at the first and the last harvest are summarized. During the first 33 days,

Table 2 The Relative Competition Coefficients (k_{af}) of *Arrhenatherum elatius* with Respect to *Festuca rubra* after 33 and 110 Days

	Unfertilized		Fertilized	
	33	110	33	110
Shoot competition	1.0	1.1	1.0	1.9
Shoot + root competition	1.6	1.9	1.7	3.2

only competition for soil factors appeared to have been important. The competitive advantage of *Arrhenatherum* during this period is probably caused by its higher initial growth rate correlated with its heavier seeds. At the last harvest above-ground competition also appears to determine dry weight production, especially under nutrient-rich conditions. In the experiment with the four hayfield species we measured a very clear difference between the vertical distribution of the leaf area of the two groups of species (Fig. 6). *Arrhenatherum* and *Alopecurus* show a rather homogeneous vertical distribution of photosynthetic area, whereas *Festuca* and *Anthoxanthum* have most of their leaf area in the lower 15 cm. So, after some time *Arrhenatherum* is able to overtop *Festuca*, which is reflected in the increased values of the relative competition coefficient in the fertilized treatments. These results suggest an important positive feedback between above-ground and below-ground competition.

In the competition between *Molinia* and *Erica* in the short term, the fast-growing species wins both under nutrient-poor and under nutrient-rich conditions. The major difference between unfertilized and fertilized series is that the relative competition coefficient shows an increasing deviation from unity with increasing nutrient supply. This phenomenon

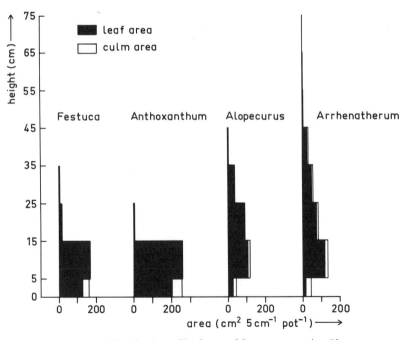

Figure 6 The vertical distribution of leaf area of four grass species. Plants were grown in pots under unfertilized conditions and harvested after 16 weeks. After Berendse and Elberse (1989).

was observed in the competition experiment with *Arrhenatherum* and *Festuca* and in many other short-term experiments as well (e.g., Van den Bergh, 1968; Berendse, 1982, 1983). Nevertheless, our analysis in which the losses of nutrients are taken into account explains why, in the long term, the fast-growing species wins under fertile conditions, whereas, in nutrient-poor environments, the slow-growing species is able to replace the other species if it has a lower relative nutrient requirement.

The question that arises now is, which plant properties determine the short-term competitive ability of the plant under the various environmental conditions and how are these plant features affected by an increasing nutrient supply and increasing biomass of the plant's neighbors? In Table 3 we give a by no means exhaustive list of features that may contribute to the competitive ability of a plant. We distinguished between plant features that affect the plant's ability to capture resources and features that determine the efficiency with which captured resources are converted into biomass production. Two essentially distinct groups of features that affect the rate of resource capture are the initial size of plant or seed and the plasticity of plant growth. The initial size of seed or plant determines to a large extent initial growth rates and the plant's

Table 3 Plant Features Contributing to the Competitive Ability of a Plant in Grassland and Heathland Communities

Features that affect the ability to capture resources
    Shoot : root ratio
    Leaf area ratio, specific leaf area
    Vertical distribution of leaf area
    Specific root length
    Root hair density and root hair length
    Active root surface : root weight ratio
    Degree of mycorrhizal infection
Features that affect the efficiency of converting captured resources into biomass production
    Allocation of carbohydrates and nutrients into chloroplasts
    Respiration costs of tissue synthesis: fraction of assimilated carbon dioxide that can be converted into growth
Initial size or weight
    Embryo and endosperm weight
    Timing of germination
    Phenology: response to temperature and day length during early spring
    Carbohydrates stored in roots or stubbles
    Photosynthetic area remaining after mowing
Plasticity of plant growth
    With respect to shoot : root ratio, leaf area ratio, specific root length, root hair density, etc.
    Response of root growth on localized nutrient supplies

ability to deplete soil resources in an early stage or to overtop its neighbors. For instance, the ability of a plant to respond rapidly to a locally increased availability of nutrients by a strong root proliferation at this place (Drew and Saker, 1978; De Jager, 1979) may be expected to contribute to its competitive ability as well. Most of the plant features listed in Table 3 (e.g., shoot : root ratio, leaf area ratio, root hair density) are known to be strongly affected by the supply of nutrients and the degree of shading (Brouwer, 1962a,b). So, the different parameters measuring the competitive ability of the plant should be expected to change with increasing nutrient supply. Grime (1979) suggested that species with higher potential growth rates generally show higher degrees of plasticity. Such behavior might contribute to the increasing deviation of the relative competition coefficient from unity with increasing nutrient supply. A further analysis of these effects is needed to obtain a sufficient insight into the interdependence between relative competition coefficient and nutrient supply.

VI. The Trade-Off between Different Adaptive Features

A matter that has attracted considerable attention during the last years is whether there is an evolutionary trade-off between plant features that are advantageous under nutrient-poor conditions and features that serve as adaptations to nutrient-rich environments (Grime, 1979; Chapin, 1980; Tilman, 1988). In this section, we first present a simple example where plant features that are advantageous under fertilized conditions are disadvantageous under unfertilized conditions and vice versa. Thereafter, we define a few new parameters that are used as a tool in a further, more detailed analysis of the consequences of some physiological and morphological plant traits.

In the competition experiment with *Festuca rubra* and *Arrhenatherum elatius* the unfertilized treatment was continued after the last harvest (at 110 days) for another 183 days. Although *Arrhenatherum* was winning in the first period of the experiment (particularly, in the fertilized treatment; cf. Table 2), in the second period *Festuca* replaced *Arrhenatherum* in the nutrient-poor series. The more homogeneous distribution of leaf area in *Arrhenatherum* (Fig. 6), which contributes to its competitive ability under nutrient-rich circumstances, has the negative consequence that this species loses more nitrogen by regular clipping than *Festuca* does. During the whole experimental period *Arrhenatherum* lost 64% of the nitrogen that it had absorbed by clipping and above-ground mortality, whereas *Festuca* lost 29% of the nitrogen that it had taken up. The hayfields in which we study the dynamics of these species are mown twice a year. So, losses by mowing may strongly affect the success of these species under field conditions. We may conclude that, in the competition between *Arrhenatherum* and *Festuca,* plant features that enable the population to succeed under fertilized conditions have a negative effect on the success of the population in the nutrient-poor series. Here there appears to be a simple trade-off between plant features that are favorable in nutrient-rich habitats and those that are favorable under less fertile conditions.

In order to analyze the adaptation of plant populations to habitats with different nutrient availabilities, it is appropriate to define a parameter that measures the efficiency with which the acquired nitrogen is used for carbon assimilation or dry matter production. Such a parameter should include two components: (1) the instantaneous rate of dry matter production per unit nitrogen in the plant, i.e., the nitrogen productivity; and (2) the period during which the acquired nitrogen can be used for carbon assimilation, i.e., the mean residence time of nitrogen in the plant. We proposed to define the nitrogen-use-efficiency (NUE) as the product of the nitrogen productivity (A) and the mean residence time ($1/L_n$; Berendse and Aerts, 1987):

$$NUE = A/L_n \tag{7}$$

The NUE defined in such a way measures the dry weight that can be produced per unit nitrogen that has been absorbed (e.g., g DW g^{-1} N). This definition has been proposed previously by other authors (Hirose, 1975; Boerner, 1984), but they did not distinguish between the two components, which is essential in our analysis. In order to take into account the interdependence between the carbon and nitrogen balances of the plant, we define a second parameter that measures the efficiency with which the assimilated carbon is used for the acquisition of nitrogen. The nitrogen-acquisition-efficiency (NAE) can be defined by an analogous formulation:

$$NAE = A'/L_c \tag{8}$$

where A' is the nitrogen absorption per unit of carbon or dry weight in the plant (g N g^{-1} DW day^{-1}) and $1/L_c$ is the mean residence time of carbon or dry matter (e.g., day). If we simplify the functioning of the plant to the extreme limit, we may state that the plant needs nitrogen for the assimilation of carbon and that it needs carbon for the assimilation of nitrogen. The two interdependent differential equations for the dynamics of the amounts of carbon (C) and nitrogen (N) in a single plant are given by:

$$dC/dt = AN - L_c C \tag{9a}$$

$$dN/dt = A'C - L_n N \tag{9b}$$

It can now be seen that, under steady-state conditions

$$\text{NUE} \times \text{NAE} = 1 \tag{10}$$

which implies an important constraint to the various combinations that are possible between values of NUE and NAE. Tilman (1988) stresses the importance of the trade-off between the allocation to above-ground plant parts and that to below-ground parts, which is one of the main components of the trade-off between NUE and NAE. As a matter of fact, there is an obvious allometric relationship between the allocation to above-ground and below-ground plant parts. Most plant species show a strong phenotypic response to an increase in nitrogen or water supply by an increased allocation to the above-ground plant parts (Brouwer, 1962a,b; Chapin, 1980). However, it is not clear whether plant species are generally adapted to nutrient-poor environments by a genetically determined higher allocation to roots. Boot (unpublished results) found a higher allocation to root biomass in *Molinia* than in *Erica*. In the pot experiment with different grass species of hayfields, we also found a higher relative allocation to the roots in the species that are characteristic of relatively nutrient-rich sites, whereas the phenotypic response upon fertilization showed the opposite trend (cf. Chapin, 1980).

Below we focus on the relationship between the two different components of the NUE: the nitrogen productivity and the mean residence time. It is not easy to measure the NUE of plant populations under field conditions, especially if root productivity and nitrogen losses by root mortality are included. In 1982 and 1983 we measured production and mortality of both above-ground and below-ground plant parts in populations of *Erica tetralix* and *Molinia caerulea* that occurred under similar environmental conditions (Berendse *et al.*, 1987a,b). Combining data of these two years and assuming that in both species 50% of the nitrogen in dying roots is withdrawn and redistributed, we calculated preliminary values of NUE, A, and $1/L_n$; these are summarized in Table 4. The large difference between the nitrogen productivity of *Erica* and *Molinia* is, to a

Table 4　Calculated Values of the Nitrogen Productivity (A), the Mean Residence Time ($1/L_n$), and the Nitrogen-Use-Efficiency (NUE) of *Erica tetralix* and *Molinia caerulea*[a]

	A	$1/L_n$	NUE
Erica tetralix	23.9	4.3	102.8
Molinia caerulea	94.2	1.4	131.9

After Berendse and Aerts (1987).

[a] Units of measurement: A: g dry matter g^{-1} N yr^{-1}; $1/L_n$: yr; NUE: g dry matter g^{-1} N.

large extent, canceled out by the difference between the mean residence times. This observation suggests that there is a trade-off between plant features that lead to a high value of A and features that lead to a high value of $1/L_n$. So, let us consider plant parameters that have opposite effects on nitrogen productivity and mean residence time. One of the main features in this context is the relative investment into the photosynthetic system. At the end of the growing season in 1982 48% of the nitrogen in *Molinia* plants had been invested in leaves and culms, whereas just 12% of the nitrogen in *Erica* plants was present in the leaves. Because of the different seasonal dynamics of these species, the differences over a whole year will be smaller, but it appears that there are quantitative differences between *Molinia* and *Erica* in the allocation of nitrogen and carbon into green plant parts. This difference between *Erica* and *Molinia* is mainly caused by the carbon and nitrogen that is needed in *Erica* for current year stem production and secondary stem growth and not by differences in allocation to roots. Because of the inherently low mortality of stems (Table 5), this investment of nitrogen into woody tissues increases the mean residence time of nitrogen in the plant, whereas investment of nitrogen in photosynthetic tissues decreases it because these tissues have a relatively short life span as compared to other plant parts.

Another difference between the two plant species that might affect both nitrogen productivity and mean residence time is the organochemical composition of plant tissues. The life span of *Erica* leaves is about four times as long as that of *Molinia* leaves (Table 5). The longer life span of *Erica* leaves and stems is possible because of their relatively high lignin content. The relatively high lignin concentrations in *Erica* have two consequences: (1) The carbon (and nitrogen) that is invested into lignin (and compounds that are connected with it) cannot be invested into the chloroplasts. So, higher lignin concentrations may have a negative effect on

Table 5 Life Span, Lignin Concentration, and Costs of Biosynthesis of Tissues in *Erica tetralix* and *Molinia caerulea*

	Erica	*Molinia*
Life span (yr)		
Leaves	1.3	0.35
Stems	5.7	—
Lignin concentration (%)		
Leaves	33	24
Stems	46	—
Costs of biosynthesis (g glucose/g dry matter)		
Shoot	1.80	1.41
Roots	1.69	1.40

the nitrogen productivity. (2) Lignin is expensive in terms of the quantity of substrate that is required for the supply of carbon skeletons and energy. For the biosynthesis of 1 g lignin 2.07 g glucose is required, whereas the costs involved with the synthesis of 1 g carbohydrates such as cellulose or pectin are just 1.17 g glucose (Penning de Vries *et al.*, 1974). Recently, Vertregt and Penning de Vries (1987) introduced a rapid method to determine biosynthesis costs. They presented a linear relationship between oxidation level of the carbon, which is reflected in the carbon content of the organic matter, and the quantity of glucose needed for the growth of one unit biomass. Using this relationship, we calculated weighted means of the costs of biosynthesis of 1 g above-ground biomass, consisting of different plants parts, and 1 g below-ground biomass (Table 5). *Erica* clearly shows a lower dry matter conversion efficiency than *Molinia* (56–59% and 71%, respectively). This difference may lead to values of the nitrogen productivity that are about 25% higher in *Molinia* than in *Erica*. Although this difference may significantly affect rates of dry matter production, the difference in investment in photosynthetic tissues seems to be quantitatively more significant.

Plants that are real generalists with respect to nutrient availability should combine a long mean residence time of nitrogen in the plant with a high potential growth rate. However, in many cases, these two sets of properties cannot be combined by the plant because of a number of morphological or physiological constraints. Many plant features that reduce nitrogen losses (e.g., low leaf/stem ratio, long life span of leaves, synthesis of defensive compounds) may lead at the same time to lower growth rates per unit nitrogen in the plant. Nitrogen and carbon allocated to stems, structural elements in leaves, or secondary compounds such as alkaloids or cyanogenic glycosides cannot be invested into the photosynthetic apparatus and high concentrations of lignins or phenolic compounds that enable a long life span of plant tissues lead inevitably to increased costs of biosynthesis.

The picture that emerges from the theory and the data that we have presented is that plant properties that enable the plant to conserve the nutrients that it has acquired are important adaptive features in nutrient-poor environments. In the two examples that we analyzed these features had quantitatively significant, negative effects on maximum growth rate or competitive ability under fertile conditions. It seems that low potential growth rates do not have any advantage in themselves, but are just a negative side effect of features that enable the plant to survive in nutrient-poor environments. This picture fits very well into the conceptual framework of Grime (1979) and Chapin (1980). It is striking that, in the two examples that we dealt with, completely different groups of plant characteristics were involved. In both cases, however, our theory might contribute to a better understanding of the adaptation of plant populations to habitats with different nutrient availabilities.

Acknowledgments

We thank J. P. van den Bergh, W. G. Braakhekke, T. J. de Jong, P. G. L. Klinkhamer, J. Pastor, D. Tilman, and two anonymous referees for their stimulating comments on the manuscript. Berendse thanks R. Aerts, R. Boot, H. Lambers, and M. J. A. Werger for providing a stimulating environment in which some of the ideas presented here have been developed.

References

Aerts, R., and Berendse, F. (1988). The effect of increased nutrient availability on vegetation dynamics in wet heathlands. *Vegetatio* **76**, 63–69.

Berendse, F. (1981). "Competition and Equilibrium in Grassland Communities," thesis. University of Utrecht, The Netherlands.

Berendse, F. (1982). Competition between plant populations with different rooting depths. III. Field experiments. *Oecologia* **53**, 50–55.

Berendse, F. (1983). Interspecific competition and niche differentiation between *Anthoxanthum odoratum* and *Plantago lanceolata* in a natural hayfield. *J. Ecol.* **71**, 379–390.

Berendse, F. (1985). The effect of grazing on the outcome of competition between plant populations with different nutrient requirements. *Oikos* **44**, 35–39.

Berendse, F., and Aerts, R. (1984). Competition between *Erica tetralix* L. and *Molinia caerulea* (L.) Moench as affected by the availability of nutrients. *Acta Oecol./Oecol. Plant.* **5**, 3–14.

Berendse, F., and Aerts, R. (1987). Nitrogen-use-efficiency: A biologically meaningful definition? *Funct. Ecol.* **1**, 293–296.

Berendse, F., and Elberse, W. Th. (1989). Competition and nutrient losses from the plant. *In* "Causes and consequences of variation in growth rate and productivity of higher plants" (H. Lambers *et al.*, eds.). SPB Academic Publishing Co., The Hague (in press).

Berendse, F., Beltman, B., Bobbink, R., Kwant, R., and Schmitz, M. (1987a). Primary production and nutrient availability in wet heathland ecosystems. *Acta Oecol./Oecol. Plant.* **8**, 265–279.

Berendse, F., Oudhof, H., and Bol, J. (1987b). A comparative study on nutrient cycling in wet heathland ecosystems. I. Litter production and nutrient losses from the plant. *Oecologia* **74**, 174–184.

Boerner, E. J. (1984). Foliar nutrient dynamics and nutrient use efficiency of four deciduous tree species in relation to site fertility. *J. Appl. Ecol.* **21**, 1029–1040.

Brouwer, R. (1962a). Distribution of dry matter in the plant. *Neth. J. Agric. Sci.* **10**, 361–376.

Brouwer, R. (1962b). Nutritive influences on the distribution of dry matter in the plant. *Neth. J. Agric. Sci.* **10**, 399–408.

Campbell, C. A., Paul, E. A., Rennie, D. A., and McCallum, K. J. (1967). Applicability of carbon-dating method of analysis to soil humus studies. *Soil Sci.* **104**, 217–224.

Chapin, F. S., III (1980). The mineral nutrition of wild plants. *Annu. Rev. Ecol. Syst.* **11**, 233–260.

De Jager, A. (1979). Localized stimulation of root growth and phosphate uptake in *Zea mais* L. resulting from restricted phosphate supply. *In* "The Soil Root Interface" (J. L. Harley and R. S. Russell, eds.), pp. 391–403. Academic Press, London.

De Wit, C. T. (1960). On competition. *Agric. Res. Rep.* **66.8**, 1–82.

De Wit, C. T. (1961). Space relationships within populations of one or more species. *Symp. Soc. Exp. Biol.* **15**, 314–329.

Drew, M. C., and Saker, L. R. (1978). Nutrient supply and the growth of the seminal root system in barley. III. Compensatory increase in growth of lateral roots and in rates of phosphate uptake in response to a localized supply of phosphate. *J. Exp. Bot.* **29**, 435–451.

Elberse, W. Th., Van den Bergh, J. P., and Dirven, J. G. P. (1983). Effects of use and mineral supply on the botanical composition and yield of old grassland on heavy-clay soil. *Neth. J. Agric. Sci.* **31**, 63–88.

Grime, J. P. (1979). "Plant Strategies and Vegetation Processes." Wiley, Chichester, England.

Hirose, T. (1975). Relations between turnover rate, resource utility and structure of some plant populations: A study in the matter budgets. *J. Fac. Sci.* **11**, 355–407.

Hui-jun, J., and Ingestad, T. (1984). Nutrient requirements and stress response of *Populus simonii* and *Paulownia tomentosa*. *Physiol. Plant.* **62**, 117–124.

Ingestad, T. (1979). Nitrogen stress in Birch seedlings. II. N, P, K and Mg nutrition. *Physiol. Plant.* **45**, 149–157.

Ingestad, T., and Kähr, M. (1985). Nutrition and growth of coniferous seedlings at varied relative nitrogen addition rate. *Physiol. Plant.* **65**, 109–116.

Kruijne, A. A., De Vries, D. M., and Mooi, H. (1967). Bijdrage tot de oecologie van de Nederlandse graslandplanten. *Agric. Res. Rep.* **696**, 1–65.

Mennema, J., Quené-Boterenbrood, A. J., and Plate, C. L. (1980). "Atlas van de Nederlandse Flora. I. Uitgestorven en Zeer Zeldzame Planten." Kosmos, Amsterdam.

Pastor, J., Aber, J. D., McClaugherty, C. A., and Melillo, J. M. (1984). Aboveground production and N and P cycling along a nitrogen mineralization gradient on Blackhawk Island, Wisconsin. *Ecology* **65**, 256–268.

Penning de Vries, F. W. T., Brunsting, A. H. M., and Van Laar, H. H. (1974). Products, requirements and efficiency of biosynthesis: A quantitative approach. *J. Theor. Biol.* **45**, 399.

Spitters, C. J. T. (1979). Competition and its consequences for selection in barley breeding. *Agric. Res. Rep.* **893**, 1–268.

Tilman, D. (1982). "Resource Competition and Community Structure." Princeton Univ. Press, Princeton, New Jersey.

Tilman, D. (1984). Plant dominance along an experimental nutrient gradient. *Ecology* **65**, 1445–1453.

Tilman, D. (1988). "Plant Strategies and the Structure and Dynamics of Plant Communities." Princeton Univ. Press, Princeton, New Jersey.

van den Bergh, J. P. (1968). An analysis of yields of grasses in mixed and pure stands. *Agric. Res. Rep.* **714**, 1–71.

van den Bergh, J. P. (1979). Changes in the composition of mixed populations of grassland species. *In* "The Study of Vegetation" (M. J. A. Werger, ed.), pp. 59–80. Junk, The Hague, The Netherlands.

Vermeer, J. G. (1985). The effect of nutrient addition and lowering of the water table on shoot biomass and species composition of a wet grassland community (*Cirsio-Molinietum* Siss. et de Vries, 1942). *Acta Oecol./Oecol. Plant.* **7**, 145–155.

Vermeer, J. G., and Berendse, F. (1983). The relationship between nutrient availability, shoot biomass and species richness in grassland and wetland communities. *Vegetatio* **53**, 121–126.

Vertregt, N., and Penning de Vries, F. W. T. (1987). A rapid method for determining the efficiency of biosynthesis of plant biomass. *J. Theor. Biol.* **128**, 109–119.

7

Mechanisms of Plant Competition for Nutrients: The Elements of a Predictive Theory of Competition

David Tilman

I. Introduction

John Harper, in reviewing competition studies, asserted that, although there was an extensive literature demonstrating the existence of competition and the effects of various environmental variables on competition, "it is very doubtful whether such experiments have contributed signifi-

Perspectives on Plant Competition. Copyright © 1990 by Academic Press, Inc. All rights of reproduction in any form reserved.

cantly either to understanding the mechanism of 'competition' or to generalizing about its effects" (Harper, 1977, p. 369). The main cause of this failing, I believe, has been the paucity of experimental and theoretical studies that were designed explicitly to study the underlying mechanisms of competition.

Most studies have focused on the phenomenon of competition. Field experiments have tested for the existence or the strength of competition in various habitats or have determined if a species is a superior or inferior competitor in a particular habitat. Similarly, most theories of plant competition, such as those based on the Lotka–Volterra model, the de Wit approach, or other density-based models, have amounted to little more than elaborations of the definition of competition as a process in which an increase in the density of one species leads to a decrease in the density, growth rate, or yield of another. These phenomenological theories have little predictive power (Tilman, 1987a). Within the confines of density-based theories, it is impossible to use information collected on monocultures of two species to make an a priori prediction of the outcome of their interactions when growing together. Rather, the parameters that describe the effects of interspecific competition can only be determined by fitting the model to *the observed results of a competition experiment*. Thus, these density-based models, which are still a mainstay of plant ecology, are useful mainly as a posteriori descriptors that demonstrate the existence of competition. Although this was an important task, it is now abundantly clear that interspecific competition is a major force, but by no means the only force, in habitats ranging from natural communities (e.g., Schoener, 1983; Connell, 1983; Tilman, 1987a) to highly disturbed agricultural ecosystems (e.g., Radosevich and Holt, 1984).

Although density-based models may describe the phenomenon of competition, and thus may seem general and simple, their simplicity disappears as soon as they are applied to multispecies communities. If the density-based phenomenological approach is to be used to predict the dynamics of multispecies competition, it is necessary both to study each species growing by itself (to determine its carrying capacity and maximal growth rate) and to study all possible pairs of all species on all trophic levels (e.g., Tilman, 1977, 1982, 1987a; Bender *et al.*, 1984; Keddy, this volume). If a community contained y species, this would require $(y^2 + y)/2$ experiments, with each experimental study being sufficiently well-replicated so as to estimate parameters accurately. Thus, 210 experiments would be needed for a community containing 20 species, and 820 would be needed for a 40-species community. The result of such an immense effort would be a model that was potentially capable of predicting the dynamics of multispecies interactions, but only in the habitat in which the parameters were determined.

I suggest that there is a simpler, more general, and potentially much more predictive approach that can be taken, and that is to study the mechanisms of interspecific interaction (e.g., Rapport, 1971; MacArthur, 1972; Schoener, 1971, 1986; Tilman, 1976, 1982, 1987a, 1988; Pulliam, 1985, 1986; Werner, 1984). Most interspecific interactions are consumer–resource interactions. It is disheartening that competition, an interaction in which *several species* consume the same resource or resources, is discussed by many ecologists as being conceptually different from herbivory or predation, interactions in which *one species* consumes one or more resources. A mechanistic approach can be markedly simpler than a phenomenological approach because there are many fewer consumer–resource linkages, and thus many fewer parameters to estimate, than there are pairs of species. For instance, for the old-field plant communities at Cedar Creek Natural History Area, nutrient addition experiments (Tilman, 1988) have shown that nitrogen is the main limiting soil resource. If a field contained 20 species, only 20 experiments (one set of nitrogen-limited monocultures per species) presumably would be required to predict, a priori, the outcome of multispecies competition for nitrogen.

Even with several limiting resources, a consumer–resource approach would be much simpler than a phenomenological approach. First, few plant habitats have been found to have more than three or four limiting resources, even though the habitats may contain several hundred species (Grubb, 1977). Second, organisms face tradeoffs, such as between their ability to acquire and use one resource versus their ability to acquire and use another, or between resource use at one temperature versus resource use at another temperature. Such tradeoffs would constrain the values of parameters and cause parameters to be correlated. For instance, almost all the freshwater diatoms studied have an inverse correlation between their competitive ability for phosphate (as determined by R^*; see Tilman, 1982, and Section III) and that for silicate (Tilman, 1982). This inverse correlation, which explains much of their distribution along environmental gradients (Tilman *et al.,* 1982), is presumably the result of an unavoidable tradeoff caused by the allocation of energy and materials to the acquisition and efficient utilization of phosphate versus to the acquisition and utilization of silicate. The proportion allocated to each function would determine the phosphate to silicate ratio at which a species would be a superior competitor. Terrestrial plants face similar tradeoffs between the ability to compete for a limiting soil resource versus light (e.g., Mooney, 1972; Tilman, 1988), between the ability to compete for a resource versus resistance to herbivory (Gulmon and Mooney, 1986; Bazzaz *et al.*, 1987), and so on. These tradeoffs, made unavoidable because they are based on the pattern of allocation, may greatly simplify mechanistic models by reducing the range of potential parameters and by causing parameters to be correlated. Major, broad, repeatable patterns in community composition and diversity, and in the morphology, physiology, and life history of species, are likely caused by the commonality of unavoidable tradeoffs and by patterns of resource limitation and other environmental constraints (Tilman, 1982, 1988, 1989). The ecology of the future, I predict, will consist largely of experimental and theoretical studies that explicitly consider environmental constraints and organismal tradeoffs.

II. Plant Competition

There are two major mechanisms of plant competition: resource competition and interference (mainly allelopathic?) competition. Resource competition can be further subdivided into competition for soil resources and competition for light. If plants are competing for limiting soil resources, theory can be developed that can potentially predict the dynamics and outcome of their interactions. This theory would use information on the resource dependence of the growth and reproduction of each species, the dynamics of resource supply, and the nutrient consumption rates of the species, much as was done by Tilman (1976, 1982) and Sommer (1985, this volume). Such theory is mechanistic because it directly includes the intermediate compound, the resource, which is the entity by which one individual plant affects another plant. Comparable approaches can be developed for allelopathy or any other mechanism of interspecific interaction. A model of the mechanisms of allelopathy would have to include the rate at which individuals made the allelopathic compound; the effects of the compound on growth, survival, and/or reproduction; and the rate of loss or decay of the compound. Any mechanistic theory should directly include the entity or means whereby an individual plant influences the survival, growth, and/or reproduction of plants of its own and other species.

If ecology is to achieve the ability to predict the dynamics and outcome of interactions in multispecies communities, which we must achieve if we are to manage the Earth's resources and preserve its ever dwindling biotic diversity, our approach must be more mechanistic than at present. However, the more complex a model becomes, the more difficult it is to estimate its parameters and the more likely are its predictions to be incorrect because of the compounding of sampling errors for its parameters. The optimal level of mechanistic detail needed to address a given question can only be determined empirically. The divergences of opinion seen among chapters in this book (cf. chapters by Goldberg, Keddy, Silander and Pacala, and Firbank and Watkinson in this volume with this chapter) illustrate that much more research is needed to determine this level. Perhaps the ideas presented in this chapter can aid us in this effort.

There are numerous reasons to assert that plant competition may be complex. Plants are morphologically, physiologically, and genetically complex (e.g., Givnish, 1986, and papers therein). Plant resources are spatially and temporally patchy. A plant's ability to acquire resources, and the amounts needed for its survival, growth, and reproduction, are influenced by mutualists, pathogens, predators, and herbivores (Louda *et al.*, this volume). The dynamics of resource supply, such as the rate of recycling of nitrogen, are influenced by the traits of the competing plants, thus introducing a potentially confounding feedback loop (e.g., Pastor *et al.*, 1984; Vitousek *et al.*, 1987). Given such complexity, how should we proceed in our attempts to understand plant competition? Must a theory of plant competition explicitly include all these factors? Or, can simpler models be useful?

All models are abstractions. They represent an attempt to make simplifying assumptions and then determine the logical implications of those simplifying assumptions. Simpler models are often powerful not because they explicitly include all the relevant parameters and processes, but because many processes and parameters can be summarized (or abstracted) in a few parameters, if they are correctly estimated. Schaffer has called this the process of ecological abstraction:

> Accordingly, when the empiricist fits data to equations describing the growth rates of particular species, he has, in a sense, 'abstracted' these species from a more complex matrix of interactions in which they are embedded. Nevertheless, because the species studied, as opposed to the variables in the abstracted equations, continue to interact with the remaining, unspecified components of the ecosystem, the parameter values obtained perforce reflect, in part, the species and interactions omitted from the model (1981, p. 383).

A model explicitly includes some variables of great interest, and it abstracts or summarizes all other variables in the ways that they influence the model's parameters. Models are developed so that the parameters believed to be of greatest importance are explicitly included. Other factors are included, indirectly, to the extent that the parameters of the model can abstract them. This raises another potential problem with density-based phenomenological models of competition. Their failure to include directly even the simplest mechanisms of resource competition means that they may have great difficulty abstracting any of the higher order complexity of resource competition.

III. Mechanisms of Nutrient Competition

Let us now consider five models of nutrient competition that range from a highly abstracted model to more realistic models of the mechanisms of plant nutrient competition. I restrict this to nutrient competition because models of nutrient competition can be analytically tractable, thus allowing direct comparisons of the predictions of different models. In contrast, the mathematical complexity caused by including a vertical light gradient and individual plant heights in models of plant competition for light (e.g., Givnish, 1982; Tilman, 1988) means that such models are only soluble analytically in special cases.

By comparing these five models, we can see how various plant traits determine nutrient competitive ability, as well as how simple models can abstract more complex ones. Moreover, these models suggest that a single, empirically observable number, R^*, may integrate the total effect of all plant traits on nutrient competitive ability. R^* may provide a simple but general and powerful way to predict the outcome of interspecific competition for nutrients.

A. Nutrient-Dependent Growth and Competition: Model 1

Perhaps the simplest mechanistic model of competition for a limiting soil nutrient is a model that uses the total, integrated effect of the limiting nutrient on plant population dynamics, and the effect of plant growth on the availability of the nutrient. Just such a model (Tilman, 1976) correctly predicted the outcome of plant competition for phosphate and silicate. The plants, though, were freshwater algae, and many ecologists readily accepted that such a simple model, because it was physiologically "realistic" for algae, could correctly predict the outcome of algal competition, but doubted that it could predict the outcome of competition among morphologically more complex organisms, such as vascular plants. However, a second model, which was physiologically more realistic, made less accurate predictions, perhaps because, with more parameters, there was more total error in their estimation (Tilman, 1977).

In its most general form, this simple model states that the per unit biomass rate of change of a population (which is $dB/dt \times 1/B$, or dB/Bdt, and is often called the relative growth rate, or RGR) depends on the difference between its resource-dependent net growth function, $f(R)$, and its loss rate, m. Here m is assumed to be both resource and density independent. Any resource dependence of loss is included in $f(R)$, because $f(R)$ gives the net effect of resources on the relative growth rate. There is an equation for each species, i, that states:

$$\text{rate of biomass change} = \text{growth} - \text{loss}$$

or

$$dB_i/B_i dt = f_i(R) - m_i \tag{1}$$

The dynamics of the growth-limiting resource, R, depend on the difference between the resource supply function, $y(R)$, and resource consumption summed over all species:

rate of resource change = supply rate − sum of consumption rates

or
$$dR/dt = y(R) - \sum_{i=1}^{n} [Q_i B_i f_i(R)] \qquad (2)$$

where Q_i is the nutrient content per unit biomass of species i and n is the total number of consumer species. The consumption expression thus multiplies the amount of new biomass produced during an instant, $B_i f_i(R)$, by the nutrient content per unit biomass of that species, Q_i, to obtain the total consumption rate per species. This is then summed over all species to give the total rate of nutrient consumption.

This general model may be solved to determine which species should persist and which should be driven to competitive exclusion once population and resource dynamics reach equilibrium (O'Brien, 1974; Tilman, 1976, 1977; Hsu et al., 1977). When this is done, it is found that the critical parameter is R^*. Each species has its own R^*. R^* is the level to which the concentration of the available form of the limiting resource is reduced by a monoculture of a species once that monoculture has reached equilibrium, i.e., once it has attained its carrying capacity. Expressed another way, R^* is the resource concentration at which the growth rate of a species equals its loss rate and the uptake rate of the species equals the rate of nutrient supply to the habitat. Thus, R^* is the concentration of available resource that a species requires to survive in a habitat. If the concentration were greater than R^*, the species' population size would increase. If it were lower, population size would decrease. R^* also measures the effect of a species on the limiting resource, and thus on its competitors. The lower the R^* of a species, the better is its competitive ability for the limiting resource. If all species are limited by the same nutrient, the species with the lowest R^* is predicted, at equilibrium, to displace all competitors (O'Brien, 1974; Tilman, 1976, 1977; Hsu et al., 1977). From Eq. (1), it can be calculated that R_i^* is

$$R_i^* = f_i^{-1}(m_i) \qquad (3)$$

where f^{-1} is the inverse function of f.

To make this more concrete, let us consider a particular model. Experimental studies have shown that the resource-dependent growth function, $f_i(R)$, is often a saturating function that monotonically approaches a maximal value as R increases. A common form for the growth function is

$$f_i(R) = r_i R/(R + K_i) \qquad (4)$$

which is the Monod (1950) model. In this model, r_i is the maximal, resource-saturated rate of growth per unit biomass of species i (i.e., RGR_{max}) and K_i is the resource concentration at which species i attains a per unit biomass growth rate equal to half of its maximal growth rate (Table 1). Substituting this for $f(R)$ gives

Model 1: $$R_i^* = m_i K_i / (r_i - m_i) \qquad (5)$$

This equation provides some significant insights into the process of competition for a limiting nutrient. First, it illustrates that the competitive ability of a species depends on m, i.e., on herbivory, disease, and other sources of loss (O'Brien, 1974; Louda *et al.*, this volume). When m is about equal to r, a decrease in the loss rate causes a large increase in competitive ability (i.e., a large decrease in R^*), but m has much less an effect on competitive ability when m is much less than r. Similarly, when r is about equal to m (and r must be greater than m for a plant to survive), small increases in r cause large increases in competitive ability. Because R^* scales linearly with K, a decrease in the half saturation constant leads to a comparable decrease in R^* and increase in competitive ability. The values of r, m, and K that lead to optimal nutrient competitive ability depend on the tradeoffs among these traits, and will be discussed later.

Table 1   Definitions of Variables Used in the Equations for R^* for Models 1–5[a]

Model 1
  r: maximal growth rate (RGR_{max})
  K: half-saturation constant for nutrient-limited *growth*
  m: loss rate (all causes)
Model 2
  r: maximal growth rate (RGR_{max})
  h: the minimal tissue nutrient concentration required for plant survival
  k: the half-saturation constant for Michaelis–Menton nutrient uptake
  m: the rate of loss of plant tissues (all causes; time^{-1})
  v: the maximal rate of nutrient uptake per unit plant biomass
Model 3
  The same parameters as for Model 2, and
  b: the proportion of plant biomass in root
Model 4
  The same parameters as for Model 2, except that m is deleted, and
  s: the rate at which plant tissues are lost through shedding of senescent parts
  c: the total loss rate for all sources of loss other than those included in s (i.e., to plant consumers and to death); thus $s + c$ of Model 4 is similar to m of Models 2 and 3
  q: the proportion of plant tissue nutrients that are *lost from the plant* when tissues are shed
Model 5
  The same parameters as for Model 4, except q is deleted, and
  M: the concentration of nutrient in senescent tissues when they are shed

[a] See Appendix for details of Models 2–5.

Model 1 is simple. It could be argued that it omits essential elements of plant biology that are directly related to competitive ability. For instance, the model includes neither nutrient-dependent uptake rates or growth rates that depend on tissue nutrient concentration, nor plant resorption of nutrients before senescent tissues are shed, nor any distinction between roots and shoots. Such traits have been commonly observed in plants that dominate nutrient-poor habitats (Chapin, 1980), and are thought to be important determinants of nutrient competitive ability (Berendse and Elberse, this volume). Does their absence from Model 1 mean it is incapable of predicting the outcome of interspecific competition for nutrients?

To explore this question, let us consider several more complex models of plant nutrient competition, and compare them with the simple model presented above. The mathematical details of these models and further discussion of the models are confined to the Appendix. Only their predictions for the equilibrial outcomes of competition are presented in the main text.

B. More Complex Models of Nutrient Competition: Models 2–5

Consider Model 2, which explicitly includes Michaelis–Menten resource-dependent nutrient uptake, variable tissue nutrient concentration, and growth that depends on tissue nutrient concentration (see Appendix). Its parameters are defined in Table 1. When this model is solved, it is found that the competitive dominant, at equilibrium, is the species with the lowest R^*, where R^* is

Model 2: $$R^* = rhkm/[v(r - m) - rhm] \tag{6}$$

[All the variables in Eq. (6) refer to traits of one species or to a cohort of genetically identical individuals. Thus, R^*, and all the variables, could be subscripted with an i, for species i. I have not done this to increase readability.] As before, R^* depends on all the model parameters (see Appendix and Table 1). R^* directly incorporates the response of a species to resource levels (h and r are parameters describing how growth depends on tissue nutrient levels), the effects of a species on the limiting resource (v and k are nutrient uptake parameters), and the effect of various sources of loss, herbivory, and mortality (m) on competitive ability. Competitive ability is increased (i.e., R^* is decreased) by traits that increase r and v and by traits that decrease m, h, and k.

Model 2 may be further modified. For instance, leaf and root biomass could be explicitly included. This would allow for the distinctly different functional roles of roots and leaves. When this is done, and the model is solved to determine the equilibrium outcome of competition for a single limiting resource, the resulting $R*$ for this model, Model 3 (see Appendix and Table 1), is

Model 3: $R* = rhkm/[vb(r - m - rb) - rhm]$ (7)

Here, b is the proportion of plant biomass in root (below-ground biomass). The remainder of plant biomass is assumed to be leaf, but this assumption could easily be modified. For ease of comparison with Model 2, the maximal uptake rate, v, is still expressed as uptake rate per unit of total plant biomass. $R*$ again depends on all of the model parameters, and the formula for $R*$ explicitly incorporates the effects of these parameters on predicted competitive ability. Comparison of Eqs. (6) and (7) illustrates how Model 2, which did not explicitly include root and leaf biomass, is a simplification of Model 3. $R*$ has a minimum value for an intermediate value of b (root allocation). This optimal level of allocation to root depends on the other plant traits.

Model 2 may also be modified by adding (1) nutrient resorption from tissues that are being shed, (2) species-specific rates of tissue shedding (i.e., different degrees of root longevity and of leaf evergreenness), and (3) nutrient resupply that depends on the quantity of the litter that a plant produces. The resulting model, Model 4 (see Appendix and Table 1), has an $R*$ of

Model 4: $R* = rhk(c + sq)/[v(r - c - s) - rh(c + sq)]$ (8)

Here, q is the proportion of plant nutrients lost when senescent tissues are shed, s is the rate at which they are shed, and c is the rate of loss and death to all other causes (herbivory, disease, etc.). Note that tissue loss to senescence is only harmful if the tissues contain nutrients, and the harm increases with the amount of nutrient lost. Tissue loss to herbivores would be even more harmful than indicated in Eq. (8) if herbivores chose the most nutrient-rich plant tissues, which is often the case (Louda *et al.*, this volume).

Model 4 assumes that a plant removes a fixed proportion of the limiting nutrient from its senescent tissues before they are shed as litter. An alternative formulation, slightly more complex mathematically, assumes that there is a particular level to which a plant reduces tissue nutrient concentration before tissues are shed. This formulation, called Model 5 (see Appendix and Table 1), gives a value for $R*$ of

Model 5: $R* = rhk[c + sM(r - c - s)/rh]/$

$$\{v(r - c - s) - rh[c + sM(r - c - s)/rh]\}   (9)$$

Although this equation may seem cumbersome, inspection of its terms illustrates that is closely related to the equations for $R*$ of Models 2, 3, and 4. Note that the major difference between this equation and Eq. (8) is that the q of Eq. (8) is replaced by $M(r - c - s)/rh$. This latter term is just the proportion of plant nutrient lost when tissues are shed. Equation (9) further illustrates the quantitative relationship between plant traits and predicted competitive ability. It shows, as did the preceding equations, that a single, empirically observable number, R^*, can summarize the effects of plant traits on competitive ability.

IV. Plant Traits and Nutrient Competitive Ability

These five different models of plant nutrient competition make qualitatively similar predictions as to the effects of various plant traits on nutrient competitive ability. Consider the equations for R^* derived from the models. Because a species becomes a better nutrient competitor by having a lower R^*, these equations can be used to determine how plant traits affect competitive ability. As already discussed, the traits included in the equation for R^* should be interdependent because of allocation-based tradeoffs. For a plant to change one trait in a way that increases its competitive ability, it must also change some other trait in a way that harms competitive ability. A thorough analysis of the suite of plant traits that maximizes nutrient competitive ability depends on the explicit tradeoffs between the various plant traits. These tradeoffs are not yet well quantified. They, themselves, should be predictable if explored in a mechanistic manner using theory that explicitly includes the underlying allocation processes (Tilman, 1988). Although this means that it is not yet possible to make rigorous analytical predictions of optimal plant traits, the equations for R^*, and the realization that there are tradeoffs, can provide some qualitative insights into the plant traits that are likely to lead to superior nutrient competitive ability. For the discussion below, let us consider Eq. (8), which gives R^* for Model 4. The other models lead to similar predictions.

1. *v, the maximum rate of nutrient uptake per unit plant biomass.* For values of v close to the lowest value that just allows a species to survive in a habitat, an increase in v leads to a large decrease in R^*. However, for larger values of v, changes in v have almost no effect on R^*. Thus, assuming that higher maximal rates of nutrient uptake have a cost, plants that are superior nutrient competitors should have relatively low maximal rates of nutrient uptake, rates that are just a few times greater than the minimal rate required to survive in the habitat. Chapin (1980) reviewed various physiological and morphological traits of plants in relation to the nutrient status of the habitats in which they were most abun-

dant in nature. He found that root absorption capacity (his V_{max}, my v) was usually higher in species from nutrient-rich habitats than in those from infertile habitats, and said that "plants have not adapted to nutrient stress through the evolution of an enhanced capacity to extract minerals from soil" (Chapin, 1980, p. 240).

2. *r, the maximal rate of vegetative growth (RGR_{max}).* For a species to survive in a habitat, it has to have an r that is at least greater than its total rate of loss from the shedding of senescent tissues, herbivores, pathogens, death, etc. These are summarized in $c + s$. For values of r that are just slightly greater than the total loss rate ($c + s$), slight increases in r lead to large decreases in R^*, and thus to large increases in competitive ability. However, once the maximal growth rate, r, is more than about 2 or 3 times the total loss rate, R^* is fairly insensitive to further increases in r. Thus, the maximum growth rate of a species, once it is sufficiently large to assure that a species can survive in a habitat in the absence of interspecific competition, is not a major determinant of competitive ability. Moreover, increases in r have definite costs. An increase in r requires decreased allocation to all structures other than leaves and to all functions other than photosynthesis (Monsi, 1968; Tilman, 1988). The increase in nutrient competitive ability gained from an increase in r, once r is much greater than the total loss rate, is unlikely to compensate for the decrease in competitive ability caused by lower root biomass or less efficient nutrient use. Hence, plants that are superior nutrient competitors should have low maximal growth rates. Chapin (1980, p. 244) found that "the predominance of ecotypes and species with inherently low relative growth rates in infertile habitats has been noted in graminoids, forbs, and woody species." If it is assumed that the plants that dominate nutrient-poor habitats are superior nutrient competitors, then the data reviewed by Chapin are consistent with the theoretical predictions made above. However, the usual explanations for the prevalence of low-growth-rate plants in nutrient-poor habitats (see Chapin, 1980) differ from the explanation suggested above. I suggest that there is no direct advantage associated with a lower maximal growth rate, but rather that a lower maximal growth rate is an unavoidable cost of allocation to other traits (such as high root biomass, herbivory defenses, leaf and root longevity, efficient nutrient utilization) that cause a net increase in nutrient competitive ability.

3. *s, the rate of loss via senescence and c, the rate of loss to other causes.* When the total loss rate, $c + s$, is about equal to the maximal growth rate, a decrease in either s or c causes a much more rapid than linear decrease in R^*. Slight decreases in the rate at which plant tissues are shed, such as by slight increases in leaf and root longevity, cause large increases in nutrient competitive ability. Similarly, slight decreases in the rate of herbivory

cause large increases in competitive ability. Thus, plants gain competitive ability for nutrients by increasing root and leaf longevity and by minimizing rates of herbivory, parasitism, and disease. However, the magnitude of the advantage associated with these traits decreases as $c + s$ becomes much less than r. Although longevity of roots and leaves and resistance to herbivory are predicted for plants that are good nutrient competitors, the magnitude of these traits must depend on the costs of these traits versus their benefits. The prevalence of evergreens in infertile habitats (Chapin, 1980), and the high proportional allocation to defense in plants common in infertile habitats (Coley *et al.*, 1985; Coley, 1987) suggest that the costs of tissue longevity and of herbivore defense are low compared to their benefits in infertile habitats. Indeed, the defensive compounds of plants from infertile habitats have low costs, once cost is amortized over the life of a leaf (Coley *et al.*, 1985).

4. *h, the minimal tissue nutrient content.* The minimal tissue nutrient content, h, is a measure of the efficiency with which a species can grow with low tissue nutrient levels. Lower values of h always lead to increased competitive ability, but the magnitude of the advantage decreases as h becomes smaller. Compared to plants of fertile habitats, in which light is likely to be limiting, plants of low-nutrient habitats should have lower minimal tissue nutrient concentrations. Chapin (1980) did not find such a pattern when comparing the results of short-term nutrient-limited growth experiments, but did find it when comparing tissue nutrient concentrations in plants growing in undisturbed natural habitats.

5. *q, the proportion of tissue nutrient lost when tissues are shed.* The equation for R^* from Model 4 predicts that a decrease in the proportion of tissue nutrient lost when senescent tissues are shed should lead to an increase in nutrient competitive ability. Thus, efficient nutrient retranslocation from senescent leaves should be favored in low-nutrient habitats, though the extent of such nutrient conservation will depend on both this benefit and its costs. Interestingly, Eq. (8) suggests that the actual minimization would be on qs (the product of q and s), which is the rate of nutrient loss caused by senescence. Although some nutrient resorption can be accomplished at a minimal cost, a portion of the nutrients in plant tissue is in refractory compounds. As such, after a point, it may be less costly for a plant to minimize nutrient loss by decreasing s rather than by decreasing q. This further supports the importance of leaf and root longevity as traits that would increase nutrient competitive ability.

6. *k, the half-saturation constant for nutrient uptake.* R^* depends on k in a simple, linear manner. A decrease in k will lead to a comparable decrease in R^*, and thus to a linear increase in competitive ability. Consistent with this, plants of low-nutrient habitats "generally have a lower V_{max} and

perhaps a lower apparent K_M of nutrient absorption than species from fertile habitats" (Chapin, 1980, p. 240).

In total, resource competition theory, when interpreted in the context of likely tradeoffs among plant traits, suggests that superior nutrient competitors should have relatively low maximal growth rates, long-lived roots and leaves, the ability to grow with low tissue nutrient concentrations, high resistance to herbivory, a low maximal rate of nutrient uptake, and high efficiency of nutrient uptake at low concentrations. The similarity between these predictions and the frequently observed traits of plants from nutrient-poor habitats (Chapin, 1980) suggests that the types of models presented above may be useful summaries of the mechanisms of nutrient competition and that nutrient competition may be a major force structuring communities on nutrient-poor soils. Furthermore, this suggests that the equation for R^* that can be derived from a given model of plant resource use and growth can define the quantitative role of plant traits in determining plant nutrient competitive ability. As such, the equation for R^* may prove to be a useful way to integrate the total effect of a suite of ecophysiological traits on plant fitness.

V. Predicting the Outcome of Nutrient Competition

All five models predict that there is a single number associated with each species that can be used to predict the equilibrial outcome of multispecies nutrient competition. This number is R^*. R^* incorporates the effects of all the traits of a plant on its resource competitive ability. R^*, however, is not just a theoretical construct. *R^* is an empirical entity that can be directly observed in the field.* It is the concentration to which the limiting nutrient would be reduced by an equilibrial monoculture of a given species. (To demonstrate that nutrient *reduction* is occurring, it is only necessary to compare the R^* of a species with the concentration of that nutrient in otherwise identical plots that had been kept free of all plant growth.) Thus, if *Poa pratensis* were grown in monoculture on a soil for which it was limited by nitrogen, but not by any other resource, the concentration of available nitrogen (the sum of ammonium and nitrate) in the monoculture, once monoculture biomass had reached a plateau, would be its R^*. (For perennial plants, this will require several years of growth, and might be best determined by averaging nutrient concentrations over the growing season. Further, it is imperative that the nutrient concentrations measured by the actual soil solution concentrations of nutrient that are immediately available for plant uptake. For example, exchangeable ammonium, i.e., ammonium bound to cation-exchange sites in the soil, is not immediately available for plant uptake, and should not be part of

R^*.) The R^* thus measured would directly incorporate how plant morphology, ecophysiology, litter production, loss, etc. interacted to determine nutrient competitive ability.

Nutrient reduction is the direct mechanism whereby one plant influences the growth of another during nutrient competition. The R^* of a species is a direct measure of the magnitude of this nutrient reduction. Thus, the simplest and most accurate way to *predict* the outcome of nutrient competition may be to measure directly the R^*'s of the competing species. These R^*'s represent the primary mechanism of plant competition for nutrients.

The plant traits that determine the R^* of a species represent the next level of mechanistic detail. As discussed above, R^* depends on all plant traits, including the maximal growth rate, leaf and root longevity, nutrient resorption, the efficiency of growth at low nutrient concentrations, pattern of allocation to roots and leaves, the maximal uptake rate, and the efficiency of uptake at low nutrient concentrations. Thus, although the R^* concept is simple, the biology leading to a particular R^* need not be simple.

The models presented here illustrate that R^*, when directly observed, should be considered a summary variable that synthesizes the effects of species on resources and of resources on species. A species has to have an amount of resource of at least R^* in a habitat for it to maintain a stable population. If it is maintaining a stable population, it will reduce the concentration of the limiting resource down to R^*. When this occurs, its population biomass will be such that its total rate of nutrient uptake equals the rate of nutrient supply. Thus, the effect of a resource on a species, and the effect of a species on a resource (see Goldberg, this volume), are the same thing, as long as they are measured as R^*.

As a test of the R^* concept for terrestrial plants, we grew *Agrostis scabra, Agropyron repens, Schizachyrium scoparium,* and *Andropogon gerardi* for 3 years in replicated monocultures on low-nitrogen soils and directly observed their R^*'s for nitrogen (Tilman and Wedin, pers. obs.). We found that *Schizachyrium scoparium* and *Andropogon gerardi* had significantly lower R^*'s for nitrogen than *Agrostis scabra,* but that *Agrostis* and *Agropyron* did not differ significantly. Pairwise competition experiments between *Agrostis* and each of the other three species revealed that, after 3 years of growth in replicated, two-species gardens, *Schizachyrium scoparium* and *Andropogon gerardi* both displaced *Agrostis scabra,* independent of initial population densities, just as predicted using their observed R^*'s. Moreover, *Agrostis* and *Agropyron,* which did not differ significantly in their R^*'s, co-occurred for the 3 years. In total, these results provide strong support for the R^* hypothesis.

VI. Abstraction versus Complexity

In discussing ecological abstraction, Schaffer (1981) noted that abstraction occurs because the parameters of a model, if fit to real-world observations, can incorporate aspects of the biology of an organism that are not directly stated in the model. This suggests that a simple model, such as Model 1, could be useful if it were appropriately abstracted. Because R^* is the critical parameter, R^* should be directly observed, and other parameters estimated in relation to the observed R^*. One way to do this would be to grow a plant species to equilibrium in replicated monocultures in the field on a nutrient-limited soil, but to vary loss rate and determine the R^* associated with each loss rate. This could then be used to estimate other model parameters (K and r as defined in Table 1). Q would be measured directly. The parameters thus obtained would depend not just on the processes directly included in the model, but on other more complex processes, such as those included in Models 2–5. This occurs because the measured R^*'s, which would be used in estimating these parameters, depend on these processes.

If the most complex models presented here are the most realistic, the others must be considered approximations. Inspection of the equations for R^* reveals how these approximations are made. For instance, Model 2 does not explicitly include nutrient conservation by plants. However, the equations for R^* show that the m of Model 2, which is the total loss rate, is analogous to $c + qs$ of Model 3. This term, $c + qs$, is the total loss rate of nutrients from a plant, with the full quota lost to herbivores (c), but with only qs lost when tissues are shed. If the parameters of Model 2 were estimated using R^* and other data from equilibrial monocultures, the m of Model 2 should be the total *nutrient* loss rate, $c + qs$, of Model 4. A similar process of abstraction could occur when any simpler model is used to describe a more complex process. This suggests that competition models that explicitly include resources may have an advantage over those that do not because the former may be better able to abstract many of the complexities of the mechanisms of nutrient competition.

Because the R^* measured for a species is a summary variable determined by numerous plant and habitat traits, the R^* concept, and simple models of resource competition in general, cannot be rejected just because they do not explicitly include some particular aspect of plant biology that is believed to be "important." Clearly, models are constrained from making any predictions about an omitted aspect of plant biology. However, they need not be constrained from making predictions as to the dynamics or outcome of competition, for the empirically measured variables incorporated into a model are summaries that may abstract much of the higher-level complexity not explicitly included in it.

Clearly, the $R*$ concept itself is a simplification. It applies to equilibrial populations, and is only capable, in theory, of predicting the long-term, equilibrial outcome of interspecific competition. However, it can be expanded to deal with added habitat complexity. For instance, $R*$ is likely to depend on various aspects of the physical environment, such as pH, temperature, and humidity (see Grace, this volume; Tilman *et al.*, 1981). These are physical factors and can be distinguished from resources because physical factors are not consumed. The effect of a physical factor on the long-term, equilibrial outcome of competition can be predicted, in theory, by knowing the dependence of the $R*$'s of each species on the physical factor (Tilman, 1982). Moreover, a given habitat may have more than one limiting resource. If this is so, the $R*$'s for both resources could be simultaneously measured for various ratios of the resources, providing resource-dependent growth isoclines (Tilman, 1980, 1982). These isoclines could be used, in theory, to predict the outcome of competition. These modifications are based on the assumption that competitive interactions in nature tend toward equilibrium, i.e., that the actual pattern observed in natural, relatively undisturbed habitats can be predicted by equilibrial versions of theory. Alternatively, interactions may not go toward equilibrium, in which case it would be possible to predict the long-term, but nonequilibrial, outcome of competition by modifying the theory of resource competition to include the dependence of growth rate on the frequency and magnitude of temporal variance in resource availability (e.g., Armstrong and McGehee, 1980; Levins, 1979). Even though it is obvious that the natural world is not at equilibrium, it may be a useful abstraction to ask which of the broad-scale patterns of nature can be explained by simpler models solved for their equilibrium predictions. For either approach, short-term transient dynamics, such as successional dynamics after a disturbance, could be predicted explicitly using the underlying dynamic models.

Nature is complex. On the surface, at least, there are few ecologists who would argue with this assertion. Indeed, some ecologists revel in its complexity, and may even believe that it is impossible to develop mechanistic, predictive theories of nature. Some of these individuals are ardent empiricists, and have made major contributions to ecology through their collection and analysis of empirical data. However, at times I wonder if some of them may have forgotten that empiricism, like theory, is based on a series of simplifying assumptions. By choosing what to measure and what to ignore, an empiricist is making as many simplifying assumptions as does any theoretician or experimentalist. Although nature seems complex, with every advance in our understanding of nature, nature becomes less complex, for we explain heretofore disparate patterns using a few simple concepts.

The next major advance in studies of plant competition is likely to be the development and testing of predictive theories of the dynamics and outcome of multispecies competition. In this chapter, I have suggested that this may be most rapidly accomplished by studying the mechanisms whereby one plant influences the growth rate of another, and by developing models that explicitly include these mechanisms. For what may prove to be a large class of plant–plant interactions, the major, primary mechanism is resource reduction. Theory suggests that a single, empirically observable number, R^*, is the measure of resource reduction that can directly predict the equilibrial outcome of competition for a single limiting nutrient. This number is an integrator of the numerous traits that determine a plant's ability to compete for a limiting nutrient. The potential validity and generality of the concepts presented in this chapter must still be determined via rigorous field experimentation and observation.

I have not discussed competition for light in this chapter. At the present time, at least, light competition is conceptually more complex than nutrient competition. We do not yet have either rigorous theoretical predictions or experimental results that indicate that a single number, analogous to R^*, can predict the outcome of competition for light. A fuller understanding of light competition remains a major challenge.

VII. Appendix

This appendix presents the differential equations for Models 2–5, and briefly discusses their ecological assumptions. Many of the variables used in these models are defined in Table 1. The remainder are defined as they are presented in the following text. For all models, the subscript i refers to species i. The total number of species is n.

Model 2: Nutrient Uptake and Variable Nutrient Stores

This model assumes that the rate of biomass change (per unit biomass) depends on the difference between growth [$g(Q)$, which depends on tissue nutrient concentration] and loss (m, assumed to be density independent). Here, B_i is the quantity of living plant biomass (g/m^2), R is the concentration of the available form of the limiting nutrient in the habitat, and Q_i is the tissue concentration of the limiting nutrient for species i.

$$dB_i/B_i dt = g_i(Q_i) - m_i \tag{10}$$

Tissue nutrient concentration, Q, depends on the difference between the uptake rate [$u_i(R)$, which depends on external nutrient concentration] and the rate at which tissue nutrient concentration is diluted because of growth.

$$dQ_i/dt = u_i(R) - Q_i g_i(Q_i) \tag{11}$$

R, the concentration of the nutrient in the habitat, depends on the difference between nutrient supply, $y(R)$, and nutrient uptake summed over all n species.

$$dR/dt = y(R) - \sum_{i=1}^{n} B_i u_i(R) \tag{12}$$

When these equations are solved for equilibrium (for $dB_i/B_i dt = dQ_i/dt = dR/dt = 0$), the critical parameter determining which species will be the superior nutrient competitor is R_i^*, which is obtained from Eqs. (10) and (11). Each species has its own value for R_i^*.

$$R_i^* = u_i^{-1}[m_i g_i^{-1}(m_i)] \tag{13}$$

where u^{-1} and g^{-1} are the inverse functions of u and g. Note that, when there is a single limiting resource, the supply function, $y(R)$, has no effect on which species is the competitive dominant, as long as all can survive in the habitat in the absence of interspecific competition.

Equation (13) may be made more obvious by substituting in explicit functions for u and g. The ones I use are the Michaelis–Menten function for nutrient uptake, $u(R) = vR/(R + k)$, and Droop's (1974) function for growth, $g(Q) = r(1 - h/Q)$. Because $u^{-1}(x) = kx/(v - x)$ and $g^{-1}(z) = rh/(r - z)$, this gives

$$R_i^* = r_i h_i k_i m_i/[v_i(r_i - m_i) - r_i h_i m_i] \tag{14}$$

Model 3: Root Uptake, Nutrient Stores, and Leaf Photosynthesis

In this variation on Model 2, biomass production (photosynthesis) is a leaf process and nutrient acquisition is a root process. Thus, the rate of biomass production by the entire plant depends on the production rate by the leaves and the proportion of the total plant biomass that is in leaves (which, here, is l_i). Similarly, the rate of change of plant tissue nutrient concentration depends on the rate of uptake by the roots, and the dilution of this uptake as it spread throughout the entire plant. This depends on b_i, proportional root biomass. This gives equations much like those of Model 2, except they include l_i and b_i:

$$dB_i/B_i dt = l_i g_i(Q) - m_i \tag{15}$$

$$dQ_i/dt = b_i u_i(R) - Q_i g_i(Q_i) \tag{16}$$

$$dR/dt = y(R) - \sum_{i=1}^{n} B_i u_i(R) \qquad (17)$$

$$R_i^* = u_i^{-1}[(m_i/b_i l_i)g_i^{-1}(m_i/l_i)] \qquad (18)$$

If, for simplicity, it is assumed that a plant is either leaf or root, i.e., that $l_i + b_i = 1$, and if the same explicit functions for $g(Q)$ and $u(R)$ are used as above, then

$$R_i^* = r_i h_i k_i m_i/[v_i b_i(r_i - m_i - r_i b_i) - r_i m_i h_i] \qquad (19)$$

Model 4: Nutrient Conservation, Litter Production, and Variable Stores

Model 2 may be modified in a different manner by giving a plant the ability to conserve nutrients, such as occurs when plants withdraw a portion of the nutrients in leaves and roots before they are shed, often at the end of the growing season. To do this, the equation for biomass dynamics is modified to replace m with two loss terms, with s being the rate of tissue loss to shedding of senescent tissues and c the loss rate to all other causes, such as consumption by a herbivore or death.

$$dB_i/B_i dt = g_i(Q) - s_i - c_i \qquad (20)$$

The equation for the rate of change of the tissue concentration must be modified to include nutrient resorption from tissues that are being shed. The term that does this is $Q_i s_i p_i$, where p_i is the proportion of nutrient in the tissue that is reabsorbed before it is shed.

$$dQ_i/dt = u_i(R) + Q_i s_i p_i - Q_i g_i(Q_i) \qquad (21)$$

Note that loss of plant parts without any resorption (i.e., with $p_i = 0$) does not change the nutrient *concentration* in a plant. The greater the resorption, the greater would be the increase in internal nutrient concentration associated with a given rate of loss of biomass. Also, herbivory or death (c) does not influence average plant tissue nutrient concentration, because these processes are assumed to act independent of tissue nutrient concentration. It would be easy to modify this model to include herbivores that preferentially consumed more nutrient-rich tissues. Such a modification would make herbivory have an even greater negative impact on competitive ability than modeled here, and thus would favor greater production of defensive compounds in the more nutrient-rich tissues of a plant.

The next equation follows the dynamics of the litter produced by species i. Litter dynamics control nutrient resupply to the plants. The dynamics of litter, dL/dt, depend on the rate at which litter is produced by the shedding of plant parts (sB) and on the rate at which litter decays:

$$dL_i/dt = s_iB_i - j_iL_i \tag{22}$$

This assumes that litter decay is a negative exponential process, with j, the rate of decay, being constant.

At equilibrium, these equations predict the outcome of nutrient competition will be determined by R^*, where

$$R_i^* = u_i^{-1}\{(c_i + q_is_i)[g_i^{-1}(s_i + c_i)]\} \tag{23}$$

If the functions for $u(R)$ and $g(Q)$ used for Models 2 and 3 are substituted into this expression, the following equation is derived for R^*:

$$R_i^* = r_ih_ik_i(c_i + s_iq_i)/[v_i(r_i - c_i - s_i) - r_ih_i(c_i + s_iq_i)] \tag{24}$$

Another equation is needed to make this be a complete dynamic model, but this equation does not influence R^*. The fourth equation follows the dynamics of the available (mineral) form of the limiting soil resource, R. R dynamics depend on the difference between nutrient supply and consumption. As litter decays, mineral nutrients contained in litter are released into the environment. Litter decay provides nutrients at a rate that depends on the nutrient content of the litter and its decay rate. However, the nutrient content of litter is variable, dependent on when it was produced. Thus, it is necessary to follow various litter classes, making it impossible to write a differential equation for this process. This equation, which does not influence R^*, does determine the equilibrial biomass of the competitive dominant, but not its identity, as long as all plants compete for a single limiting nutrient. However, if there were two limiting resources, such as light and nitrogen, the processes included in this equation could have a major effect on competitive interactions.

Model 5: Improved Nutrient Conservation

Model 4 may be modified to have a plant conserve tissue nutrients in a different manner. For this modification, let us assume that a plant reduces tissue nutrient concentration down to a constant level of M before the tissues are shed. This leads to the following equations.

$$dB_i/B_idt = g_i(Q) - s_i - c_i \tag{25}$$

$$dQ_i/dt = u_i(R) + s_i(Q_i - M_i) - Q_ig_i(Q_i) \tag{26}$$

$$dL_i/dt = s_iB_i - j_iL_i \tag{27}$$

Because this model assumes that all litter produced by a particular species has a nutrient content of M, resource dynamics are easily modeled:

$$dR/dt = \sum_{i=1}^{n} [j_iM_iL_i] + \sum_{i=1}^{n} [c_iB_iQ_i] - \sum_{i=1}^{n} [B_iu_i(R)] \tag{28}$$

At equilibrium, the equations of Model 5 predict that the outcome of competition will be determined by R^*, where

$$R_i^* = u_i^{-1}[s_iM_i + s_ig_i^{-1}(s_i + c_i)] \tag{29}$$

When the functions for $u(R)$ and $g(Q)$ used for Models 2 and 3 are substituted into this expression, the following equation is derived for R^*:

$$R_i^* = r_ih_ik_i[c_i + s_iM_i(r_i - c_i - s_i)/r_ih_i]/$$

$$\{v_i(r_i - c_i - s_i) - r_ih_i[c_i + s_iM_i(r_i - c_i - s_i)/r_ih_i]\} \tag{30}$$

Models 4 and 5 suggest that the litter feedback effect will influence the equilibrial biomass of the dominant species, but not influence the identity of the dominant (except in the trivial case in which litter feedback prevents the survival of the dominant in the absence of competition). However, this prediction hinges on the assumption that the nutrient will remain the only limiting factor. If this model were expanded to include competition for another resource, such as light, the litter feedback effect could strongly influence the identity of the competitive dominant because litter feedback would change the relative availability of the two limiting resources. Moreover, multiple stable equilibria might result from litter feedback effects. For instance, a good nutrient competitor with poorly mineralized litter, if that species were initially abundant, could drive a soil to low levels of available nitrogen that would favor it. Similarly, a good light competitor that produced rapidly mineralized litter, if it were initially abundant on an identical soil, could drive a soil to have high nutrient supply rates, high plant standing crop, and low light intensities, and thus exclude the superior nutrient competitor.

VIII. Summary

Classical, density-based studies of plant competition have demonstrated its existence in nature, but have not led to a general theory capable of predicting the dynamics and outcome of plant competition. Such generality and predictive power require theory that explicitly includes the mechanisms of competition. Five different models of the mechanisms of nutrient competition, including several highly complex and realistic models, all state that the critical parameter that should be capable of predicting the outcome of competition is R^*. R^* is a summary variable that incorporates the effects of a plant's ecophysiological, morphological, and life history traits, as well as such habitat characteristics as the intensity of herbivory and of other sources of loss or death. R^* is the concentration in the soil to which a limiting nutrient is reduced by an equilibrial monoculture of a species. As such it is directly empirically observable. The importance of $R*$ is also intuitive. Nutrient reduction is the mechanism whereby one plant inhibits another, when both are limited by the same nutrient. $R*$ quantifies this nutrient reduction.

The equation describing the dependence of $R*$ on plant traits predicts that a superior nutrient competitor should have a low maximal growth rate, long-lived roots and leaves, the ability to grow with low tissue nutrient concentrations, a low maximal rate of nutrient uptake, a high efficiency of uptake at low soil nutrient concentrations, and high resistance to herbivory. These predicted traits agree well with Chapin's (1980) summary of the traits of plants that dominate nutrient-poor habitats. In total, this suggests that a single, empirically observable number, $R*$, may both integrate the effects of plant traits on competitive ability and be capable of predicting the outcome of interspecific competition for a limiting nutrient.

Acknowledgments

I thank all the participants at the conference, especially Deborah Goldberg, Steve Pacala, Jim Grace, and Paul Keddy, for the free and open discussions we had. These have had a major impact on this chapter. I am indebted to David Wedin, Tania Vincent, and Jim Grace for their extensive comments on an earlier version of this manuscript. I thank the National Science Foundation (NSF/BSR-8811884) and the Andrew Mellon Foundation for supporting this work.

References

Armstrong, R. A., and McGehee, R. (1980). Competitive exclusion. *Am. Nat.* **115,** 151–170.

Bazzaz, F. A., Chiariello, N. R., Coley, P. D., and Pitelka, L. F. (1987). Allocating resources to reproduction and defense. *BioScience* **37,** 58–67.

Bender, E. A., Case, T. J., and Gilpin, M. E. (1984). Perturbation experiments in community ecology: theory and practice. *Ecology* **65,** 1–13.

Chapin, F. S., III (1980). The mineral nutrition of wild plants. *Annu. Rev. Ecol. Syst.* **11,** 233–260.

Coley, P. D. (1987). Interspecific variation in plant anti-herbivore properties: The role of habitat quality and rate of disturbance. *New Phytol.* **106,** 251–263.

Coley, P. D., Bryant, J. P., and Chapin, F. S., III (1985). Resource availability and plant antiherbivore defense. *Science* **230,** 895–898.

Connell, J. (1983). On the prevalence and relative importance of interspecific competition: Evidence from field experiments. *Am. Nat.* **122,** 661–696.

Droop, M. R. (1974). The nutrient status of algal cells in continuous culture. *J. Mar. Biol. Assoc. U.K.* **54,** 825–855.

Givnish, T. J. (ed.) (1986). "On the Economy of Plant Form and Function," 696 pp. Cambridge Univ. Press, Cambridge, England.

Givnish, T. J. (1982). On the adaptive significance of leaf height in forest herbs. *Am. Nat.* **120,** 353–381.

Grubb, P. (1977). The maintenance of species richness in plant communities: The importance of the regeneration niche. *Biol. Rev.* **52,** 107–145.

Gulmon, S. L., and Mooney, H. A. (1986). Costs of defense and their effects on plant productivity. *In* "On the Economy of Plant Form and Function" (T. J. Givnish, ed.), pp. 681–696. Cambridge Univ. Press, Cambridge, England.

Harper, J. L. (1977). "Population Biology of Plants," 892 pp. Academic Press, London.

Hsu, S. B., Hubbell, S. P., and Waltman, P. (1977). A mathematical theory for single-nutrient competition in continuous cultures of microorganisms. *S.I.A.M. J. Appl. Math.* **32,** 366–383.

Levins, R. (1979). Coexistence in a variable environment. *Am. Nat.* **114,** 765–783.

MacArthur, R. H. (1972). "Geographical Ecology: Patterns in the Distribution of Species," 269 pp. Harper & Row, New York.

Monod, J. (1950). La technique de culture continue: theorie et applications. *Ann. Inst. Pasteur* **79,** 390–410.

Monsi, M. (1968). Mathematical models of plant communities. *In* "Functioning of Terrestrial Ecosystems at the Primary Production Level" (F. E. Eckardt, ed.). Vaillant-Carmanne, Liege, Belgium.

Mooney, H. A. (1972). The carbon balance of plants. *Annu. Rev. Ecol. Syst.* **3,** 315–346.

O'Brien, W. J. (1974). The dynamics of nutrient limitation of phytoplankton algae: A model reconsidered. *Ecology* **55,** 135–141.

Pastor, J., Aber, J. D., McClaugherty, C. A., and Melillo, J. M. (1984). Above ground production and N and P cycling along a nitrogen mineralization gradient on Blackhawk Island, Wisconsin. *Ecology* **65,** 256–268.

Pulliam, H. R. (1985). Foraging efficiency, resource partitioning, and the coexistence of sparrows. *Ecology* **66,** 1829–1836.

Pulliam, H. R. (1986). Niche expansion and contraction in a variable environment. *Am. Zool.* **26,** 71–79.

Radosevich, S. R., and Holt, J. S. (1984). "Weed Ecology: Implications for Vegetation Management." Wiley, New York.

Rapport, D. J. (1971). An optimization model of food selection. *Am. Nat.* **105,** 575–578.

Schaffer, W. M. (1981). Ecological abstraction: The consequences of reduced dimensionality in ecological models. *Ecol. Monogr.* **51,** 383–401.

Schoener, T. W. (1971). Theory of feeding strategies. *Annu. Rev. Ecol. Syst.* **2,** 369–404.

Schoener, T. W. (1983). Field experiments on interspecific competition. *Am. Nat.* **122,** 240–285.

Schoener, T. W. (1986). Alternatives to Lotka–Volterra competition: Models of intermediate complexity. *Theor. Pop. Biol.* **10,** 309–333.

Sommer, U. (1985). Comparison between steady state and non-steady state competition: Experiments with natural phytoplankton. *Limnol. Oceanogr.* **30,** 335–346.

Tilman, D. (1976). Ecological competition between algae: Experimental confirmation of resource-based competition theory. *Science* **192,** 463–465.

Tilman, D. (1977). Resource competition between planktonic algae: An experimental and theoretical approach. *Ecology* **58,** 338–348.

Tilman, D. (1980). Resources: A graphical–mechanistic approach to competition and predation. *Am. Nat.* **116,** 362–393.

Tilman, D. (1982). "Resource Competition and Community Structure." Princeton Univ. Press, Princeton, New Jersey.

Tilman, D. (1987a). The importance of the mechanisms of interspecific competition. *Am. Nat.* **129,** 769–774.

Tilman, D. (1987b). Secondary succession and the pattern of plant dominance along experimental nitrogen gradients. *Ecol. Monogr.* **57,** 189–214.

Tilman, D. (1988). "Plant Strategies and the Dynamics and Structure of Plant Communities." Princeton University Press, Princeton, New Jersey.

Tilman, D. (1989). Discussion: Population Dynamics and Species Interactions. *In* "Perspectives in Ecological Theory" (J. Roughgarden, R. May, and S. Levin, eds), pp. 89–100. Princeton University Press, Princeton, New Jersey.

Tilman, D., Kilham, S. S., and Kilham, P. (1982). Phytoplankton community ecology: The role of limiting nutrients. *Annu. Rev. Ecol. Syst.* **13,** 349–372.

Tilman, D., Mattson, M., and Langer, S. (1981). Competition and nutrient kinetics along a temperature gradient: An experimental test of a mechanistic approach to niche theory. *Limnol. Oceanogr.* **26,** 1020–1033.

Vitousek, P. M., Walker, L. R., Whiteaker, L. D., Mueller-Dombois, D., and Matson, P. A. (1987). Biological invasion by *Myrica faya* alters ecosystem development in Hawaii. *Science* **238,** 802.

Werner, E. E. (1984). The mechanisms of species interactions and community organization in fish. *In* "Ecological Communities: Conceptual Issues and the Evidence" (D. R. Strong, Jr., D. Simberloff, L. G. Abele, and A. B. Thistle, eds.), pp. 360–382. Princeton Univ. Press, Princeton, New Jersey.

8

Allelopathy, Koch's Postulates, and the Neck Riddle

G. Bruce Williamson

I. Introduction

A weird creature came to a meeting of men,
Hauled itself in to the high commerce
Of the wise. It lurched with one eye,
Two feet, twelve hundred heads,
A back and belly—two hands, arms,
Shoulders—one neck, two sides.
Untwist your mind and say what I mean.[1]

Throughout the decade of the 1960s, Muller (1965, 1966, 1969; Muller *et al.,* 1964; McPherson and Muller, 1969) developed techniques to study allelopathy in the chaparral of southern California. Methodically moving from one type of evidence to another, from one species to next, from hydrophobic to hydrophilic compounds, from soft to hard chaparral, his team of researchers and students sought evidence to support the hypothesis that shrubs release compounds that inhibit the germination, growth, and development of potential competitors. The motivation for Muller's research arose from his field observations that few if any herbaceous plants grew among the chaparral shrubs, although they flourished on the same soils immediately outside the bare zones or halos surrounding the shrubs.

After a decade of research, Muller's results were disquieting for many reasons (Muller, 1969). First, lacking chemical expertise, he seemed to postulate a spectre whose effects could be seen but whose identity and mode of action were often unknown. Second, the spectre haunted the prevailing view that interactions among plants occurred as resource competition—the utilization of common, limited resources. Third, the spectre had begun to appear in other plant communities with additional advocates in the scientific community, most notably allelochemical inhibition of nitrification and slowing of succession in temperate grasslands and oak forests (Rice, 1964, 1965, 1968; Rice and Pancholy, 1972, 1973, 1974). Allelopathy as a field of research had grown, albeit sluggishly, to the stage where it required integration into general ecological theory. Despite Muller's (1969) own review of allelopathy, more critical adjudications were to be declared by researchers outside the field.

If Muller's achievements could be gauged by the seniority of his judges, then accolades would have prevailed. However, in the end a split decision was rendered. The first to speak was Whittaker (1969, 1970), who hailed the discovery of chemical interactions between plant species:

> It is reasonable to judge that allelopathic effects are common and that the observed cases stand out from a background of more widespread, less conspicuous effects on plant growth and populations. . . . Allelopathy may consequently be of widespread significance in plant communities. (Whittaker and Feeney, 1971, pp. 757–758).

What had been the steady advance of a few investigative teams was now superseded by a rash of ventures by ecologists and phytochemists. Journal articles on the topic tripled in the year following Whittaker and Feeney's (1971) article in *Science,* never again to recede to previous levels of productivity (Fig. 1). Other touchstones of scientific progress emerged. In the same year, the National Academy of Sciences (1971) sponsored a working conference on allelopathy. In 1974 the *Journal of Chemical Ecology* appeared, a further legitimization of allelopathy as one of the elements of the developing investigations into secondary compounds. In the same year, no less significant was the publication of *Allelopathy* (Rice, 1974), which provided a ready review for neophytes in the field.

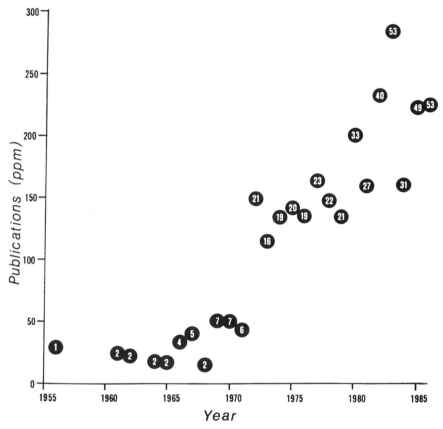

Figure 1 Annual growth of studies in allelopathy as measured by the number of journal articles listed by *Bioabstracts* relative to total annual listings (in publications per million or ppm); absolute number of articles on allelopathy shown inside each point.

Almost simultaneous with the exaltation from Whittaker came experimental evidence supporting an alternative explanation for the bare zones around chaparral shrubs. Bartholomew (1970) demonstrated that halos around some shrubs in the soft chaparral might be maintained by herbivory from animals inhabiting the shrubs. Successive attempts to test the relative importance of allelopathy versus herbivory in both the development and the maintenance of bare zones showed that chemical inhibition occurred in some species but not in others and that herbivory of unexclosed seedlings could be as intense as 100% (Halligan, 1973, 1975, 1976; Christensen and Muller, 1975a,b). Regrettably, the herbivory exclosures were never maintained long enough to determine if seedlings present would become reproductive adults or whether they would succumb to allelochemics. Nevertheless, their presence was accepted generally as conclusive proof of the role of herbivory. Muller was accused of having overrated the importance of allelopathy, and his conclusions were deemed premature and oversimplified, if not altogether fictitious (Bartholomew, 1970; Halligan, 1973). Surprisingly, these criticisms did little to dampen activity in allelopathy, which prospered under the flourishing aegis of chemical ecology and the blessing from Whittaker (1969, 1970) (Fig. 1).

In the latter half of the decade a second adjudication on allelopathy was rendered, and again the magnitude of the impact conformed to the seniority of the judge. Harper (1975, 1977) gave allelopathy a less than enthusiastic welcome into plant population biology:

> The history of research into toxic interactions has included many *causes celebres:* startling claims that have subsequently been abandoned in favour of some other interpretation (1975, p. 370).

Equally critical, but perhaps more important to the future of allelopathy as a science was the censure of its methods. Stowe (1979) determined that classical, unadulterated bioassays—whole plant extracts, foliar washes, and decomposing litter—caused inhibition of germination or growth, even when the plants were not suspected of allelopathy. His findings, in essence, were that positive results were more often the case than not—a conclusion casting a shadow over many prior studies. Furthermore, documented cases of inhibition caused by the pH or osmotic potential of extracts, without reference to any phytotoxins contained therein, added credence to the notion that positive effects in bioassays might be coincidental (Anderson and Loucks, 1966; Bell, 1974; Reynolds, 1975a,b). In allelopathy, the null hypotheses appeared even more susceptible to challenge than in other fields of ecological research where challenges arose at the end of the last decade (e.g., MacFayden, 1975; Futuyma, 1975; Smith, 1976; Strong, 1980; Simberloff, 1980; Pielou, 1981; Feinsinger *et al.,* 1981; Lewin, 1983a,b). Bioassays needed to become more realistic in their simulation of field conditions and controls more rigorous in their exclusion of alternative hypotheses.

In this air of skepticism, productivity measured by publications stabilized during the latter half of the 1970s (Fig. 1). Then, the decade of the 1980s witnessed a renewed resurgence in allelopathic research. While the basic sciences had indulged in the luxury of critique, compelling examples of chemical inhibition in agriculture and forestry stimulated further investigations (Patrick *et al.,* 1963, 1964; Guenzi and McCalla, 1966a,b; Webb *et al.,* 1967; Tukey, 1969, 1971; Einhellig and Rasmussen, 1973, 1978, 1979; Putnam and Duke, 1978). Applied scientists moved to the forefront of basic research in allelopathy, as evidenced by their attempts to reckon with criticisms of the discipline (Putnam and Duke, 1978; Fuerst and Putnam, 1983) and by their productivity (see

Rice, 1984), funding (U.S. Department of Agriculture, 1985), and organization.[2]

In contrast, basic science remained mired in disbelief. While Whittaker's flirtations with chemical ecology were brief, Harper's opposition became so doctrinaire that a general ecology text that he coauthored a decade later would contain only six lines devoted to the topic (Begon *et al.*, 1986). In this climate of apprehension, the proof required in allelopathic investigations became more stringent than that demanded in other areas of ecology: experimental as well as correlative evidence, field as well as laboratory tests, and elimination of all reasonable alternative hypotheses through perfectly controlled designs. In the face of such scrutiny, Harper counseled acquiescence:

> Demonstrating this [toxicity in the field] has proved extraordinarily difficult—it is logically impossible to prove that it doesn't happen and perhaps nearly impossible to prove absolutely that it does (1977, p. 494).

II. Koch's Postulates: A Neck Riddle?

Studies of allelopathy are riddled with a difficulty common to all studies of interactions between species: the conditions in which toxins act are also the conditions in which all other interactions, such as competition and herbivory, occur. Therefore, the most fundamental problem in research design is how to separate the allelopathic component from other factors. The solution, usually formulated by analogy to Koch's postulates regarding the cause of a disease, is enticing: most simply, proof of allelopathy requires application of the suspected compounds under natural field conditions, when the plant that produces the compounds is absent or removed, to ascertain if the symptoms or suspected effects are recreated in other plants. This apparent solution has been formulated by both proponents and opponents of allelopathy (Harper, 1975, 1977; Fuerst and Putnam, 1983; Putnam and Tang, 1986).

However, the microbial analogy is not altogether appropriate for several reasons. First, chemicals do not reproduce, whereas cells do. Therefore, introduction of a few cells of a microorganism into a host may be enough to generate a disease and its symptoms. But this is not so with an allelochemical, which must be introduced continually in the exact dosage that it is released from its source plant. The active compound cannot simply be applied in the field as an innoculum to reproduce itself to the natural levels to spawn a disorder. Second, chemicals not only do not reproduce but they degrade, and, in allelopathy, the degradation products are often believed to be the inhibitory compounds (Putnam and Duke, 1978; Tanrisever *et al.*, 1987). In such cases, locating the source of the toxin becomes more difficult, and attributing any adaptive significance to its toxicity, more precarious. To further confound matters, production of toxins via degradation might be considered to be adaptive because it reduces potential autotoxicity. Third, among microbes there is a one-to-one correspondence between disease and organism, but to date all cases of alleged allelopathy appear to involve a complex of chemicals which interact synergistically (Rasmussen and Einhellig, 1977; Putnam and Tang, 1986; Einhellig, 1986). Therefore, the analogous reinoculation requires knowledge of the chemical complex, the concentration of each component, and the mechanism of release of each component.

The advantage to be derived from the application of Koch's postulates is clear—one final experiment in which the allelopathic plant is absent but its putative phytotoxins are present, so any effects measured must be due to the toxins, not to all alternative interactions. Aside from the fact that the requisite research skills and knowledge span the gamut of biological sciences from the ecological to chemical, the task is extraordinarily difficult, if not impossible. The suspected allelochemicals must be isolated, identified, then quantified on a rate release basis, purchased or synthesized, and finally reapplied simultaneously in the field at natural rates over a prolonged period of time. What are the conditions for such an experiment? The answer is more elusive than the solution to a neck riddle.[1]

Unsurprisingly then, application of Koch's postulates in allelopathic studies has been more by inference than by fact. To date, no such proof of allelopathy has been forthcoming because few, if any, tests have been performed (Barnes and Putnam, 1987). As if he foresaw the entanglement of Koch's postulates, Muller (1953) summed up the problem elegantly:

> The natural habitat . . . is far too intricate a system of influences and factors, physical and biological, to hope that there may be found a single factor controlling the complicated life of a perennial species. An explanation when it is arrived at, will be at least as intricate as the situation it seems to describe (p. 59).

Despite its idealism, the analogy to Koch's postulates provides a template for constructive experimental design, namely, suspected allelotoxins should be applied in the field as naturally as possible. However, such applications are neither the genesis nor the culmination of an allelopathic investigation, but merely one step in the scientific method. In fact, such applications will normally involve many different trials in experiments designed to determine dosage effects, synergisms, and the importance of chemical inhibition relative to resource competition and herbivory under different conditions.

III. Some Obligations in Allelopathic Research

Interestingly, outside of allelopathy, the field of ecology has been more generous in accepting experimental demonstration of effects as proof of a process without elucidation of the underlying mechanisms (Connell, 1983 and this volume; Tilman, 1987). For example, demonstration of resource competition rarely requires identification of the limiting resource, yet allelopathy is incredible without identification of the phytotoxin. Early in his research, Muller (1969) recognized this disparity in the evaluative criteria for competitive and allelopathic investigations:

> That the burden of proof lies upon the proponent of biochemical inhibition and that he is obliged to show that competition for some necessary factor of the environment is not the cause of apparent inhibition. This obligation the experimenter gladly assumes (p. 348).

There is a logical justification for the uneven application of criteria. If allelopathy is more than coincidental, i.e., if there has been selection for chemical inhibition, then in many situations allelopathy may have evolved in response to resource competition. In such cases, allelopathy will be associated with current or past resource competition, so allelopathic research is somewhat obligated to test for resource competition alongside tests for biochemical inhibition.

This conclusion is not to suggest that allelopathy can only emerge from resource competition. Too often, in fact, studies of allelopathy do not rigorously address the mechanism of natural selection for production of allelochemicals. To state that the selective advantage is the inhibition of potential competitors is a relatively diffuse, meaningless assertion. The specific advantage gained under specific environmental conditions must be defined in order to generate testable hypotheses. Why has allelopathy evolved in some cases of resource competition but not others? And what conditions other than resource competition may lead to allelopathy?

Every ecological process has particular aspects that must be addressed and investigated in order to test if and how that process is operating in nature. Refuting resource competition is one such obligation of any interpretation of the dissociation of two plant species based on chemical inhibition. A second and rather obvious obligation particular to allelopathic investigations is the isolation and identification of the putative phytotoxins (Muller, 1969; Fuerst and Putnam, 1983). A third particular aspect is determination of how autotoxicity is minimized.

To gain a selective advantage from production of allelotoxins, a species must inhibit its potential competitors more than it inhibits itself. What mechanisms exist to reduce or preclude autotoxicity? Autotoxicity is one of the more contradictory aspects of allelopathy (Newman, 1978; Stowe, 1979). A species that produces, packages, and releases chemical inhibitors is likely to impact its own growth and development negatively. To assume arbitrarily that it will have evolved resistance to its own toxins is contradictory to the notion that the presumed target species will not have evolved such resistance (Newman, 1978). Furthermore, since the chemicals are presumed to originate in the source plant, they are likely to be most concentrated and most damaging there, unless mechanisms exist to reduce autotoxicity. Consequently, it behooves the proponent of allelopathy to determine the mechanisms whereby autotoxicity is reduced or circumvented.

I conclude with a brief discussion of (1) the selective advantage for allelopathy with tests to refute resource competition, and (2) the avoidance of autotoxicity based on research in upland plant communities in the Southeastern Coastal Plain. Readers interested in more detail on the chemical moieties are referred to previous works (Tanrisever *et al.*, 1987, 1988; Fischer *et al.*, 1987; Williamson *et al.*, 1989a). Parallels to California's chaparral and herbaceous communities suggest possible generalization.

IV. Sand Pine Scrub: The Coastal Plain Chaparral?

Two different plant communities are intermingled on the upland, well-drained sands of the Southeastern Coastal Plain: (1) the scrub characterized by sand pine [*Pinus clausa* (Chapm. ex Engelm.) Vasey ex Sarg.], with a dense shrub cover but no herbaceous ground cover, and (2) the sandhill dominated by longleaf pine (*P. palustris* Mill.) with a complete graminoid ground cover but few shrubs (Chapman, 1932; Laessle, 1958). Fire is a frequent feature of the sandhill, burning the deciduous surface fuels every 3–8 years (Williamson and Black, 1981). However, scrub is evergreen and burns infrequently, about once every 50 years or once per generation of the dominant sand pine, which regenerates via serotinous cones (Harper, 1914; Richardson, 1977). The differences between scrub and sandhill are remarkably parallel to the differences between California chaparral shrubs and adjacent grasslands (Table 1). Additionally, the sharpness of the ecotone between scrub and sandhill is reminiscent of the bare zones between chaparral and grasslands. The correspondence is inexplicable on the basis of soils (mainly clays in California versus sands in the Southeast), or on the basis of climate (dry in California, with precipitation concentrated in the winter, versus wet in the Southeast, with precipitation concentrated in the summer).

Table 1 Comparison of Southeastern Coastal Plain Scrub and
Sandhill Communities to California Chaparral and
Grassland Communities

Community		
Southeastern Coastal Plain	Scrub	Sandhill
California	Chaparral	Grassland
Physiognomy		
Ground cover	None	Complete
Shrub cover	Very dense	Very sparse
Fuel traits		
Surface litter, quantity	Low	High
Surface litter, quality	Compressed	Loose, aerated
Crown litter, quantity	High	None
Foliage phenology	Evergreen	Deciduous
Fire traits		
Frequency	20–50 yr	3–8 yr
Type	Crown	Surface
Suspected of allelopathy	Yes	No

The prevailing explanation for the existence of scrub and sandhill vegetation types in Florida has been differences in soil nutrients, although attempts to find differences have been more popular than fruitful (Harper, 1914; Webber, 1953; Kurz, 1942; Laessle, 1958, 1968; Kalisz and Stone, 1984; Richardson, 1985). Furthermore, the addition of fertilizers on plots in the scrub produces neither greater germination nor greater growth of seedlings than on control plots (Richardson, 1985).

Surface fires keep scrub species out of the sandhill. In the absence of surface fires in the sandhills, the woody species from the scrub will colonize sandhill sites (Laessle, 1958; Veno, 1976; Myers, 1985). They grow well in the sandhills, often faster than in the scrub, but ultimately surface fires sweeping through the graminoid ground cover kill them (Veno, 1976; Hebb, 1982). Myers (1985) has suggested that different fire regimes maintain scrub and sandhill. Allelopathy may have evolved to reduce fire risk.

A. The Selective Advantage of Biochemical Interference

Generally, surface fires move through the sandhills until they encounter islands or strands of scrub vegetation which extinguish them by changes in fuel and live vegetation. Webber (1935) called the scrub "a firefighting machine." However, on rare occasions fire spreads into the scrub, where it explodes into the crowns, destroying all above-ground woody vegetation. Subsequent regeneration occurs through dormant seeds in the soil, serotinous cones, or resprouting. We have been investigating the hypothesis that shrubs of the early successional scrub community produce chemicals to inhibit germination and growth of the grasses and pines which provide the fuel for surface fires that otherwise would kill the shrubs. This hypothesis offers a very specific mechanism for the selective advantage of allelopathy, namely, fuel control to prevent damage from surface fires.

To test for allelopathy, a field transplant experiment was designed to control for competition (resource utilization) but still allow potential allelopathic interactions. Plugs of wiregrass (*Aristida stricta* Michx.) were removed from the sandhill with a golf green cupcutter and inserted into sections of PVC pipes, 10 cm diameter by 20 cm long, open at both ends. The grass was pruned to the soil surface and monitored in a greenhouse until new growth emerged. Then, 20 pipes with healthy grass shoots were transplanted into each of 8 field plots, 4 scrub sites, and 4 sandhill sites. In the field, the plants remained in the sandhill soil within the pipes but were exposed to water and gas exchange through the ends of the pipes.

In addition, plugs of sandhill soil without plants were extracted from the sandhill and inserted in the PVC pipes. Then each pipe was planted with 10 seeds of slash pine, longleaf pine, or sand pine. After the seedlings emerged, the pipes were placed in the ground at the scrub and sandhill field sites. Pine seedlings were protected from herbivory by a cone of hardware cloth for several months. After 2 months, each pipe was thinned to 3 pine seedlings to reduce root competition.

A number of provisions were made to preclude resource competition between the native vegetation and the transplants. The plots, 1 × 20 m, were oriented in an east–west direction to ensure maximum exposure to sunlight. Additionally, any vegetation overhanging the plots was pruned. In the plots vegetation was pruned at the soil surface, and the plot perimeter was root pruned by inserting a spade to a depth of 30 cm. All pruning was repeated every 2 weeks. Finally, the pipes were rotated 180° every 2 weeks to ensure that no roots grew into or out of the pipes. During periods of drought, plants were watered with 500 ml per pipe, but only once in any 2-week period.

After 16 months, plants grown in the scrub exhibited only 62% of the dry weight of those in the sandhill for longleaf pine ($p = 0.01$), 35% for slash pine ($p = 0.03$) and 56% for sand pine ($p = 0.03$) (Table 2). The height of pines in the scrub was 55% of those in the sandhill for sand pine ($p = 0.002$), and 72% for slash pine, but the latter difference was not statistically significant (Table 2). No height measurements were recorded for longleaf pine seedlings, which were in the stemless, "grass" stage of development. Furthermore, the wiregrass plugs in the scrub weighed only 57% of those in the sandhill ($p = 0.05$) (Table 2). In the sandhill 95% (74/78) of the grass plugs survived, while in the scrub only 76% (53/70) survived.

Table 2 Sizes (Mean ± Standard Deviation) from
16-Month Transplant Experiment of Pines and
Wiregrass Grown in Scrub and Sandhill Sites[a]

Species	Scrub	Sandhill	Ratio
Dry weight (g)			
Longleaf pine	1.04 ± 0.12	1.69 ± 0.56	0.62**
Slash pine	0.49 ± 0.08	1.39 ± 1.36	0.35*
Sand pine	0.28 ± 0.05	0.50 ± 0.12	0.56*
Wiregrass	4.48 ± 1.71	7.87 ± 0.76	0.57*
Meristem height (cm)			
Slash pine	3.82 ± 0.63	5.31 ± 1.53	0.72
Sand pine	4.17 ± 0.88	7.53 ± 0.78	0.55*

[a] Scrub and sandhill means that are significantly different are indicated on the ratio by * if $p \leq 0.05$ and by ** if $p \leq 0.01$.

The experiment attempted to control resource competition, but there are alternative interpretations for the observed differences. One can argue that, if toxins moved into the pipes in the scrub, then nutrients could have moved out. Also, we recorded differences (2°C) in soil temperatures, another factor that might have caused the growth differences. However, the allelopathic conclusion from this experiment is strengthened by other results showing seasonality in inhibition, the identity of the toxins, and transport mechanisms to the soil (e.g., Richardson, 1985; Tanrisever *et al.*, 1988; Williamson and Richardson, 1988; Williamson *et al.*, 1989a).

The same mechanism, protection from fire, may have resulted in selection for allelopathy by shrubs in the chapparal. Frequent burning converts chaparral to grassland (Kay, 1960; Biswell, 1974) and successive fires may result in exceptionally high mortality of the dominant chaparral shrubs, which seem to regenerate well after fire at long intervals (Zedler *et al.*, 1983). Preliminary investigations of another pair of proximate communities, high elevation grassland and shrubland (paramo) in Central America, the former with frequent surface fires and the latter with infrequent crown fires, also implicate allelopathy by fire-sensitive paramo shrubs (Williamson *et al.*, 1986).

B. Mechanisms of Avoiding Autotoxicity

Perhaps the best mechanism for avoiding autotoxicity entails the production of inhibitors from plant products after they are removed from the source plant. Various avenues exist to execute such external production. Many compounds appear to be associated with decaying litter, although the exact mechanisms have not been elucidated for most systems (Putnam and Duke, 1978; Grace, 1983; Carter and Grace, 1986). Second, microbes associated with the host plant's soil may produce the inhibitors (Kaminsky, 1981). Finally, relatively innocuous compounds released from the host may degrade to form more toxic products. For example, *Ceratiola ericoides* in the Coastal Plain scrub community releases the novel compound ceratiolin in foliar runoff of rainwater. Ceratiolin is relatively inactive but undergoes degradation to produce hydrocinnamic acid which is highly active, inhibiting seed germination and radicle growth of grasses in concentrations of less than 60 ppm (Tanrisever *et al.*, 1987; Williamson *et al.*, 1989b). The degradation occurs only when ceratiolin is in aqueous solution and is accelerated by exposure to light, heat, and acid—conditions intrinsic to the upland sandy soils where *Ceratiola* occurs. The compound is released from both fresh foliage and from litter, and bioassays of the latter suggest greater toxicity than the former (Richardson, 1985; Williamson and Richardson, 1988). In the field, the species exhibits remarkable halo, devoid of other plants, both when it is in the scrub (Fig. 2a) and when it is colonizing disturbed sites (Fig. 2b). The latter implies that herbivory is not the cause of halos because the shrubs are too small to harbor any rodents (Bartholomew, 1970; Halligan, 1973, 1975, 1976).

In contrast to toxins produced external to the source plant, inhibitors produced on the source plant may be sequestered to reduce autotoxicity. Inhibitors produced on source plants include monoterpenes, one of the largest classes of compounds invoked as allelopathic agents; many monoterpenes are extremely toxic, nonpolar, volatile compounds (Muller *et al.*, 1964; Muller, 1965; Muller and del Moral, 1966; Fischer, 1986; Tanrisever *et al.*, 1988).

In the Coastal Plain scrub, two endemic mints, *Conradina canescens* and *Calamintha ashei,* produce an extensive array of monoterpenes (de la Peña, 1985; Tanrisever *et al.*, 1988). These compounds are released by glandular trichomes on the leaf surfaces together with copious quantities of ursolic acid (5–10% dry wt of leaves), a weak biological detergent. In this way, the monoterpenes are isolated from the leaf tissue but trapped within a thin layer of cuticle (Fig. 3). Water dripped over fresh foliage will leach the monoterpenes from the leaf surface into the soil solution.

The role of ursolic acid is unknown, although it appears to form micelles with some monoterpenes (Fischer, 1986; Williamson *et al.*, 1989b), possibly serving three different functions. First, it can lower the vapor pressure of the monoterpenes and reduce their volatility. Second, it can increase the solubility of the monoterpenes in water, thereby allowing them to leach from the plant in foliar runoff water. Third, once in the soil, the micelle may facilitate entry of the monoterpenes into target seeds by causing leaks in cell membranes.

Figure 2 (a) *Ceratiola ericoides* growing at the scrub ecotone with a halo; (b) *C. ericoides* colonizing a disturbed site.

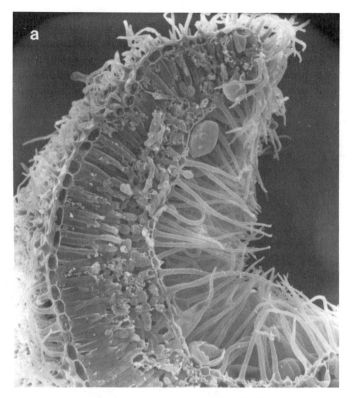

Figure 3   (a) Cross section of a leaf of *Conradina canescens,* showing secretory trichomes in a maze of nonsecretory hairs. (b) Cross section through a secretory trichome, revealing secretory cells beneath the fold of the cuticle which harbors the solution of monoterpenes.

Homologous secretory trichomes on chaparral plants, especially the mints (*Salvia* spp.), probably compartmentalize their monoterpenes; however, the delivery or transport systems appear to be different. In California, where rain is sparse, the monoterpenes of *Salvia* volatilize on hot days, settle on the soil, and are adsorbed on clay particles until they contact lipophilic seed or seedling membranes (Muller and del Moral, 1966). In Florida, where rain is copious, the sandy soils are not lipophilic like the clays of California, and the potential for volatile monoterpenes to settle on the soil is remote when surface temperatures commonly exceed 50°C (Richardson, 1985). Therefore, the aqueous leaching of monoterpenes in micelles may be a more effective delivery mechanism in Florida.

A third mechanism for minimizing autotoxicity is the production of a root system well below the soil surface. Then, inhibitors released onto

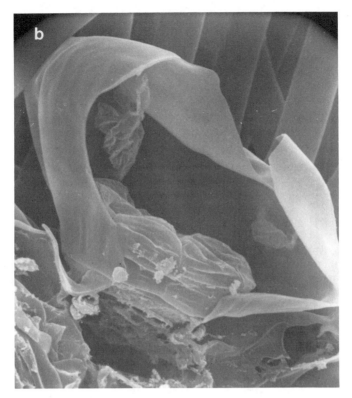

Figure 3   *(Continued)*

the surface will contact other species roots in the upper soil horizon before contacting roots of the source species. By the time compounds move into the lower soil strata, their solutions are likely more dilute and they may have been degraded by microorganisms. For example, *Ceratiola* exhibits a thick (8–20 cm) layer of roots, which begins about 10 cm below the soil surface. In contrast, invading grasses have dense fibrous roots concentrated near the soil surface.

V. Summary

The study of allelopathy has developed in size and scope over the past three decades since pioneering efforts by Muller. Growth of the discipline has been somewhat erratic as critics have expressed extreme opinions, alleging that allelopathy was both everywhere and nowhere. Such doctrinaire judgments may have left plant ecologists more opinionated on allelopathy than on other forms of interactions among plants, when in reality most of the criticisms of chemical inhibition can be leveled as squarely at resource competition. The future of allelopathy, and perhaps competition as well, seems to lie in the study of specific mechanisms of plant–plant interactions. From such studies both generalities and exceptions will become evident.

Parallels between California chaparral and Southeastern Coastal Plain scrub provide compelling evidence for allelopathy, despite the edaphic and climatic differences between the regions. The common feature is the proximity of two communities in each region with different fire regimes, one with frequent surface fires, the other with infrequent crown fires.

Acknowledgments

This material is based on work supported by the Cooperative State Research Service, U.S. Department of Agriculture under agreement No. 88-33520-4077 of the Competitive Research Grants Program for Forest and Rangeland Renewable Resources. SEM photos were prepared by Sharon Mathews.

Notes

1. In less rational times, a condemned man's last chance to save his neck was to be able to ask a riddle that could be solved by neither his judge nor his executioners. Solutions to neck riddles were necessarily obscure (Taylor, 1949), and the answer to the one here is the "One-eyed Seller of Garlic" (Williamson, 1982).
2. The first national public conference, the North American Symposium on Allelopathy, was sponsored in 1982 by the Illinois Agricultural Experimental Stations, Departments of Forestry and Agronomy at the University of Illinois, the North Central Forest Experimental Station, and the U.S. Forest Service. In contrast, the earlier working conference on allelopathy held in 1971 was sponsored by the National Academy of Sciences and included equal representations of basic and applied scientists.

References

Anderson, R. C., and Loucks, O. L. (1966). Osmotic pressure influence in germination tests for antibiosis. *Science* **152,** 771–773.

Barnes, J. P., and Putnam, A. R. (1987). Role of benzoxazinones in allelopathy by rye (*Secale cereale* L.). *J. Chem. Ecol.* **13,** 889–906.

Bartholomew, B. (1970). Bare zone between California shrub and grassland communities: The role of animals. *Science* **170,** 1210–1212.

Begon, M., Harper, J. L., and Townsend, C. R. (1986). Ecology: Individuals, populations and communities. Sinauer, Sunderland, Massachusetts.

Bell, D. T. (1974). The influence of osmotic pressure in tests for allelopathy. *Trans. Ill. State Acad. Sci.* **67,** 312–317.

Biswell, H. H. (1974). The effects of fire on chaparral. *In* "Fire and Ecosystems" (T. T. Kozlowski and C. E. Ahlgren, eds.), pp. 321–364. Academic Press, New York.

Carter, M. F., and Grace, J. B. (1986). Relative effects of *Justicia americana* litter on germination, seedlings, established plants of *Polygonum lapathifolium. Aquat. Bot.* **23**, 341–349.

Chapman, H. H. (1932). Is longleaf type a climax? *Ecology* **13**, 328–334.

Christensen, N. L., and Muller, C. H. (1975a). Relative importance of factors controlling germination and seedling survival in *Adenostema* chaparral. *Am. Midl. Nat.* **93**, 71–78.

Christensen, N. L., and Muller, C. H. (1975b). Effects of fire on factors controlling plant growth in *Adenostema* chaparral. *Ecol. Monogr.* **45**, 29–55.

Connell, J. H. (1983). On the prevalence and relative importance of interspecific competition; evidence from field experiments. *Am. Nat.* **122**, 661–696.

de la Pena, A. C. (1985). "Terpenoids from *Conradina canescens* (Labiatae) with Possible Allelopathic Activity," M.S. thesis, 64 pp. Louisiana State Univ., Baton Rouge, Louisiana.

Einhellig, F. A. (1986). Mechanisms and modes of action of allelochemicals. *In* "The Science of Allelopathy" (A. R. Putnam and C. S. Tang, eds.), pp. 171–188. Wiley, New York.

Einhellig, F. A., and Rasmussen, J. A. (1973). Allelopathic effects of *Rumex crispus* on *Amaranths retroflexus*, grain sorghum and field corn. *Am. Midl. Nat.* **90**, 79–86.

Einhellig, F. A., and Rasmussen, J. A. (1978). Synergistic inhibitory effects of vanillic and p-hydroxybenzoic acids on radish and grain sorghum. *J. Chem. Ecol. 4*, 425–436.

Einhellig, F. A. and Rasmussen, J. A. (1979). Effects of three phenolic acids on chlorophyll content and growth of soybean and grain sorghum seedlings. *J. Chem. Ecol.* **5**, 815–824.

Feinsinger, P., Whelan, R. J., and Kiltie, R. A. (1981). Some notes on community composition: Assembly by rules or by dartboards? *Bull. Ecol. Soc. Am.* **62**, 19–23.

Fischer, N. H. (1986). The function of mono and sesquiterpenes as plant germination and growth regulators. *In* "The Science of Allelopathy" (A. R. Putnam and C. S. Tang, eds.), pp. 203–218. Wiley, New York.

Fischer, N. H., Tanrisever, N., de la Peña, A., and Williamson, G. B. (1987). The chemistry and allelopathic mechanisms in the Florida scrub community. *Proceedings, 1987 Plant Growth Regulator Society of America* **14**, 192–208.

Fuerst, E. P., and Putnam, A. R. (1983). Separating the competitive and allelopathic components of interference: Theoretical principles. *J. Chem. Ecol.* **9**, 937–944.

Futuyma, D. J. (1975). Competition and the structure of bird communities (a review). *Q. Rev. Biol.* **50**, 217.

Grace, J. B. (1983). Autotoxic inhibition of seed germination by *Typha latifolia:* An evaluation. *Oecologia* **59**, 366–369.

Guenzi, W. D., and McCalla, T. M. (1966a). Phenolic acids in oats, wheat, sorghum, and corn residues and their phytotoxicity. *Agron. J.* **58**, 303–304.

Guenzi, W. D., and McCalla, T. M. (1966b). Phytotoxic substances extracted from soil. *Soil Sci. Soc. Am. Proc.* **30**, 214–216.

Halligan, J. P. (1973). Bare areas associated with shrub stands in grasslands: The case of *Artemisia californica. BioScience* **23**, 429–432.

Halligan, J. P. (1975). Toxic terpenes from *Artemisia californica. Ecology* **56**, 999–1003.

Halligan, J. P. (1976). Toxicity of *Artemisia californica* to four associated herb species. *Am. Midl. Nat.* **95**, 406–421.

Harper, J. L. (1975). Allelopathy (a review). *Q. Rev. Biol.* **50**, 493–495.

Harper, J. L. (1977). "Population Biology of Plants." Academic Press, New York.

Harper, R. M. (1914). Geography and vegetation of northern Florida. *Annu. Rep. Fla. State Geol. Surv.* **6**, 163–391.

Hebb, E. A. (1982). Sand pine performs well in the Georgia–Carolina sandhills. *South. J. Appl. For.* **6**, 144–147.

Kalisz, P. J., and Stone, E. L. (1984). The longleaf pine islands of the Ocala National Forest, Florida: A soil study. *Ecology* **65**, 1743–1754.

Kaminsky, R. (1981). The microbial origin of the allelopathic potential of *Adenostema fasciculatum* H & A. *Ecol. Monogr.* **51**, 365–382.

Kay, B. L. (1960). Effect of fire on seed forage species. *J. Range Manage.* **13**, 31–33.

Kurz, H. (1942). Florida sand dunes and scrub, vegetation and geology. *Fla. Geol. Surv., Geol. Bull.* **23**, 1–154.

Laessle, A. M. (1958). The origin and successional relationship of sandhill vegetation and sand-pine scrub. *Ecol. Monogr.* **28**, 361–387.

Laessle, A. M. (1968). Relationships of sand pine scrub to form shorelines. *Q. J. Fla. Acad. Sci.* **30**, 269–286.

Lewin, R. (1983a). Santa Rosalia was a goat. *Science* **221**, 636–639.

Lewin, R. (1983b). Predators and hurricanes change ecology. *Science* **221**, 737–740.

MacFayden, A. (1975). Some thoughts on the behaviour of ecologists. *J. Anim. Ecol.* **44**, 351–363.

McPherson, J. K., and Muller, C. H. (1969). Allelopathic effects of *Adenostoma fasciculatum*, Chamise, in the California chaparral. *Ecol. Monogr.* **39**, 177–198.

Muller, C. H. (1953). The association of desert annuals with shrubs. *Am. J. Bot.* **40**, 53–60.

Muller, C. H. (1965). Inhibitory terpenes volatilized from *Salvia* shrubs. *Bull. Torrey Bot. Club* **92**, 38–45.

Muller, C. H. (1966). The role of chemical inhibition (allelopathy) in vegetational composition. *Bull. Torrey Bot. Club* **93**, 332–351.

Muller, C. H. (1969). Allelopathy as a factor in ecological process. *Vegetatio* **18**, 348–357.

Muller, C. H., and del Moral, R. (1966). Soil toxicity induced by terpenes from *Salvia leucophylla*. *Bull. Torrey Bot. Club* **93**, 332–351.

Muller, C. H., Muller, W. H., and Haines, B. L. (1964). Volatile growth inhibitors produced by shrubs. *Science* **143**, 471–473.

Myers, R. L. (1985). Fire and the dynamic relationship between Florida sandhill and sand pine scrub vegetation. *Bull. Torrey Bot. Club* **112**, 241–252.

National Academy of Sciences, U.S. (1971). "Biochemical Interactions among Plants" (U.S. National Committee for International Biological Programs, eds.). Washington, D.C.

Newman, E. I. (1978). Allelopathy: Adaptation or accident? *In* "Biochemical Aspects of Plant and Animal Coevolution" (J. B. Harborne, ed.), pp. 327–342. Academic Press, London.

Patrick, Z. A., Toussoun, T. A., and Snyder, W. C. (1963). Phytotoxic substances in arable soils associated with decomposition of plant residues. *Phytopathology* **53**, 152–161.

Patrick, Z. A., Toussoun, T. A., and Koch, L. W. (1964). Effect of crop residue decomposition products on plant roots. *Annu. Rev. Phytopathol.* **2**, 267–292.

Pielou, E. C. (1981). The usefulness of ecological models: A stock-taking. *Q. Rev. Biol.* **56**, 17–31.

Putnam, A. R., and Duke, W. B. (1978). Allelopathy in agroecosystems. *Annu. Rev. Phytopathol.* **16**, 431–451.

Putnam, A. R., and Tang, C. S. (1986). Allelopathy: State of the science. *In* "The Science of Allelopathy" (A. R. Putnam and C. S. Tang, eds.), pp. 1–19. Wiley, New York.

Rasmussen, J., and Einhellig, F. (1977). Synergistic inhibitory effects of p-coumaric and ferulic acids on germination and growth of grain sorghum. *J. Chem. Ecol.* **3**, 197–205.

Reynolds, T. (1975a). Characterization of osmotic restraints on lettuce fruit germination. *Ann. Bot.* **39**, 791–796.

Reynolds, T. (1975b). pH restraints on lettuce fruit germination. *Ann. Bot.* **39**, 797–805.

Rice, E. L. (1964). Inhibition of nitrogen-fixing and nitrifying bacteria by seed plants. I. *Ecology* **45**, 824–837.

Rice, E. L. (1965). Inhibition of nitrogen-fixing and nitrifying bacteria by seed plants. II. Characterization and identification of inhibitors. *Physiol. Plant.* **18,** 255–268.

Rice, E. L. (1968). Inhibition of nodulation of inoculated legumes by pioneer plant species from abandoned fields. *Bull. Torrey Bot. Club* **95,** 346–358.

Rice, E. L. (1974). "Allelopathy." Academic Press, New York.

Rice, E. L. (1984). "Allelopathy," 2nd ed. Academic press, Orlando, Florida.

Rice, E. L., and Pancholy, S. K. (1972). Inhibition of nitrification by climax ecosystems. *Am. J. Bot.* **59,** 1033–1040.

Rice, E. L., and Pancholy, S. K. (1973). Inhibition of nitrification by climax ecosystems. II. Additional evidence and possible role of tannins. *Am. J. Bot.* **60,** 691–702.

Rice, E. L., and Pancholy, S. K. (1974). Inhibition of nitrification by climax ecosystems. III. Inhibitors other than tannins. *Am. J. Bot.* **61,** 1095–1103.

Richardson, D. R. (1977). Vegetation of the Atlantic Coastal Ridge of Palm Beach county, Florida. *Fl. Sci.* **40,** 281–330.

Richardson, D. R. (1985). "Allelopathic Effects of Species in the Sand Pine Scrub of Florida," Ph.D. dissertation. Univ. of South Florida, Tampa, Florida.

Simberloff, D. (1980). A succession of paradigms in ecology: Essentialism to materialism and probabilism. *Synthese* **43,** 3–39.

Smith, F. E. (1976). Ecology: Progress and self-criticism. *Science* **192,** 546.

Stowe, L. G. (1979). Allelopathy and its influence on the distribution of plants in an Illinois old-field. *J. Ecol.* **67,** 1065–1085.

Strong, D. R., Jr. (1980). Null hypotheses in ecology. *Synthese* **43,** 271–285.

Tanrisever, N., Fronczek, F. R., Fischer, N. H., and Williamson, G. B. (1987). Ceratiolin and other flavonoids from *Ceratiola ericoides*. *Phytochemistry* **26,** 175–179.

Tanrisever, N., Fischer, N. H., and Williamson, G. B. (1988). Calaminthone and other menthofurans from *Calamintha ashei;* their germination and growth regulatory effects on *Schizachyrium scoparium* and *Lactuca sativa*. *Phytochemistry,* **27,** 2523–2526.

Taylor, A. (1949). The varieties of riddles. *In* "Philologica: The Malone Anniversary Studies" (T. A. Kirby and H. B. Woolf, eds.), pp. 1–8. Johns Hopkins University Press, Baltimore, Maryland.

Tilman, D. (1987). The importance of the mechanisms of interspecific competition. *Am. Nat.* **129,** 769–774.

Tukey, H. B., Jr. (1969). Implications of allelopathy in agricultural plant sicence. *Bot. Rev.* **35,** 1–16.

Tukey, H. B., Jr. (1971). Leaching of substances from plants. *In* "Biochemical Interactions among Plants" (U.S. National Committee for International Biological Programs, eds.), pp. 25–32. Nat. Acad. Sci., Washington, D.C.

U.S. Department of Agriculture (1985). Competitive research grants program for forest and rangeland renewable resources for fiscal year 1985. *Fed. Regist.* **50,** 16524–16526.

Veno, P. A. (1976). Successional relationships of five Florida plant communities. *Ecology* **57,** 498–508.

Webb, L. J., Tracey, J. G., and Haydock, K. P. (1967). A factor toxic to seedlings of the same species associated with living roots of the non-gregarious subtropical rain forest tree, *Grevillea robusta*. *J. Appl. Ecol.* **4,** 13–25.

Webber, H. J. (1935). The Florida scrub, a fire-fighting association. *Am. J. Bot.* **22,** 344–361.

Whittaker, R. H. (1969). The chemistry of communities. *In* "Biochemical Interactions among Plants" (U.S. National Committee for IBP, eds.), pp. 10–18. Nat. Acad. Sci., Washington, D.C.

Whittaker, R. H. (1970). The biochemical ecology of higher plants. *In* "Chemical Ecology" (E. Sondheimer and J. B. Simeone, eds.), pp. 43–70. Academic Press, New York.

162   *G. Bruce Williamson*

Whittaker, R. H., and Feeney, P. P. (1971). Allelochemics: Chemical interactions between species. *Science* **171,** 757–770.

Williamson, C. B. (1982). "A Feast of Creatures: Anglo–Saxon Riddle-Songs. Univ. Pennsylvania Press, Philadelphia, Pennsylvania.

Williamson, G. B., and Black, E. M. (1981). High temperatures of forest fires under pines as a selective advantage over oaks. *Nature (London)* **293,** 643–644.

Williamson, G. B., and Richardson, D. R. (1988). Bioassays for allelopathy: Measuring treatment responses with independent controls. *J. Chem. Ecol.* **14,** 181–187.

Williamson, G. B., Schatz, G. E., Alvarado, A., Redhead, C. S., Stam, A. C., and Sterner, R. W. (1986). Effects of repeated fires on tropical paramo vegetation. *Trop. Ecol.* **27,** 62–69.

Williamson, G. B., Fischer, N. H., Richardson, D. R., and de la Pena, A. (1989a). Chemical inhibition of fire-prone grasses by the fire-sensitive shrub, *Conradina canescens. J. Chem. Ecol.* **15,** 1567–1577.

Williamson, G. B., Richardson, D. R., and Fischer, N. H. (1989b). Allelopathic mechanisms in fire-prone communities. *In* "Frontiers of Allelochemical Research" (S. J. H. Rizvi, ed.), in press. Nijhoff, Dordrecht, The Netherlands.

Zedler, P. H., Gautier, C. R., and McMaster, G. S. (1983). Vegetation change in response to extreme events: The effect of short interval between fires in California chaparral and coastal scrub. *Ecology* **64,** 809–818.

II

The Role of Competition in Community Structure

9

On the Effects of Competition: From Monocultures to Mixtures

L. G. Firbank   **A. R. Watkinson**

I. Introduction

Competition can be defined as "an interaction between individuals brought about by a shared requirement for a resource in limited supply, and leading to a reduction in the survivorship, growth, and/or reproduc-

Perspectives on Plant Competition. Copyright © 1990 by Academic Press, Inc. All rights of reproduction in any form reserved.

tion of the individuals concerned" (Begon *et al.*, 1986). The first part of this definition focuses on the causes of competition—resources are required by different individuals, which are shared out among them. The second part focuses on the effects on population dynamics—the reduction in the contribution made by individuals to future generations when they are brought together.

Competition needs to be studied at a variety of different levels, from the physiological mechanisms of competition to its role in determining the structure and dynamics of plant communities. For population ecologists, the effects of competition on the numbers of births, deaths, immigrants, and emigrants are of greater importance than the underlying mechanisms. This emphasis on effects enables us to consider any negative–negative interaction between individuals as being due to competition (Odum, 1959; Williamson, 1972), thus avoiding the need to verify the mechanism of the interaction for every single situation. It is the aim of this chapter to analyze the dynamics of intra- and interspecific competition in an agricultural context. Experiments in agriculture and forestry provide much of the best information on the effects of competition and allow us to explore how crowding as determined by the age, size, and type of neighbors influences the performance of individual plants, the development of the stand, and the dynamics of interspecific interactions.

II. Competition within Monocultures

Much of our understanding of how competition affects the growth of plants within monocultures arises from the series of papers entitled "Intraspecific Competition among Higher Plants" written during the 1950s and 1960s by Kira, Hozumi, Shinozaki, Yoda, and colleagues in Japan. Their careful observations rank among the outstanding achievements in competition studies to date, and their work foreshadows much of our current understanding of the effects of competition. They identified three major effects of intraspecific competition at high densities in monocultures of plants. At high density the mean size of surviving plants is reduced (the competition–density effect), the probability of survival is reduced (self-thinning), and the size structure of the population is altered. Furthermore, they discovered that the first two effects obey clear, definable mathematical relationships.

A. The Competition–Density Effect

The first paper in the series was published by Kira *et al.* (1953), who studied the effects of density on regularly spaced plants. They assumed, as do virtually all subsequent studies under controlled conditions, that "total plant weight . . . [may be] . . . taken as the measure of competition; namely, the smaller the mean plant weight, the more intense the competition within the population . . . [is] . . . considered to be." The results from a range of experiments showed that "mean values of both fresh and dry weight became significantly higher with decreasing density. In the earlier stages, however, a certain lower limit of density was found below which mean weight no more increased. So far as indicated by total plant weight, no competition was operative below the limit. This minimum density for competition fell with time."

When the mean weights per plant were plotted against density on a log–log graph (Fig. 1), they found that, except at low densities where there was no competition, mean weight declined linearly with increasing

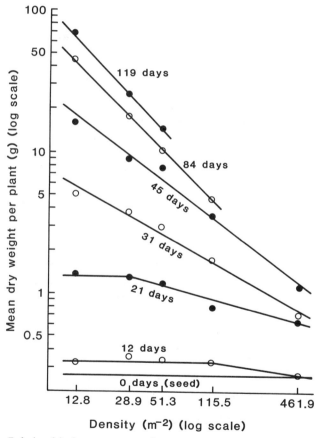

Figure 1   Relationship between mean dry weight per plant (g) and density (m^{-2}) at different times during growth in soybean populations. Redrawn from Kira *et al.* (1953).

density, suggesting to them the model

$$wN^c = K \tag{1}$$

where w is the mean weight per plant, N is the density of plants, and c and K are constants which vary according to growth stage and experimental conditions. The constant c starts at a value of 0 in experiments established from seeds, but increases with time to a value of approximately 1.

Equation (1) cannot be considered to be a satisfactory description of the competition–density effect as it does not describe yield at low densities. In a later paper of the series, Shinozaki and Kira (1956) proposed a second model, also used by de Wit (1960), which describes mean yield per plant at all densities at a given time

$$w^{-1} = AN + B \tag{2}$$

where w is again mean weight per plant and N is density; A and B are constants. This reciprocal equation was derived from an assumption of logistic growth by the component plants. Unfortunately, this assumption resulted in the workers ignoring the need for the power term c in Eq. (1) so clearly shown in the earlier paper. The new model assumes that at high densities mean yield per plant is inversely proportional to density, in other words, total yield per unit area is independent of density. This assumption is referred to as the "law of constant final yield."

Not all monocultures obey this assumption, however, which was relaxed again by Bleasdale and Nelder (1960) and subsequently by Watkinson (1980) by including another power term:

$$w = w_m(1 + aN)^{-b} \tag{3}$$

where w_m estimates the mean weight of isolated plants, a estimates the area required by a plant to grow to w_m, and b describes the efficiency of the use of resources by the population. The parameter b is often needed to provide an adequate fit to data, and is affected by the efficiency of resource utilization of individual plants and also by the size structure of the stand (Firbank and Watkinson, 1985a). As time progresses, the parameter w_m increases with time in a stand along with a and b (Watkinson, 1984). Competition–density equations can also describe the yield of plant parts as well as whole plants, in which case the parameter estimates subsume the allometric relationships between the sizes of the parts to the sizes of the whole plants (Watkinson, 1980; Spitters, 1983a), which may vary in different environments (Morris and Myerscough, 1987).

Equation (3) may be fitted to data by using a nonlinear regression technique in which the sum of squared residuals is minimized by iteration; log–log transformation usually gives the most even distribution of the residuals. Maximum likelihood methods may also be used. Unlike Cousens *et al.* (1988), who used a wide range of unreplicated densities, Spitters (1983b) and Mead and Curnow (1983) recommend that the densities should be replicated. However, mortality invariably causes the densities at harvest to depart from those intended, and if mean yield per plant is regressed against desired densities, large errors and biases may be introduced (unless yield per unit area is used as the response variable). The experimental design and degree of replication should depend on the purposes of the experiment, and where possible, should be guided by the probable choice of model and by data from previous experiments.

A typical mean yield–density response is shown in Fig. 2, which uses data from an experiment reported by Firbank *et al.* (1984). Data from this experiment will also be used to illustrate several other aspects of competition in this chapter. In October 1982, *Bromus sterilis* was sown in monocultures and mixtures with constant density of winter wheat at

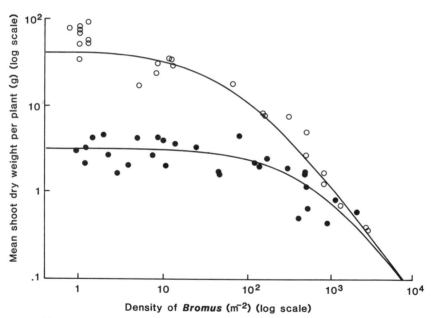

Figure 2 The relationship between mean shoot dry weight per plant and density at harvest for *Bromus sterilis* in the presence (closed circles) and absence (open circles) of a constant sown density of winter wheat. The curves show Eqs. (3) and (10) fitted to the monoculture and mixtures data, giving the equations $w = 42.6(1 + 0.02N)^{-1.22}$ for the monocultures and $w = 42.6[1 + 0.02(N + 1.3N_c)]^{-1.22}$, where w is the mean weight per plant of *Bromus*, N is the density of *Bromus*, and N_c is the density of the crop. Taken from Firbank *et al.* (1984).

densities of 1, 10, 100, and 1000 plants m^{-2}. There were 12 replicates at each density and a permanent quadrat was set up in each replicate. The *Bromus* plants within each quadrat were individually tagged, and the leaf and tiller number of each was recorded at approximately 3-week intervals. These plants were harvested at the end of the experiment, along with all wheat plants within sample quadrats on all the wheat-containing plots.

B. Self-Thinning

The second effect of density on monocultures is that mortality increases at high densities even in the absence of other factors. Yoda *et al.* (1963) showed that such mortality can be described by the relationship (as reparameterized by Watkinson, 1980)

$$N_s = N_i(1 + mN_i)^{-1} \qquad (4)$$

where N_i and N_s are the initial and final densities, respectively, and $1/m$ represents the asymptotic value of N_s as N_i tends to infinity. The parameter m can be estimated using nonlinear regression on data from a wide range of initial densities, some of which should be very high indeed.

Not surprisingly, the value of m decreases with time in a plant stand; as the plants grow in size, there is room for fewer of them. Yoda *et al.* (1963) found that once a population had reached the maximum density, mortality occurred in such a way that

$$w = cN_s^{-k} \qquad (5)$$

where w is mean weight per plant, c is a constant which varies from species to species, and k has been widely reported to take the value of approximately 3/2 for a wide range of species (Yoda *et al.*, 1963; White, 1980; Westoby, 1984). Such a population is said to be self-thinning (Yoda *et al.*, 1963). The relationship between Eqs. (4) and (5) is clarified by realizing that, in a self-thinning population, N_s must equal m^{-1} (Firbank, 1984), i.e.,

$$w = cm^k \qquad (6)$$

It is notoriously easy to confuse the self-thinning effect with the competition–density effect. The competition–density effect refers to the relationship between mean plant size and density of stands grown under the same conditions and inspected at one time, whereas self-thinning describes how density declines as mean yield per plant increases in a single stand as time progresses. This distinction is clarified using the diagram which Yoda and colleagues used for this purpose (Fig. 3).

Weller (1987a) has discussed several problems in estimating the values of c and k. It is difficult to justify any regression technique when the two

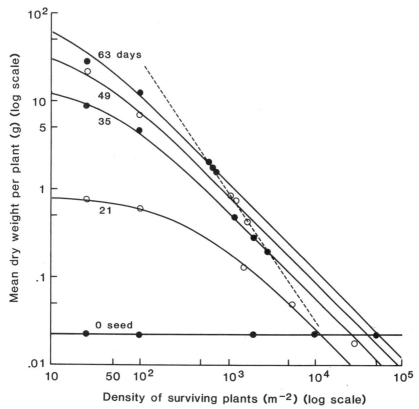

Figure 3 Competition–density curves (solid lines) of plants at different ages and the self-thinning line (dotted line) in even-aged monocultures of buckwheat, showing the interrelation between the two. Redrawn from Yoda *et al.* (1963).

variables, yield and density, vary concomitantly—there is no independent variable. An alternative solution is to use principle components analysis to determine the slope and intercept of the thinning line (Mohler *et al.*, 1978). Furthermore, because the thinning line is approached asymptotically by a population (see, e.g., Fig. 3), it is difficult to decide at what point the population has actually reached the line. Finally, many workers have failed to test whether the slopes of their thinning lines agree with a $-k$ value of $-3/2$. Indeed, Weller (1987a) has recalculated thinning gradients and has found that many data sets show significant differences from this value, and that others, previously quoted as being examples of self-thinning, fail to show any correlation between yield per unit area and density. Given these problems in estimating thinning gradients, one should bear in mind that the resulting values are indeed only estimates!

C. Variation in Plant Size

So far, we have discussed only the response of the "mean" plant to density. Intraspecific competition, however, also affects the variation that occurs among individual plants within a population. In a typically detailed series of studies, Koyama and Kira (1956) found that the frequency of individual plant weight is usually normal at the seed and early seedling stage, but thereafter the distribution of plant weights becomes progressively skewed, with few large individuals and many small ones. The rate of change from normal to skewed distributions varies greatly between populations, but is usually increased at high densities. In addition, Ford (1975) found that a bimodal size distribution may be generated (although the appearance of bimodality does depend on the selection of classes in the size histogram; there are no formal tests for bimodality) and that, during self-thinning, the frequency distribution becomes less skewed again, as some of the smaller plants die.

Recent workers have tended to stress changes in size inequality rather than skewness (van Andel *et al.*, 1984; Weiner and Solbrig, 1984; Biere, 1987; Schmitt, Eccleston, and Ehrhardt, 1987; Hara, 1986a, 1988). For example, Weiner and Thomas (1986) have calculated Gini coefficients for size hierarchies from a range of previously published experiments. Fourteen out of 16 of these show increases in inequality at higher densities. Such increases are also found in Knox's (1987) analysis of *Pinus taeda* stands and in the *Bromus* experiment described earlier. Initially, all of the *Bromus* plants on each plot had one leaf, and so in terms of leaf number there was no variability within the population. However, as time progressed, differences between plants became apparent and the Gini coefficient increased. There was substantial variation between plots of the same sowing density, but when the values were combined there was also a consistent trend showing an increase in variability with increased sowing density (Table 1A).

There are a limited number of possible causes of variation in plant size in an even-aged stand lacking herbivores and pathogens, which are excellently reviewed by Benjamin and Hardwick (1986). Essentially, they are variation in seed size, in depth of burial and preemergence growth to cause variation in emergence time, size at emergence, relative growth rate, and duration of growth. The changing frequency distributions observed by Koyama and Kira (1956), Ford (1975), and others result in part from different relative growth rates among the individual plants after emergence. These rates may differ either because of factors intrinsic to the plants, the action of herbivores and pathogens, or in response to competition from neighboring plants. Before looking at the possible mechanisms of these changes in relative growth rate, we shall look at experiments designed to investigate the role of competition in the growth of individual plants.

Table 1   The Variability of Plant Size with Density and Time in Populations of
Bromus sterilis Grown in the Presence and Absence of Wheat[a]

	Monitor Date							
	1982	1983						Weight at
Density	8/12	6/1	25/1	15/2	7/3	29/3	19/4	Harvest
A. Wheat absent								
1	—	—	—	—	—	—	—	0.276
10	0	0.031	0.105	0.145	0.191	0.192	0.297	0.274
100	0	0.048	0.095	0.129	0.193	0.207	0.229	0.344
1000	0	0.096	0.162	0.152	0.221	0.244	—	0.470
Means	0	0.058	0.121	0.142	0.202	0.214	0.263	0.363
B. Wheat present								
1	—	—	—	—	—	—	—	0.409
10	0	0.035	0.081	0.126	0.190	0.186	0.206	0.301
100	0	0.056	0.108	0.131	0.191	0.181	0.189	0.286
1000	0	0.066	0.144	0.161	0.151	0.170	0.212	0.312
Means	0	0.052	0.111	0.139	0.177	0.179	0.202	0.300

[a] Leaf number was used as the measure of plant size (Gini coefficient), except at the final harvest where shoot weight was used. Each value is the mean of between 3 and 11 plots. See text for details.

1. The Analysis of the Yields of Individual Plants   Since the average plant in monoculture responds to density, it follows that individual plants should respond to the number of neighboring plants. In particular, it might be expected that the relative growth rates of plants will be functions of the space available to individual plants, since this will affect the availability of resources (Benjamin and Hardwick, 1986). Firbank and Watkinson (1987) identified four classes of models defining the space available to an individual plant, and fitted examples from three of these to data from an experiment in which *Triticum aestivum* and *Agrostemma githago* were grown in monocultures and mixtures. In the first analysis, individual plant weights were related to the areas of their Voronoi polygons; these are constructed to include the area closer to the focal plant than to any other plant (e.g., Mead, 1966; Mithen *et al.*, 1984; Sutherland and Benjamin, 1987). In the second analysis, following Schellner *et al.* (1982) and Mack and Harper (1977), plant size was related to the number of plants within a given radius of each plant. Four radii were used. In the third model, following Weiner (1982), an estimate of competitive pressure was devised for each plant, which took into account the proximity of all plants in a plot as well their sizes. Major statistical problems were identified with fitting the third model, for just as the neighboring plants affect the weight of the focal plant, the focal plant in turn affects them. As Cormack (1979) and Ford and Diggle (1981) point out, such models should *not* be used unless the model can be tested against replicate data sets. In addition, if there has been any density-dependent mortality, one might also find that individual plant weight affects the number of neighbors and the Voronoi polygon area.

These problems did not affect our conclusions—typically only 20% of the variation in individual plant yield could be explained by any of these methods of analysis. Nor are these results unusual. Similar levels of explained variation are reported by, among others, Daniels (1976), Liddle *et al.* (1982), and Schellner *et al.* (1982). Interestingly, high correlations between individual plant weight and measures of available area within a population are consistently found amongst desert and dune plants (e.g., Yeaton and Cody, 1976; Nobel, 1981) where competition may not be for light (e.g., Nobel and Franco, 1986). Obviously, if very wide ranges of densities are used, then the relationship between mean size per plant and density becomes increasingly apparent and explains a greater proportion of the variation in plant size. The models of Silander and Pacala (1985) simply reaffirm this competition–density effect, without exploring the causes of the variation in plant size about this relationship.

2. The Effects of Plant Age   One important reason why measures of available area often fail to explain variability in plant sizes is that even a slight variation in emergence time can affect yield, either by altering the time available for growth, by giving the earlier emerging plants a competitive advantage, or by virtue of some cohorts of seedlings encountering more favorable environmental conditions for survival and growth than others (Benjamin and Hardwick, 1986). Studies by Black and Wilkinson (1963) and by Ross and Harper (1972), among others (see Miller, 1987) commonly report a negative correlation between plant weight at harvest and emergence time. Unequivocal evidence that these relationships are due to competition, rather than to the other possible causes, is much rarer (see Benjamin and Hardwick, 1986).

The *Bromus* experiment does supply such evidence (Table 2). The plants were sown at four densities, between which there were no significant differences in the age structure of seedlings. At the higher densities, the later-emerging plants were smaller than plants of the same age at low densities. In addition, when the relationship between plant size and sowing density was investigated for the separate cohorts, it was apparent that the weights of the later-emerging plants were affected more than those of the early-emerging plants. These results are exactly as one would expect if competition at the high densities was restricting the growth of the younger plants. The interaction between plant emergence time (as expressed by the variable *cohort:* cohort 1 plants had emerged before December 8, 1982; cohort 2 plants had emerged between then and 6

Table 2 Analysis of the Effects of Emergence Time and Density for Individual Plants of *Bromus sterilis* Grown at Four Sowing Densities

A. Summary of regression analysis of $\log_{10}$(individual plant weight at harvest in grams) against cohort for each density[a]

Density (m^{-2})	Intercept	Slope	r^2 (adj for d.f.)	Prob. Level
1	1.89	−0.027	0.055	NS
10	1.41	−0.070	0.094	<0.001
100	1.15	−0.217	0.298	<0.001
1000	0.342	−0.170	0.327	<0.001

B. The relationship between $\log_{10}$(individual plant weight in grams) and sowing density for plants in different cohorts[a]

Cohort	Intercept	Slope	r^2 (adj for d.f.)	Prob. Level
1	1.93	−0.536	0.566	<0.001
2	1.81	−0.588	0.754	<0.001
3	1.81	−0.602	0.726	<0.001
4 and 5	1.74	−0.640	0.914	<0.001
6 and over	1.68	−0.844	0.852	<0.001

[a] Cohort 1 plants are the oldest; they had emerged before the first monitor on December 8; cohort 2 plants emerged before January 6; the other cohort dates also correspond to the monitor dates given in Table 1 (see text for details). Log-transformation was used, as it gives an even spread of residuals.

January 1983; the other cohorts correspond to the monitor dates given in Table 1) and density is confirmed by fitting the simple model

$$\log w = a + b* \log N + c* \text{cohort} + d* \text{cohort}* \log N \qquad (7)$$

where w is individual plant weight (g), N is the sowing density (m^{-2}), and a, b, c, and d are parameters, to give the highly significant relationship

$$\log w = 1.860 − 0.426 \log N − 0.078 \text{cohort}* \log N \qquad (8)$$

($r^2 = 0.702$, $n = 333$, rss $= 43.40$; all parameters significant at $p < 0.001$) in which the cohort term, parameter c, is not significantly different from zero and so is excluded from the analysis. The significant interaction between age and density is clearly indicated.

D. Toward an Understanding of Plant Competition within Monocultures

Competition among plants can be investigated at a number of levels, from the physiological responses of individual plants to the effects of competition on the dynamics of whole communities. So far in this chapter, we have confined ourselves to describing the effects of competition on population structure. In this section, we examine how the competition–density effect, self-thinning, and size inequality might be generated as a result of the process of competition.

One can imagine a plant as taking light, water, and nutrients from a circular area around it. If the size of the plant is directly related to this zone, then it follows that, as the plant grows, then so will its zone of resource depletion. If the resource depletion zones of two plants overlap and resources are in limited supply, then competition will occur for the resources within the overlap area. If neither plant has a competitive advantage, the resources will be shared equally and competition may be regarded as being two-sided. However, it might more generally be expected that the plant which has emerged first and/or has the fastest growth rate will reduce the amount of light, water, and nutrients available to competing plants, thereby suppressing their growth. Such competition is one-sided, with those plants with the larger resource depletion zones using a disproportionate share of the resources within regions of overlap.

Simple models based on these ideas have been developed by Diggle (1976), Gates (1978), Aikman and Watkinson (1980), and Firbank and Watkinson (1985a), and are reviewed by Hara (1988). All of them show that one-sided competition gives rise to hierarchy development through time which mimics that observed in real monocultures. Diggle (1976) and Gates (1978) also showed that the assumption of one-sided partitioning of resources is necessary to mimic the bimodel size distributions apparently found in some populations (e.g., Ford, 1975).

Aikman and Watkinson (1980) went on to model size hierarchy development with time over a range of densities. By assuming that plants died once their relative growth rates became negative, and that competition was one-sided, self-thinning was generated. The self-thinning gradient was $-3/2$, because of an implicit allometric relationship for the plants between weight and resource depletion area. The time course of self-thinning was not well described by this model, as spatial relationships between plants were not made explicit, and so each plant was competing not with specific neighbors but with the "mean" of all the other plants in the stand.

Firbank and Watkinson (1985a), in contrast, related the size of each individual plant to the area of resources available to it after accounting for competition. Each plant was assigned a random position and allowed to grow and compete with its neighbors. When run at a range of different densities, this model generated realistic relationships between yield and density, regardless of the type of competition assumed and of the age structure of the population. Different age structures and assumptions about the nature of competition affected the parameter values of the competition density relationship [Eq. (3)] without affecting its suitability to describe the data. The competition density effect simply reflects that the total yield per unit area is restricted by the total amount of resources; the effect subsumes how those resources are actually divided among the individual plants. It is therefore possible, at least in principle, to reparameterize the effect in terms of those resources (e.g., Thornley, 1983) and the time available for plants to use them (e.g., Scaife *et al.*, 1987).

While the model by Firbank and Watkinson (1985a) mimicked self-thinning in a monoculture, the precise causes of the self-thinning rule are still questioned. Part of the reason for this is perhaps that the term "self-thinning" is used to refer to a process which has three distinct features, which have tended to become confused. First, self-thinning describes mortality due to competition in plant monocultures. Characteristically, it is the smaller plants in crowded areas which die, presumably after being overtopped and suppressed by their neighbors (Watkinson *et al.*, 1983; Schmitt, Eccleston, and Ehrhardt, 1987). This is the mechanism used in the models of Aikman and Watkinson (1980) and Firbank and Watkinson (1985a); in the latter model, if death follows a longer period of negative growth, self-thinning occurs at higher densities (Fig. 4).

The second feature of self-thinning is the fact that while self-thinning takes place, the total yield per unit area of the stand can continue to increase unless the resource levels are so low as to prevent further

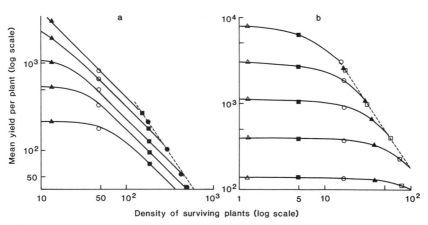

Figure 4 Competition–density curves of populations of different ages in simulated monocultures in which plants die (a) as soon as they have a negative relative growth rate and (b) after three time units of continuous negative growth. Scales are relative. Taken from Firbank and Watkinson (1985a).

growth (Lonsdale and Watkinson, 1982). It is inevitable that the thinning slope k in Eq. (8) must be stepper than -1. Weller (1987a) strongly suggests that the k value of $-3/2$ represents a typical, rather than the "true" value, perhaps varying with the allometries of the dimensions of the plants (Firbank and Watkinson, 1985a; Weller, 1987b; Norberg, 1988).

The third aspect of self-thinning is that, if all known thinning lines are plotted together, they appear to be bounded by one overall line,

$$\log_{10} w = 5 - 1.5 \log_{10} N \tag{9}$$

where w is mean weight per plant (g) and N is density (m^{-2}) (White, 1985). The differences noted above between self-thinning lines of different populations in no way invalidate this overall relationship, which applies only to the most extreme combinations of biomass and density for a given stand (White, 1985). The value of the intercept of Eq. (9) should be regarded as revealing a limit to the amount of biomass which can be supported by plant tissues (Givnish, 1986; Norberg, 1988), and the slope of $-3/2$ could well simply relate to Yoda *et al.*'s (1963) explanation that canopy mass should be proportional to the cube of the linear dimension height, whereas area of actively photosynthesizing tissue should scale as the square of this dimension (Givnish, 1986; Hardwick, 1987). Even populations undergoing two-sided competition would be subject to these constraints, and should not display combinations of yield and density beyond Eq. (9).

Like the earlier models, the Firbank and Watkinson (1985a) model mimicked the development of size hierarchies. Furthermore, because the position of each plant was known, it was possible to regress the sizes of these computer-generated plants against the various measures of neighborhood density outlined above. Unless there was virtually no variation in emergence time, or the competitive advantage to the larger plants was very small or nonexistent, these regressions were rarely significant. Even in the absence of environmental heterogeneity, it appears that the effects of local density and those of emergence time and intrinsic growth rate interact in a nonlinear manner which makes analysis very difficult (Firbank and Watkinson, 1987).

To summarize, all three major effects of density on monocultures have been modeled by assuming that plant size is a function of resource depletion area and that competition is one-sided. Such general models are not well suited for fitting to specific populations, as the parameters may be difficult or impossible to measure. However, Benjamin (1988) has developed a model of one-sided competition which can be fitted to plant stands, as can Hara's (1986a,b) models which describe hierarchy development as a function of photosynthetic rates and light interception.

III. Competition within Two-Species Mixtures of Plants

A. The Effects of Competition on Individual Plants

The behavior of mixtures of plants can be explained, at least in principle, by the behavior of the component individual plants. Plants are competing for the same resources—light, water, and nutrients—and so the performance of each plant depends on its ability to obtain and utilize those resources, just as it does in monocultures. One might expect hierarchies to form and develop in mixtures in the same way as happens in monocultures. In extremely crowded conditions, this behavior should lead to self-thinning. Also, as the total yield of the mixture depends on the amounts of resources available, one should expect mixtures to obey competition–density effects (e.g., Kira *et al.*, 1953). Nevertheless, competition within mixtures is more complex than that in monocultures as the different species may have different resource requirements, different patterns of growth, respond differently to environmental conditions, and modify the environment for each other.

Few studies have investigated the effects of competition on individual plants in mixtures. In a comparison of size hierarchy development in monocultures and two-species mixtures of wild oats (*Avena fatua*), leafless pea (*Pisum sativum* cv. Filbe), and conventional pea (*Pisum sativum* cv. Birte), Butcher (1983) found that the degree of inequality among all individuals in the mixtures appeared to be related to the ability of one species to outcompete the other. Where the two species seemed evenly matched, the degree of size inequality was similar to that observed in the monocultures. In contrast, the inequality of the conventional pea–wild oat mixture, in which the pea outgrew the oat, was considerably greater than that in either monoculture. Considering the inequality within the component species of the mixtures, Weiner (1985) found that *Lolium perenne,* the dominant species when grown with *Trifolium incarnatum,* showed less size inequality in mixture than in monoculture, whereas *Trifolium* plants, the suppressed species, usually developed a greater degree of inequality. Furthermore, plants of *Bromus sterilis* in the experiment described earlier showed less variation in size in the presence of wheat than they did in monoculture (Table 1).

Neighborhood analyses have also been attempted on individual plants from mixtures, but, just as in monocultures, a great deal of the variance of plant size frequently remains unexplained by the spatial arrangement of the plants alone (e.g., Firbank and Watkinson, 1987; Pacala and Silander, 1987). While the outcome of competition between pairs of species is often apparently dependent on relative sizes (e.g., Miller and Werner, 1987; but see Connolly, 1986), it is potentially extremely sensitive to the relative emergence times (Spitters and Aerts, 1983). There have been few experiments involving the manipulation of starting conditions on competing species, although experiments designed to elucidate the ideal time to control weed infestations, the so-called critical period of competition (reviewed by Radosevich and Holt, 1984), suggest that it is those cohorts of weeds which emerge before or with the crop which cause the greatest loss of crop yield. We clearly need more experiments that are closely controlled, along with the methods for analyzing them, if we wish to clarify the development of hierarchies in two-species mixtures.

B. Competition at the Level of the Mean Plant

Our knowledge of self-thinning in mixtures is perhaps even more meager than our knowledge of hierarchy development. It seems, however, that mixtures of plants self-thin according to Eq. (5) (Bazzaz and Harper, 1976) and yield–density combinations are bounded by the same line [Eq. (9)] as are monocultures when all component species are considered collectively (White, 1985).

The response of yield per unit area to competition in two-species mixtures has been studied much more widely because of its relevance to agriculture. If the papers by the Japanese group did much to lay the foundations of our understanding of competition within monocultures, then surely the work of de Wit pioneered the analysis of competition within mixtures of plants with his replacement series design (de Wit, 1960; de Wit and ven den Bergh, 1965).

1. Replacement Series Designs   In a replacement series, the species are grown in mixtures of varying proportions and in monocultures, always keeping the total density the same (de Wit, 1960). The results are expressed using indices such as the relative yield total (the sum of the yields in mixture of the two species expressed as proportions of their yields in monoculture), which describes how efficiently the two species use resources when grown together, and the relative crowding coefficient, which gives some indication as to the competitive abilities of the two species (de Wit, 1960; de Wit and van den Bergh, 1965). The analysis [reviewed by Hall (1974a), Harper (1977), and many others] is extremely popular, not least because it is based on small experiments.

A series of papers (e.g., DeBenedictis, 1977; Inouye and Schaffer, 1981; Jolliffe *et al.,* 1984; Firbank and Watkinson, 1985b; Connolly, 1986; Law and Watkinson, 1987) has strongly criticized this experimental design. Marshall and Jain (1969) showed that the results of the experiment and hence the values of the standard indices vary according to the total density selected, a point stressed by Firbank and Watkinson (1985b). This means that the results from a replacement series experiment cannot provide a reliable indicator of the outcome of competition over several generations (Law and Watkinson, 1987). Furthermore, the values of the indices are highly unstable, being highly dependent on the chosen experimental design (Connolly, 1986), and their statistical behavior is difficult to comprehend (see Thomas, 1970) and as a result tends to be ignored.

The replacement series is, however, extremely valuable for comparing the outcome of competition between two plants species under different conditions. Its use has led to important insights into the nature of niche differentiation (Trenbath, 1974) and differential resource use by plants. For example, Berendse (1982) demonstrated niche differentiation between species with different rooting depths, and Hall (1974b) discovered that the depressing effect of *Setaria* on *Desmodium* could be ameliorated by the addition of potassium. Other experiments have shown that pathogens can alter the outcome of competition between species (reviewed by Burdon, 1987), as can parasitic plants (Gibson, 1986), herbivores (Whittaker, 1979), and the presence or absence of mycorrhiza (Fitter, 1977). In a slightly different use of the replacement series, Cottam (1985) showed that herbivores may feed on plants according to their proportions in mixtures. Such experiments are highly informative, as they show the effects of single factors on the outcome of competition and suggest how plant interactions may be modified in the field. Nevertheless, alternative designs are available, and these include the additive design and the addition series.

2. Additive Designs   In an additive experiment, the density of one component is held constant while that of the other is varied. This design has been criticized, notably by Harper (1977; Begon *et al.*, 1986), on the grounds that total density and proportion vary together, thus confounding the effects of total density and frequency. But this is a problem only if the model used to describe the effects of competition uses these two variables; if the two equivalent variables of the density of each species are used, then the problem disappears.

The additive design lends itself to the investigation of weed–crop systems in particular. It is of course important in agriculture to quantify the effects of competition between weeds and crops. One approach is to modify the single-species competition model

$$w = w_{\mathrm{m}}(1 + aN)^{-b} \tag{3}$$

where w is the mean weight per plant, N is the density of plants, w_{m} is the weight in the absence of competition, and a and b are fitted parameters to give the model

$$w_{\mathrm{A}} = w_{\mathrm{mA}}[1 + a_{\mathrm{A}}(N_{\mathrm{A}} + \alpha N_{\mathrm{B}})]^{-b_{\mathrm{A}}} \tag{10}$$

where A and B are used to identify the two species and α is the competition coefficient which describes the average effect of an individual of species B on an individual of species A (Watkinson, 1981, after Hassell and Comins, 1976). An analogous equation can be used to describe density-dependent mortality in mixtures

$$N_{sA} = N_{iA}[1 + m_A(N_{iA} + \beta N_{iB})]^{-1} \qquad (11)$$

where N_i and N_s refer to initial and final densities, respectively, and β again is a competition coefficient. These two equations can readily be used to analyze the performance of the species which varies in density in additive experiments (Fig. 2).

In agriculture, there is usually greater interest in the yield of the species at constant density, the crop, than in the performance of the weed. In a major review of the subject, Zimdahl (1980) concluded that the relationship between crop yield and weed density is typically sigmoidal, with little change in crop yields at very low or very high weed densities. But in a more critical study, Cousens (1985) showed that the evidence actually reveals the function to be typically hyperbolic. In a comparison of 18 models he suggested that the model

$$y = y_m[1 - IN_w/(1 + IN_w/A)] \qquad (12)$$

gave a particularly good fit to the data, where y is crop yield per unit area, y_m is the weed-free yield, N_w is the density of weeds (either initial density

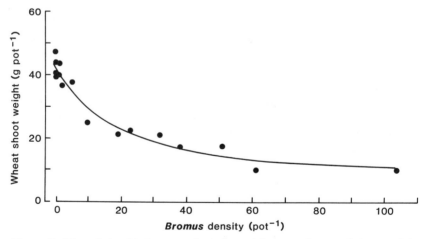

Figure 5   The relationship between shoot dry weight per unit area of wheat and density of *Bromus sterilis* in an additive pot experiment. The curve shows Eq. (12) fitted to data, giving the equation $y = 42.7[1 - 0.052N(1 + 0.052N/0.88)^{-1}]$, where y is wheat shoot weight (g per pot) and N is *Bromus* density per pot (L. G. Firbank, R. Cousens, A. M. Mortimer, and R. R. Smith, unpublished observations).

or at harvest), and I and A are fitted parameters (Fig. 5). Alternatively, Eq. (10) can be modified to describe crop yield as a function of weed density, giving the model

$$y = y_m(1 + \alpha N_w/N_c)^{-b} \tag{13}$$

where N_c is the crop density and other parameters are as previously defined [note that α tends to be correlated with b, making interpretation of these parameters much less clear than for Eqs. (3) and (10)]. Both Eqs. (12) and (13) require a wide range of weed densities in order to give good parameter estimates, and log-transformation is not usually needed.

3. Addition Series Neither the replacement nor the additive designs attempt to describe the complete range of outcomes of competition between two species. These outcomes form a response surface, and the substitutive and additive designs are restricted in that they merely take slices through that surface. A thorough understanding of the competitive interaction between pairs of species can only be achieved by growing them in a complete design which includes a wide range of frequencies and total densities.

One possible experimental approach is to replicate a replacement series design at a wide range of densities (Marshall and Jain, 1969), giving an addition series (Spitters, 1983b). Firbank and Watkinson (1985b) analyzed such an experiment, and showed that the competition models (10) and (11) could describe mean yield per plant and density at harvest for both *Agrostemma githago* and spring wheat over all densities and frequencies studied. However, Law and Watkinson (1987) made a much more intensive study of the response surface by using a far greater number of density combinations (Fig. 6). They found that, when *Phleum arenarium* and *Vulpia fasciculata* were grown together, Eqs. (10) and (11) failed to provide a good fit to the data. Their preferred model to describe the effects of the densities of species A and B on the mean weight of plants of species A was of the form

$$w_A = w_{mA}/(1 + N_A^{c_{AA}} + N_B^{c_{AB}}) \tag{14}$$

where the power terms c_{AA} and c_{AB} are fitted parameters, and the other parameters remain as previously defined. This model fits because it does not include a constant competition coefficient term; in other words, the competitive equivalence between plants of different species is allowed to vary with frequency and density. Perhaps the morphology of the plants responded in different ways to different combinations of frequency and density. If this result is typical, then it becomes even more important to realize the limitations of virtually all competition experiments carried out to date, for, without a wide range of densities and frequencies within

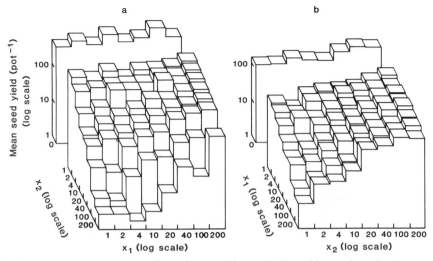

Figure 6 The response surface of competition, as indicated by seed production per pot, among (a) *Phleum arenarium* and (b) *Vulpia fasciculata* sown in monocultures and mixtures over a range of densities and frequencies. From Law and Watkinson (1987).

the design, the parameter estimates are bound to be sensitive to the densities which the experimenter happens to have selected.

IV. Forecasting the Dynamics of Monocultures and Mixtures

In both pure and applied ecology, it is often of interest to know how a plant population will behave in the future. In weed control, for example, it is highly desirable to know how a weed infestation will respond to a particular management regime and what crop yield losses will result. Rational weed management systems require adequate models of weed–crop systems which show the effects of the weed on crop yield and which can predict the size of the weed population into future years.

The arable weed *Agrostemma githago,* the corncockle, has a particularly simple lifecycle and can be used to demonstrate how one can construct a population model (Firbank and Watkinson, 1986). This once-common weed of cereal crops is a self-pollinating winter annual with a negligible seed bank. From a given density of seeds at the beginning of autumn, we can forecast the density of seeds the following autumn if we know the germination success of the seeds, the survival rate of the seedlings, and the seed production of the adult plants. For *Agrostemma* growing in spring wheat, the number of seeds germinating is a simple proportion of the density of seeds, the survival of plants is a function of their density which can be described using Eq. (11), and seed production is an allometric function of plant size, which is related to density and that of the crop according to Eq. (10). One can estimate the parameters of these functions using one year's data to give a simple predictive population model, which is described in detail in Firbank and Watkinson (1986).

Modeling the dynamics of *Agrostemma* is made easier by the fact that it is a crop mimic; its seeds used to be harvested along with the crop and subsequently resown with the crop seeds. The major form of control involved separating the weed seeds from the grain. The model shows how moderate levels of control would have little effect on weed population size, but that increasing seed cleaning efficiency has a progressively greater effect on the weed population (Fig. 7). This finding ties in well with the history of *Agrostemma*. A similar model has been developed for infestations of *Alopecurus myosuroides* to help determine the threshold weed densities above which control is economically worthwhile (Doyle *et al.*, 1986).

Most weeds are, however, more difficult to model than *Agrostemma*, as the propagules are already in the soil when the crop is sown. The relative emergence times are likely to differ from year to year (e.g., Roberts, 1984), which can markedly affect the development of the size hierarchies, thus affecting the outcome of competition. Furthermore, the recruitment, growth, and mortality of the weed may change from season to season (e.g., Reader, 1985), from place to place (Cousens *et al.*, 1988), from soil type to soil type (L. G. Firbank, R. Cousens, A. M. Mortimer, and R. R. Smith, unpublished observations), and with management practice (Pollard, 1982). Also, many weed species have complex life histories, including seed banks (e.g., *Avena fatua*) and bud banks from rhizomes (e.g., *Elymus repens*) which complicate their dynamics still further (e.g., MacDonald and Watkinson, 1981; Cousens *et al.*, 1986).

It is one thing to develop a model which describes in general terms the population dynamics of a species; it is quite another to develop a model which will predict how its numbers will change in the future. For example, when the *Agrostemma* model was modified and applied to infestations of *Bromus sterilis* in wheat, it failed to predict the weed densities in the following season (Firbank *et al.*, 1985). The data used to estimate the parameters of the model were those reported earlier in this chapter, coupled with estimates of seed survival during the following summer, i.e., 1983. Knowing the seedling densities in December 1983, it was possible to predict the densities expected in December 1984. However, the summer of 1984 was wetter than the previous year, stimulating earlier germination of the *Bromus* seeds. As a result, a higher proportion of the *Bromus* population was killed during cultivation and sowing of the

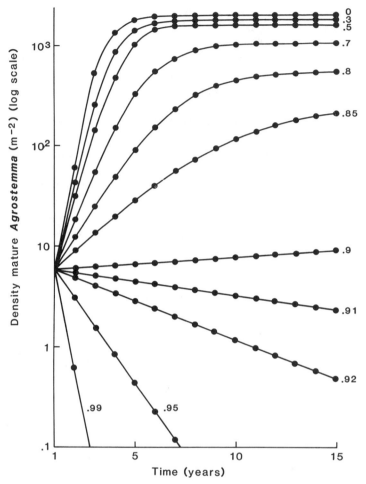

Figure 7   The effects of different levels of mortality on the dynamics of populations of *Agrostemma githago,* with an initial density of 10 seeds m^{-2} growing in continuous wheat as predicted by the population model of Firbank and Watkinson (1986). From Firbank and Watkinson (1986).

crop. A second source of error was that those seedlings which did survive were mostly established before the crop, and so were higher in the yield hierarchy than in the previous year, thus affecting the outcome of competition with the crop (Firbank *et al.,* 1985).

   If we are to predict the behavior of plant mixtures, we need to discover how the various environmental factors affect the outcome of competition. The greater the degree of precision that is required, the greater the amount of data and the more complex the model needed. We need to combine simulation modeling, tightly controlled experiments, and natural experiments (cf. Diamond, 1986). We hope to have shown that a great deal is known about the effects of competitive interactions on the population biology of monocultures and two-species mixtures, at least in agricultural systems. The next steps are to discover how the parameters of the models vary, both empirically and in terms of the underlying physiological processes, and to relate these models to the dynamics of populations within natural communities.

V. Summary

The effects of competition among plants have been studied intensively within monocultures and two-species mixtures. Within monocultures, as crowding and hence competition increase, the mean size per plant is reduced. Also, in crowded stands, self-thinning occurs and the degree of inequality among the plants increases with time.

All of these effects can be modeled by assuming that individual plants compete by preempting access to resources. The same assumption appears to hold true for competition between species in two-species mixtures. Differences in relative growth rates and relative emergence times thus have a major influence on the outcome of competition between pairs of species.

Many studies of competition in two-species mixtures are limited in scope because of the use of restricted experimental designs such as the replacement series and the additive series. A complete design, such as the addition series, is needed to describe the complete response of each species.

We show how the dynamics of competing species may be modeled, but the parameters of such models differ from place to place and from year to year. As a result, we are not yet in a position to predict the effects of competition on the dynamics of plant populations with a high degree of accuracy.

References

Aikman, D. P., and Watkinson, A. R. (1980). A model for growth and self-thinning in even-aged monocultures of plants. *Ann. Bot.* **45,** 419–427.

Bazzaz, F. A., and Harper, J. L. (1976). Relationship between plant weight and numbers in mixed populations of *Sinapis alba* (L.) Rabenh. and *Lepidium sativum* L. *J. Appl. Ecol.* **13,** 211–216.

Begon, M., Harper, J. L., and Townsend, C. R. (1986). "Ecology, Individuals, Populations and Communities." Blackwell, Oxford, England.

Benjamin, L. R. (1988). A single equation to quantify the hierarchy in plant size induced by competition within monocultures. *Ann. Bot.* **62,** 199–214.

Benjamin, L. R., and Hardwick, R. C. (1986). Sources of variation and measures of variability in even-aged stands of plants. *Ann. Bot.* **58,** 757–778.

Berendse, F. (1982). Competition between plant populations with different rooting depths. III. Field experiments. *Oecologia* **53,** 50–55.

Biere, A. (1987). Ecological significance of size variation within populations. *In* "Disturbance in Grasslands" (J. van Andel *et al.,* ed.), pp. 253–263. Junk, The Hague, The Netherlands.

Black, J. N., and Wilkinson, G. N. (1963). The role of time of emergence in determining the growth of individual plants in swards of subterranean clover (*Trifolium subterraneum* L.). *Aust. J. Agric. Res.* **14,** 628–638.

Bleasdale, J. K. A., and Nelder, J. A. (1960). Plant population and crop yield. *Nature (London)* **188,** 342.

Burdon, J. J. (1987). "Diseases and Plant Population Biology." Cambridge Univ. Press, Cambridge, England.

Butcher, R. E. (1983). "Studies on Interference between Weeds and Peas," Ph.D. thesis. Univ. of East Anglia, Norwich, England.

Connolly, J. (1986). On difficulties with replacement-series methodology in mixtures experiments *J. Appl. Ecol.* **23,** 125–137.

Cormack, R. M. (1979). Spatial patterns of competition between individuals. *In* "Spatial and Temporal Analysis in Ecology" (R. M. Cormack and J. K. Ord, eds.), pp. 151–252. Int. Publ. House, Fairfield, Maryland.

Cottam, D. A. (1985). Frequency-dependent grazing by slugs and grasshoppers. *J. Ecol.* **73,** 925–933.

Cousens, R. (1985). A simple model relating yield loss to weed density. *Ann. Appl. Biol.* **107,** 239–252.

Cousens, R., Doyle, C. J., Wilson, R. J., and Cussans, G. W. (1986). Modelling the economics of controlling *Avena fatua* in winter wheat. *Pestic. Sci.* **17,** 1–12.

Cousens, R., Firbank, L. G., Mortimer, A. M., and Smith, R. R. (1988). Variability in the relationship between crop yield and weed density for winter wheat and *Bromus sterilis. J. Appl. Ecol.* **25,** 1033–1044.

Daniels, R. F. (1976). Simple competition indices and their correlation with annual loblolly pine growth. *For. Sci.* **22,** 454–456.

DeBenedictis, P. A. (1977). The meaning and measurement of frequency-dependent competition. *Ecology* **58,** 158–166.

de Wit, C. T. (1960). On competition. *Versl. Landbouwkd. Onderz.* **66,** 1–82.

de Wit, C. T., and van den Bergh, J. P. (1965). Competition among herbage plants. *Neth. J. Agric. Sci.* **13,** 212–221.

Diamond, J. M. (1986). Overview: Laboratory experiments, field experiments and natural experiments. *In* "Community Ecology" (J. M. Diamond and T. J. Case, eds.), pp. 3–22. Harper & Row, New York.

Diggle, P. J. (1976). A spatial stochastic model of inter-plant competition. *J. Appl. Probability* **13,** 662–671.

Doyle, C. J., Cousens, R., and Moss, S. R. (1986). A model of the economics of controlling *Alopecurus myosuroides* Huds. in winter wheat. *Crop Prot.* **5,** 143–150.

Firbank, L. G. (1984). "The Population Biology of *Agrostemma githago* L.," Ph.D. thesis. Univ. of East Anglia, Norwich, England.

Firbank, L. G., and Watkinson, A. R. (1985a). A model of interference within plant monocultures. *J. Theor. Biol.* **116,** 291–311.

Firbank, L. G., and Watkinson, A. R. (1985b). On the analysis of competition within two-species mixtures of plants. *J. Appl. Ecol.* **22,** 503–517.

Firbank, L. G., and Watkinson, A. R. (1986). Modelling the population dynamics of an arable weed and its effect upon crop yield. *J. Appl. Ecol.* **23,** 147–159.

Firbank, L. G., and Watkinson, A. R. (1987). On the analysis of competition at the level of the individual plant. *Oecologia* **71,** 308–317.

Firbank, L. G., Manlove, R. J., Mortimer, A. M., and Putwain, P. D. (1984). The management of grass weeds in cereal crops, a population biology approach. *Proc. Int. Symp. Weed Biol. Ecol. Syst., 7th* pp. 375–384.

Firbank, L. G., Mortimer, A. M., and Putwain, P. D. (1985). *Bromus sterilis* in winter wheat: A test of a predictive population model. *Aspects Appl. Biol.* **9,** 59–66.

Fitter, A. H. (1977). Influence of mycorrhizal infection on competition for phosphorus and potassium by two grasses. *New Phytol.* **79,** 119–125.

Ford, E. D. (1975). Competition and stand structure in some even-aged monocultures. *J. Ecol.* **63,** 311–333.

Ford, E. D., and Diggle, P. J. (1981). Competition for light in a plant monoculture modelled as a spatial stochastic process. *Ann. Bot.* **48,** 481–500.

Gates, D. J. (1978). Bimodality in even-aged plant monocultures. *J. Theor. Biol.* **71,** 525–540.

Gibson, C. C. (1986). "The Population and Community Biology of *Rhinanthus minor* L.," Ph.D. thesis. Univ. of East Anglia, Norwich, England.

Givnish, T. J. (1986). Biomechanical constraints on self-thinning in plant populations. *J. Theor. Biol.* **119,** 139–146.

Hall, R. L. (1974a). Analysis of the nature of interference between plants of different species. I. Concepts and extension of the de Wit analysis to examine effects. *Aust. J. Agric. Res.* **25,** 739–747.

Hall, R. L. (1974b). Analysis of the nature of interference between plants of different species. II. Nutrient relations in a Nandi Setaria and Greenleaf *Desmodium* association with particular reference to potassium. *Aust. J. Agric. Res.* **25,** 749–756.

Hara, T. (1986a). Effects of density and extinction coefficient on size variability in plant populations. *Ann. Bot.* **57,** 885–892.

Hara, T. (1986b). Growth of individuals in plant populations. *Ann. Bot.* **57,** 55–68.

Hara, T. (1988). Dynamics of size structure in plant populations. *Trends in Ecology and Evolution* **3,** 129–133.

Hardwick, R. C. (1987). The nitrogen content of plants and the self-thinning rule of plant ecology: a test of the core-skin hypothesis. *Ann. Bot.* **60,** 439–446.

Harper, J. L. (1977). "The Population Biology of Plants." Academic Press, London.

Hassell, M. P., and Comins, H. N. (1976). Discrete time models for two-species competition. *Theor. Pop. Biol.* **9,** 202–221.

Inouye, R. S., and Schaffer, W. M. (1981). On the ecological meaning of ratio (de Wit) diagrams in plant ecology. *Ecology* **62,** 1679–1681.

Jolliffe, P. A., Minjas, A. N., and Runeckles, V. C. (1984). A reinterpretation of yield relationships in replacement series experiments. *J. Appl. Ecol.* **21,** 227–243.

Kira, T., Ogawa, H., and Sakazaki, N. (1953). Intraspecific competition among higher plants. I. Competition–yield–density interrelationships in regularly dispersed populations. *J. Inst. Polytech., Osaka City Univ., Ser. D* **4,** 1–16.

Knox, R. (1987). "Hypothesis Testing in Plant Community Ecology: Analysis of Long-Term Experiments and Regional Vegetational Data," Ph.D. thesis. Univ. of North Carolina, Chapel Hill, North Carolina.

Koyama, H., and Kira, T. (1956). Intraspecific competition among higher plants. VIII. Frequency distribution of individual plant weight as affected by individual plants. *J. Inst. Polytech., Osaka City Univ., Ser. D* **7,** 73–94.

Law, R., and Watkinson, A. R. (1987). Response–surface analysis of two-species competition: An experiment on *Phleum arenarium* and *Vulpia fasciculata. J. Ecol.* **75,** 871–886.

Liddle, M. J., Budd, C. S. J., and Hutchings, M. J. (1982). Population dynamics and neighbourhood effects in establishing swards of *Festuca rubra*. *Oikos* **38,** 52–59.

Lonsdale, W. M., and Watkinson, A. R. (1982). Light and self-thinning. *New Phytol.* **90,** 431–445.

MacDonald, N., and Watkinson, A. R. (1981). Models of an annual plant population with a seedbank. *J. Theor. Biol.* **93,** 643–653.

Mack, R., and Harper, J. L. (1977). Interference in dune annuals: Spatial pattern and neighbourhood effects. *J. Ecol.* **65,** 345–363.

Marshall, D. R., and Jain, S. K. (1969). Interference in pure and mixed populations of *Avena fatua* and *A. barbata. J. Ecol.* **57,** 251–270.

Mead, R. (1966). A relationship between individual plant spacing and yield. *Ann. Bot.* **30,** 301–309.

Mead, R., and Curnow, R. N. (1983). "Statistical Methods in Agriculture and Experimental Biology." Chapman and Hall, London.

Miller, T. E. (1987). Effects of emergence time on survival and growth in an early old-field plant community. *Oecologia* **72,** 272–282.

Miller, T. E., and Werner, P. A. (1987). Competitive effects and responses between plant species in a first year old-field community. *Ecology* **68,** 1201–1210.

Mithen, R., Harper, J. L. and Weiner, J. (1984). Growth and mortality of individual plants as a function of "available area." *Oecologia* **62,** 57–60.

Mohler, C. L., Marks, P. L., and Sprugel, D. G. (1978). Stand structure and allometry of trees during self-thinning of pure stands. *J. Ecol.* **66,** 599–614.

Morris, E. C., and Myerscough, P. J. (1987). Allometric effects on plant interference. *Ann. Bot.* **59,** 629–633.

Nobel, P. S. (1981). Spacing and transpiration of various sized clumps of a desert grass, *Hilaria rigida. J. Ecol.* **69,** 735–742.

Nobel, P. S., and Franco, A. C. (1986). Annual root growth and intraspecific competition for a desert bunchgrass. *J. Ecol.* **74,** 1119–1126.

Norberg, R. A. (1988). Theory of growth geometry of plants and self-thinning of plant populations: geometric similarity, elastic similarity, and different growth modes of plant parts. *Am. Nat.* **131,** 220–256.

Odum, E. P. (1959). "Fundamentals of Ecology." Saunders, Philadelphia, Pennsylvania.

Pacala, S. W., and Silander, J. A. (1987). Neighborhood interference among velvet leaf *Abutilon theophrasti*, and pigweed, *Amaranthus retrofluxus. Oikos* **48,** 217–224.

Pollard, F. (1982). A computer model for predicting changes in a population of *Bromus sterilis* in continuous winter cereals. *Proc. Br. Crop Prot. Conf.—Weeds* pp. 973–979.

Radosevich, S. R., and Holt, J. S. (1984). "Weed Ecology: Implications for Vegetation Management." Wiley, New York.

Reader, R. J. (1985). Temporal variation in recruitment and mortality for the pasture weed *Hieracium floribundum:* Implications for a model of population dynamics. *J. Appl. Ecol.* **22,** 175–183.

Roberts, H. A. (1984). Crop and weed emergence patterns in relation to time of cultivation and rainfall. *Ann. Appl. Biol.* **105,** 263–275.

Ross, M. A., and Harper, J. L. (1972). Occupation of biological space during seedling establishment. *J. Ecol.* **60,** 77–88.

Scaife, A., Cox, E. F., and Morris, G. E. L. (1987). The relationship between shoot weight, plant density and time during the propogation of four vegetable species. *Ann. Bot.* **59,** 325–334.

Schellner, R. A., Newell, S. J., and Solbrig, O. T. (1982). Studies on the population biology of the genus *Viola*. IV. Spatial patterns of ramets and seedlings in three stoloniferous species. *J. Ecol.* **70,** 273–290.

Schmitt, H. J. Eccleston, J., and Ehrhardt, D. W. (1987). Dominance and suppression, size-dependent growth and self-thinning in a natural *Impatiens capensis* population. *J. Ecol.* **75**, 651–665.

Shinozaki, K., and Kira, T. (1956). Intraspecific competition among higher plants. VII. Logistic theory of the C–D effect. *J. Inst. Polytech., Osaka City Univ. Ser. D* **7**, 35–72.

Silander, J. A., and Pacala, S. W. (1985). Neighborhood predictors of plant performance. *Oecologia* **66**, 256–263.

Spitters, C. J. T. (1983a). An alternative approach to the analysis of mixed cropping experiments. 2. Marketable yield. *Neth. J. Agric. Sci.* **31**, 143–155.

Spitters, C. J. T. (1983b). An alternative approach to the analysis of mixed cropping experiments. 1. Estimation of competition effects. *Neth. J. Agric. Sci.* **31**, 1–11.

Spitters, C. J. T., and Aerts, R. (1983). Simulation of competition for light and water in crop–weed associations. *Aspects Appl. Biol.* **4**, 467–483.

Sutherland, R. A. and Benjamin, L. R. (1987). A new model relating crop yield and plant arrangement. *Ann. Bot.* **59**, 399–411.

Thomas, V. J. (1970). A mathematical approach to fitting parameters in a competition model. *J. Appl. Ecol.* **7**, 487–496.

Thornley, J. H. M. (1983). Crop yield and planting density. *Ann. Bot.* **52**, 257–259.

Trenbath, B. R. (1974). Biomass productivity of mixtures. *Adv. Agron.* **26**, 177–210.

van Andel, J., Nelissen, H. M. J., Wattel, E., van Valen, T. A., and Wassenaar, A. T. (1984). Teil's inequality index applied to quantify population variation of plants with regard to dry matter allocation. *Acta Bot. Neerl.* **33**, 161–175.

Watkinson, A. R. (1980). Density-dependence in single-species populations of plants. *J. Theor. Biol.* **83**, 345–357.

Watkinson, A. R. (1981). Interference in pure and mixed populations of *Agrostemma githago*. *J. Appl. Ecol.* **18**, 967–976.

Watkinson, A. R. (1984). Yield–density relationships: The influence of resource availability on growth and self-thinning in populations of *Vulpia fasciculata*. *Ann. Bot.* **53**, 469–482.

Watkinson, A. R., Lonsdale, W. M., and Firbank, L. G. (1983). A neighbourhood approach to self-thinning. *Oecologia* **56**, 381–384.

Weiner, J. (1982). A neighbourhood model of plant interference. *Ecology* **63**, 1237–1241.

Weiner, J. (1985). Size hierarchies in experimental populations of annual plants. *Ecology* **66**, 743–752.

Weiner, J., and Solbrig, O. T. (1984). The meaning and measurement of size hierarchies in plant populations. *Oecologia* **61**, 334–336.

Weiner, J., and Thomas, S. C. (1986). Size variability and competition in plant monocultures. *Oikos* **47**, 211–222.

Weller, D. E. (1987a). A re-evaluation of the −3/2 power rule of plant self-thinning. *Ecol. Monogr.* **57**, 23–43.

Weller, D. E. (1987b). Self-thinning exponent correlated with allometric measures of plant geometry. *Ecology* **68**, 813–821.

Westoby, M. (1984). The self-thinning rule. *Adv. Ecol. Res.* **14**, 167–225.

White, J. (1980). Demographic factors in populations of plants. *In* "Demography and Evolution in Plant Populations" (O. T. Solbrig, ed.), pp. 21–48. Blackwell, Oxford, England.

White, J. (1985). The thinning rule and its application to mixtures of plant populations. *In* "Studies on Plant Demography: A Festschrift for John L. Harper" (J. White, ed.), pp. 291–309. Academic Press, London.

Whittaker, J. B. (1979). Invertebrate grazing, competition and plant dynamics. *In* "Population Dynamics" (R. M. Anderson, B. D. Turner, and L. R. Taylor, eds.), pp. 207–222. Blackwell, Oxford, England.

Williamson, M. H. (1972). "The Analysis of Biological Populations." Arnold, London.

Yeaton, R. I., and Cody, M. L. (1976). Competition and spacing in plant communities: The northern Mojave desert. *J. Ecol.* **57,** 37–44.

Yoda, K., Kira, T., Ogawa, H., and Hozumi, K. (1963). Self-thinning in overcrowded pure stands under cultivated and natural conditions (intraspecific competition among higher plants. XI). *J. Biol., Osaka City Univ.* **14,** 107–129.

Zimdahl, R. L. (1980). "Weed Crop Competition: A Review." Int. Plant Prot. Cent., Corvallis, Oregon.

10

Phytoplankton Nutrient Competition—from Laboratory to Lake

U. Sommer

I. Introduction

This chapter is written 10 years after the publication of Tilman's (1977) pioneering chemostat competition experiments between the diatoms *Asterionella formosa* and *Cyclotella meneghiniana*. Since then, studies to assess exploitative competition among phytoplankton have strongly expanded,

Perspectives on Plant Competition. Copyright © 1990 by Academic Press, Inc. All rights of reproduction in any form reserved.

and experiments conducted at different laboratories have yielded a largely contradiction-free body of information and an impressive agreement between experimental results and underlying theory (Tilman, 1982). Although experiments published so far used the mineral nutrients P, Si, or N as limiting resources, theory suggests that the principal findings apply to any kind of nonsubstitutable resources, including light. For steady-state conditions, the following conclusions seem warranted: (1) Only as many species can coexist as there are limiting resources. (2) If species coexist, their relative abundance is controlled by the ratio of the supply rates of the limiting resources. (3) The outcome of competition experiments can be predicted from single-species physiological parameters (in the case of dissolved nutrients, nutrient-saturated reproductive rates, half-saturation constants of reproduction, yield coefficients). (4) There are tradeoffs in competitive abilities for different resources, e.g., among most diatom species tested so far, competitive good competitors for Si are poor ones for P, and vice versa (Tilman *et al.*, 1982). If many species are inversely ranked in their competitive ability for two different resources, all these species are competitive dominants at their optimal resource ratios. (5) Results of competition experiments are robust in that, at the generic level, the experiments are repeatable from laboratory to laboratory. *Synedra* spp., for example, have been the best competitors for P when there is a nonlimiting supply of Si (Tilman, 1981; Smith and Kalff, 1983; Sommer, 1983; Kilham, 1986). (6) There is a tendency toward similar behavior of species within higher taxa. Among multispecies assemblages, for example, diatoms tend to dominate at high Si : P ratios (Sommer, 1983), blue-green algae at low N : P ratios and high temperatures, and green algae at high N : P and low Si : P ratios (Tilman *et al.*, 1986).

The theoretical interest in the apparent contradiction between predictions stemming from the concept of competitive exclusion and the species richness of phytoplankton [Hutchinson's (1961) "paradox of the plankton") has lead to a modification of the steady-state experimental design by pulsed nutrient addition or pulsed dilution. As anticipated by theory (Armstrong and McGehee, 1977), nonsteady-state experiments with periodic perturbations have shown that more species can coexist than there are limiting resources in the classic sense. This can be interpreted as fluctuations being exploited as "additional resources" (Tilman, 1982). Moreover, optimal resource ratios of some species may change as conditions change from steady state to nonsteady state (Sommer, 1985). However, qualitative trends, e.g., dominance of diatoms at high Si : P ratios, were the same as in steady-state experiments.

Despite its pronounced success in experimental research, the concept of resource competition has met considerable opposition among phytoplankton ecologists (Harris, 1986), not because of alleged flaws in competition theory or experimentation but because of supposed absence of competition *in situ*. This assumption is based on the negation of nutrient limitation of phytoplankton growth rates either because of rapid and patchy recycling of nutrients by zooplankton (Goldman *et al.*, 1979), or because of the frequency of physical perturbation (Harris, 1986). Despite some deficiencies in Goldman's arguments (Tett *et al.*, 1985), his influential paper has cast doubt on the previous matter-of-course views about nutrient limitation, and reversed the burden of proof. Low ($<k_s$) ambient nutrient concentrations are no longer viewed as providing sufficient proof of nutrient limitation. By definition (Tilman, 1982), competition for resources does not occur if the resources do not limit reproductive rates.

After a decade of continuous culture experiments, we have learned much about mechanisms of competition, but relatively little about the importance of competition *in situ*. Support for the concept of competition could in some cases be obtained if the distribution of phytoplankton species is correlated with resource ratios, or if species replacements following shifts in resource ratios conform to experimentally obtained predictions. For such patterns to emerge, it is necessary, however, that competition proceeds under relatively invariable conditions long enough to give the competitively dominant species time to become numerically dominant. Indirect evidence for nutrient competition could be obtained by proving that reproductive rates are nutrient limited and that nutrient limitation is a consequence of prior consumption by algae. For lake plankton, nutrient limitation is usually a consequence of algal consumption, because total concentrations (dissolved plus bound in organisms and their debris) even in most oligotrophic lakes are much higher than typical half-saturation constants for reproduction. Searching for resource limitation has the advantage that resource competition becomes detectable long before species composition approaches the competitive equilibrium. In this chapter, I present some of the examples for both ways to check for competition *in situ* and discuss the methodological and principal problems involved. In the final section, I address problems which arise from the fact that, in the long run, competition locally eliminates those species which are most strongly negatively affected by it.

II. Are Nutrients Limiting *in Situ*?

Some nutrients, such as phosphorus and nitrogen, can be taken up in excess relative to the immediate demand and can be stored for later use. If nutrient-rich conditions are followed by nutrient-poor ones, reproductive rates can be maintained at a higher level than predicted by steady-state equations. Reproductive rates, however, still remain a function of intracellular limiting nutrient concentrations ("cell quota"; Droop, 1973, 1983):

$$\mu = \mu_{max}(1 - q_0/q) \tag{1}$$

where μ is the reproductive rate, μ_{max} the species-specific maximal reproductive rate, q the cell quota of the limiting nutrient, and q_0 the species-specific minimal cell quota. Direct application of the Droop model to field data is usually impossible, since natural samples usually contain many different species and detritus and the two constants of the equation are species specific. Goldman *et al.* (1979) suggested a simplification to overcome this problem: normalize the cell quota to biomass instead of cell number and replace μ by the quotient μ/μ_{max} ("relative growth rate" in their terminology). The "relative growth rate" then becomes a linear function of q^{-1} or of the carbon : limiting nutrient ratio:

$$\mu_{rel} = 1 - q_0/q \tag{2}$$

If biomass-specific minimal cell quotas are similar among species (cf. Sections II,B and IV) and detritus does not deviate greatly from living organisms, seston stoichiometry becomes a direct indicator of the nutritional status. Based on culture experiments with several species of marine algae, Goldman *et al.* (1979) suggested a particulate matter C : P ratio of 106 : 1 and a C : N ratio of 6.625 : 1 as criterion values between nutrient limitation and nutrient saturation; higher ratios indicate limitation.

The most unsatisfactory aspect of seston stoichiometry as an indicator of nutrient limitation lies in the fact that it does not detect interspecific differences in the intensity of nutrient limitation. If strongly nutrient-limited algae are being outcompeted by less limited ones, the average chemical composition becomes increasingly dominated by the less limited species and the importance of competition will be progressively more underestimated. More complicated techniques, such as the measurement of alkaline phosphatase activity (Berman, 1970) or short-term uptake measurements (Lean, 1984) share this problem. Testing the growth response after nutrient addition by cell counts can be species specific and has the advantage that it permits the study of rare species. However, bioassays of this type may overestimate the intensity of nutrient limitation, because in the control bottle, algae are protected from several loss factors (sinking, grazing) and, thus, may accumulate more biomass than *in situ*, becoming more nutrient limited. Diluting the plankton sample with filtered lake water (Løvstad, 1986) can overcome this problem. If nutrient addition is sufficient to permit growth at μ_{max} in the enrichment bottles, the results of bioassays can be tested by seston stoichiometry. A linear regression of μ_{rel} (community average of μ_{rel}, weighted by biomass) should give an intercept near unity, and the slope of the regression should be an aggregated analog of the q_0 parameter of Droop's equation [Eq. (1)].

A. Examples

The data of Goldman *et al.* (1979) (all referring to the tropical oceans) showed that seston stoichiometry clustered around the Redfield ratio (C : N : P = 106 : 16 : 1). The authors concluded that nutrient limitation was unlikely and attributed this result to the patchy excretion of nutrients by zooplankton and to rapid nonsteady-state uptake of the nutrients from micropatches by algae. Despite the geographic restriction of the data set, a number of authors (e.g., Harris, 1986) took this article as definite death of the longstanding concept of nutrient limitation of algal growth. Tett *et al.* (1985), however, noted that Goldman *et al.*'s (1979) data set included phytoplankton from deep-water layers, where limitation by insufficient light could preclude nutrient limitation.

Meanwhile, seston stoichiometry data have been published which indicate nutrient limitation for a variety of limnetic and marine sites. Sakshaug *et al.* (1983) found C : P ratios of up to 357 in the Trondheimfjord (Norway) and 400 in the eutrophic lake Haugatjern (Norway), and C : N ratios of up to 14.7 in the North Sea and 12.9 in Haugatjern. In Lake Constance (FRG), maximal C : P ratios in the light-saturated layer are around 400 (Sommer, 1987), and in Esthwaite Water (England), ratios as high as 717 occur during the annual bloom of *Ceratium* (Heaney *et al.*, 1987). According to Goldman (1979), a C : P ratio of 400 indicates a μ_{rel} of 0.6 to 0.7, while a C : N ratio of 14, a μ_{rel} of about 0.25. Except for Esthwaite Water, the examples cited above show that elevated values of C : N or C : P persist only for a few weeks and are frequently interrupted by periods of nutrient saturation.

Similar results were found with enrichment bioassays during a 1.5-year study of phytoplankton from moderately eutrophic Schöhsee (FRG) (Sommer, 1988a). N or P limitation of algae was usually weak, and frequently interrupted (Fig. 1). The opposite was true for Si limitation of diatoms: throughout the entire stratified period these algae were limited strongly. The taxonomic distribution of the patterns of nutrient limitation conformed to the expectations based on competition experiments: When limitation occurred, diatoms were most commonly Si-limited, blue-green algae most commonly P-limited, and green algae most commonly N-limited. Flagellate taxa were not nutrient limited in >50% of cases; if limited, P limitation was most common, except for *Ceratium hirundinella* which was more often N-limited. The results of the bioassays

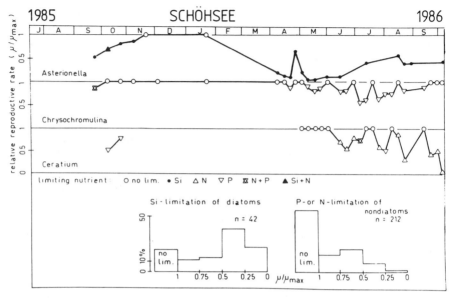

Figure 1 Nutrient limitation of Schöhsee phytoplankton as revealed by enrichment bioassays. Top three panels: Relative reproductive rates of a usually Si-limited alga (*Asterionella formosa*), a usually P-limited alga (*Chrysochromulina parva*), and a usually N-limited alga (*Ceratium hirundinella*). Bottom panel: Frequency distribution of relative reproductive rates of diatoms (if limited, Si-limited) and of nondiatomaceous algae (if limited, P- or N-limited).

were tested by comparison with C : N, C : P, and C : Si ratios of the total seston and of seston size-fractions, which were sometimes nearly detritus free. Regression of the average μ_{rel} on the cellular nutrient ratios (Table 1 in Sommer, 1988a) usually yielded good correlation coefficients, and as expected theoretically, the intercepts were around unity (from 0.93 to 1.26). Moreover, Goldman *et al.*'s (1979) criterion values were confirmed. The regressions predicted μ_{rel} values from 0.91 to 0.96 for a C : P ratio of 106 and from 0.95 to 1.00 for a C : N ratio of 6.625. However, there is a broad band of uncertainty. Up to C : P ratios of ~160 and a C : N of ~9.5, a μ_{rel} of 1 is within the 95% confidence intervals for the dependent variable.

In summary, Goldman *et al.*'s (1979) view that phytoplankton are growing at μ_{max} is no longer tenable. Limitation by N and P, which are rapidly recycled, appears discontinuous rather than continuous. *In situ* competition for these nutrients has probably more similarity with pulsed-state competition experiments (Sommer, 1985) than with steady-state experiments. Silicate limitation of diatoms, however, seems to be both stronger and more persistent, at least in thermally stratified systems. The interrupted nature of P and N competition is an important finding with respect to Hutchinson's "paradox of the plankton." After a decade of experimental competition research, Hutchinson's paradox no longer appears paradoxical. Experiments with simple deviations from steady state, such as periodic nutrient pulses (Sommer, 1985) or periodic dilution (Robinson and Sandgren, 1983; Gaedeke and Sommer, 1986), have shown that more species coexist than there are different limiting resources. If those experiments mimic lake conditions, we should expect that in Schöhsee the nonsiliceous algae (which discontinuously compete for N or P) would be more diverse than the diatoms (which more or less continuously compete for Si). This is indeed the case. Among diatoms, *Asterionella* makes up more than 90% of biomass throughout most of the growth season; among the other algae, usually at least five species exceed 10% and a single species never exceeds 50% of nondiatom biomass. In hydrographically less stable Lake Constance, where Si also shows strong variability during periods of stratification, diatoms are also quite diverse.

B. Problems and Perspectives

The use of Redfield stoichiometry for the criterion value between nutrient limitation and nutrient saturation requires some uniformity in the minimal cell quotas of the dominant phytoplankton species in a sample. Goldman (1979) found this uniformity among cultures of marine algae. Similarly, the analysis of field data (Sommer, 1988a) confirmed the usefulness of the Redfield ratio as the criterion value (see above). This contrasts with the wide scatter of q_0 values in the literature. I found 31 values for freshwater phytoplankton where the minimal cell quota was expressed on a biomass basis (Nyholm, 1977; Shuter, 1978; Gotham and Rhee, 1981; Uehlinger, 1980; Smith and Kalff, 1982; Elrifi and Turpin, 1985; Ahlgren, 1987). If more than one value for a species was found, the one with temperature closest to 20°C was taken. If biomass was expressed as cell volume, the cell volume was converted to carbon according to Rocha and Duncan (1985). The data are widely scattered and fairly log normally distributed (see frequency distribution plot in Fig. 3). The mean value ($q_0 = 1.44 \times 10^{-3}$ P/C) corresponds to a relative reproductive rate of 0.847 at the Redfield ratio. The mean plus its standard deviation (3.56×10^{-3} P/C) corresponds to a μ_{rel} of 0.938, and the mean minus standard deviation (5.8×10^{-3} P/C) to a μ_{rel} of 0.623 at the Redfield ratio. The total range of data predicts anything from no growth at all to a μ_{rel} of 0.979. If field populations were as diverse in their q_0 as those laboratory strains, biomass stoichiometry would be of no indicative value at all. The discrepancy between the wide scatter among laboratory strains and the apparent relative uniformity of field data will be discussed further in Section IV.

Even if important lake algae are physiologically more uniform than the laboratory strains tested so far, the use of community average values of cellular stoichiometry might be misleading. A mixture of P-limited algae containing surplus N with N-limited algae containing surplus P, for example, may give an overall average close to Redfield stoichiometry. Moreover, if species competing for the same nutrient have similar μ_{max} values and similar resistance against mortality, the one with the higher μ_{rel} should win the competition. Competition will lead to an increasing enrichment of less-nutrient-limited algae. Thus, the overall community average of biomass stoichiometry may seem to provide evidence for only slight nutrient limitation, despite strong nutrient limitation of the losers. In conclusion, the stoichiometry approach of Goldman (1979) is biased *against* the discovery of nutrient limitation and competition. Because nutrient limitation has been detected nevertheless, we can be the more confident of its importance as a factor prompting competitive succession.

It is important to measure cellular nutrient stoichiometry not only for plankton mixtures but for separate populations. Size fractionation alone only occasionally yields monospecific fractions; the same is true for density gradient separation (Ierland and Peperzak, 1984). A combination of the two will more often be successful. An even more promising, though more expensive, technique is the measurement of the individual cell's nutrient content by energy-dispersive X-ray analysis (EDAX). This method also provides an estimate of within-population variability of cell quota values, which may describe the patchiness of nutrient availability from an alga's perspective (Lehman, 1985).

III. Dominance of Algal Taxa in Relation to Nutrient Ratios

A. Examples

The examples presented here relate to two taxa, diatoms and blue-green algae, about which relatively much is known from competition experiments. Unfortunately, we know nearly nothing about the competitive performance of such important taxa as chrysophytes, cryptophytes, and dinoflagellates. The examples here address the following hypotheses: (1) *Asterionella* is favored over *Cyclotella* by high Si : P ratios, (2) diatoms are favored by high Si : P ratios, and (3) blue-green algae are favored by low N : P ratios. Rejection of the null hypothesis, in each case is assumed to provide evidence for the importance of competition in the determination of species composition.

In Tilman's (1977) first series of competition experiments, *Asterionella formosa* excluded *Cyclotella meneghiniana* when Si : P in the chemostat inflow medium was >90 : 1. *Cyclotella* dominated at ratios of <6 : 1. In the intermediate range, the two species coexisted in perfect equilibrium, with *Asterionella* increasing in relative abundance with increasing Si : P ratios. Analysis of field data along a river-plume to mid-lake gradient in Lake Michigan revealed that 74% of the variance in the ratio *Cyclotella* : *Cyclotella* + *Asterionella* could be explained by the logarithm of the dissolved, molar Si : P concentration ratios. The trend line for the field data seemed to agree very well with the experimental results. Later, a calculation error in the ratios was discovered (Sell *et al.*, 1984): the actual ratios were three times higher than published earlier. The positive relationship between percent *Asterionella* and log(Si : P) was retained, but the trend line for the field data was no longer the same as for the experimental data. Tilman *et al.* (1984) ascribed this discrepancy to the fact that the ratios of the supply rates of Si and P (the real determinant of competitive success) are lower than Si : P concentration ratios (the surrogate parameter). The similarity of slopes for the experimental and the field data trend lines, however, suggested a proportionality between supply rate ratios and concentration ratios.

Most phytoplankton field ecologists are probably more familiar with seasonal shifts in species composition than with shifts along horizontal gradients. The examples presented here are extracted from my own data sets from deep, nutrient-rich (P_{tot} during circulation ~3 μM) Lake Constance (Sommer, 1987) and from moderately deep, mesotrophic Schöhsee (P_{tot} during circulation ~0.7 μM). In both lakes, diatoms showed a pronounced seasonality and so did dissolved Si : P ratios (Fig. 2). There were three peaks of diatom contribution to total phytoplankton biomass in Lake Constance in 1979. The summer and the fall peaks followed peaks in dissolved Si : P. Both diatom peaks were composed of several members of the family Fragilariaceae, which are good competitors for P and have high optimal Si : P (Tilman *et al.*, 1982). The spring peak of diatoms was not preceded by elevated Si : P ratios. This peak was composed mainly of small centric diatoms whose optimal Si : P ratios are about 6 : 1, which is lower than the minimal ratios in Lake Constance. In Schöhsee, both the large vernal peak of *Asterionella formosa* and the small late summer peak (*Asterionella* and *Synedra acus*) were preceded by peaks in the concentration ratios of Si and P. Note, however, that in both lakes a time lag of 1 to 2 weeks occurred between the Si : P ratios and the relative biomass of diatoms. This was about the time needed to grow up from being almost indetectable (~10^3 colonies per liter) to be a dominant species (~10^6 to 10^7 colonies per liter). The time lags point to the nonequilibrium character of the systems. Increasing Si : P ratios shift the competitive advantage to the diatoms, which then consume increasing amounts of silicate as population densities increase. Since silicate, unlike phosphate, is not recycled by grazing zooplankton, high Si : P ratios can-

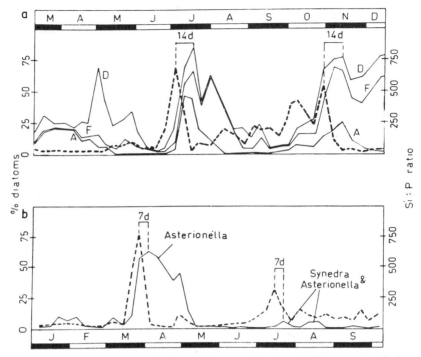

Figure 2 Temporal change in the relative contribution of diatoms to total phytoplankton biomass in Lake Constance and Schöhsee as delayed response to Si : P concentration ratios. (a) Lake Constance in 1979: thick broken line gives the Si : P concentration ratios in euphotic zone; thin lines are A, *Asterionella;* F, Fragilariaceae; D, all diatoms. (b) Schöhsee in 1986: thick broken line gives Si : P ratios in the euphotic zone; thin line, all diatoms (practically only *Asterionella* and *Synedra*).

not be maintained. A correlation between percent diatoms and Si : P ratios without time lag would be negative instead of positive. The annual maxima of concentration ratios for Si and P in both lakes were about 700 : 1; the following dominance peak by diatoms reached about 85% in Lake Constance and about 70% in Schöhsee, respectively. In multispecies steady-state experiments with Lake Constance phytoplankton, a 75% dominance by diatoms was achieved at an Si : P inflow ratio of 20 : 1 (Sommer, 1983). Under pulsed nutrient supply (Sommer, 1985), an input ratio of about 100 : 1 was required to reach that degree of dominance by diatoms. In microcosm experiments under simultaneous grazing pressure and nutrient competition (Sommer, 1988c), that degree of dominance was achieved at ratios of 600 to 700.

Smith (1983) tested the hypothesis that low N : P ratios favor blue-green algae by comparing the growth-season mean of blue-green algae contribution to total biomass with the growth-season mean of epilimnetic total nitrogen to total phosphorus ratios of 17 lakes. Lakes with $N_{tot} : P_{tot}$ of >64 : 1 (atomic ratio) had low (usually <10%) proportions of blue-greens, whereas lakes with $N_{tot} : P_{tot}$ of <64 : 1 had low to very high (nearly 100%) proportions of blue-greens. Smith concluded that low N : P ratios permitted, but do not necessarily cause, dominance by blue-greens. Deep mixing, for example, may prevent the growth of bloom-forming blue-green algae, even if the nutrient regime is favorable (Reynolds *et al.*, 1984). Despite the qualitative agreement between Smith's field data and continuous culture experiments (Tilman *et al.*, 1986), there remains an unresolved quantitative discrepancy. Smith's threshold ratio of 64 : 1 is much higher then the usually cited optimal N : P of 16 : 1 for average phytoplankton. In a survey of optimal N : P ratios of several algal species, Rhee and Gotham (1980) found a maximal value of 30 : 1 (*Scenedesmus obliquus*) and a minimal value of 7 : 1 (*Melosira binderana* = *Stephanodiscus binderanus*).

According to competition theory, bloom-forming blue-greens (many of them as N_2 fixers independent of N ions) should become dominant at ratios lower than the optimal ratios of the other algae. Total nitrogen might be a poor indicator of biologically available nitrogen, because a large proportion is usually highly refractory dissolved organic nitrogen (Liao and Lean, 1978). This might have led to an overestimate of the ratios of biologically available N and P.

B. Problems and Perspectives

The above examples show qualitative agreement between predictions based on resource ratios and observed trends in taxonomic composition, but do not show much agreement at the level of numerical values of resource ratios. This problem arises from the fact that, under natural conditions, different nutrients may cycle with different velocities. In a chemostat experiment, nutrient supply is solely determined by the external input as long as there is no internal cycling of nutrients in the culture flask. Nutrient supply rates are proportional to dissolved nutrient concentrations in the inflow medium, and dissolved nutrient concentrations in the inflow medium equal total nutrient concentration in the culture flask. Hence, nutrient supply rate ratios equal input medium ratios and total nutrient ratios in the culture. In lakes, this identity between the surrogate parameters (dissolved nutrient ratios, total nutrient ratios) and the real independent variable (supply rate ratios) cannot be guaranteed.

As long as vertical transport through the thermocline (eddy diffusion) is the only supply of nutrients to epilimnetic phytoplankton populations, supply rate ratios equal the ratios of the vertical concentration gradients. Vertical transport of nutrients by migrating organisms can already lead to substantial change in the ratios of supply rates. Even more important is the short cycle of nutrients within the epilimnion. Usually, excretion by zooplankton is a major source of dissolved nutrients for algal populations ("regenerated production," sensu Dugdale and Goering, 1967). Nitrogen is usually excreted as dissolved ammonium and urea, phosphorus as soluble reactive phosphorus (Sterner, 1989). Both are readily taken up by phytoplankton. Silicate, however, is excreted as particulate debris which dissolves slowly. Most debris of diatom origin is lost by sedimentation before dissolution (Sommer, 1988b). In the presence of grazing, it follows that Si will cycle much slower than P and N; Si : P or Si : N supply rate ratios will be lower than the concentration ratios.

Similar problems might occur with N : P ratios, if excretion and assimilation into animal biomass are not proportional to the N : P ratios in the food algae. Zooplankton stop excreting significant amounts of P if the P content of the food organisms is lower than the P content of zooplankton biomass. For *Daphnia,* this occurs at an atomic C : P ratio of $\sim$300 (Olsen *et al.,* 1986), which implies only moderate P limitation of algae. This is a positive feedback loop which tends to reinforce P limitation of algae, because the more they are P-limited, the more they are cut off from P recycling. The same mechanisms do not apply to the recycling of N by zooplankton. NH_4 is an inevitable end product of protein metabolism which is excreted irrespective of the chemical composition of the food (Sterner, 1989). These differences between P and N recycling explain the relatively abrupt shifts between N limitation (at low zooplankton densities) and P limitation (at high zooplankton densities) reported by Elser *et al.* (1988). As a consequence, the relationship between the ratios of concentrations of N and P and the ratios of supply rates is by no means simple and is influenced by complex interactions between the nutritional status of phytoplankton and of zooplankton.

The use of concentration ratios as surrogate parameters makes an implicit assumption. Concentration ratios are supposed to be proportional to, or at least positively correlated with, supply rate ratios. This assumption seems plausible, but I know of no study providing a rigorous proof. The considerations above indicate that there is also not much hope that we will ever be able to construct a "calibration curve" which unambiguously translates concentration ratios into supply rate ratios. For the relatively simple cases of Si : P ratios, the relationship between both depends on the relative importance of "fresh" (through the thermocline or from the watershed) and "regenerated" nutrient sources, and both sources undergo pronounced seasonal changes. In the case of N : P ratios, even more complex interactions have to be considered. Moreover, excretion by zooplankton does not only make the ratio of nutrient supply rates differ from the ratio of nutrient concentration, it also changes the supply mode by causing microscale (excretion plumes) and macroscale (diel feeding patterns, zooplankton abundance fluctuations) patchiness of nutrient supply, which can also influence the outcome of competition.

IV. The Role of Competition in Assembling a Lake's Species Pool

To distinguish between coexistence and competitive exclusion, phytoplankton competition experiments typically had durations of 25–50 days. This duration was needed until the experimenters felt confident about which species were being excluded and which were able to persist. Time demand for competitive exclusion was relatively uniform across experiments, and independent of dilution rates and nutrient ratios. In the steady-state multispecies competition experiments of Smith and Kalff (1983) and Sommer (1983, 1986), it took 14 ± 3 (mean ± SD) days until the last looser started to decline, and 23 ± 5 days until the winning species or winning couple of species reached 95% of total biomass. This time demand relates relatively well to the time scale for species replacements within the seasonal succession of phytoplankton. Such species replacements take place within a well-defined species pool, which in turn has been assembled over the course of years. The question remains, How important has competition been in assembling a lake's species pool? Searching for the importance of competition at a longer time scale (decades) means turning Hutchinson's paradox upside down: Why there are so few phytoplankton species in a lake?

Typically, a lake's species pool contains several hundred phytoplankton species. This number is about two orders of magnitude lower than the total number of planktonic algal species worldwide. Except for recently filled reservoirs and lakes which have recently undergone drastic environmental change, colonization history probably plays only a very minor role, for phytoplankton are very vagile. Endemism is rare and, unlike terrestrial vegetation, similar lakes in different continents tend to have the same phytoplankton species. Even slow-growing algae (generation times of 2 to 4 days) need only a small inoculum and a few years to establish their populations if the conditions are suitable. Therefore, the explanation of presence and absence has to be sought in the local conditions for survival and growth. A small but unknown portion of the absences is probably due to absolute impossibility of survival caused by physical or chemical properties of a particular lake. The rest has to be explained by a long-term negative balance between reproductive rates and mortality rates, which either excludes a previously present population or prevents the establishment of a new one. Here the question of the relative importance of competition comes into play. Admittedly, such an inquiry means searching for the "ghost of competition past" (Connell, 1980). Competitive exclusion of those species which least withstand exploitative competition eliminates just those species which most clearly show the impact of competition and makes competition less detectable. However, the "ghost of competition past" may leave sufficient tracks to be detectable.

The most direct way is the long-term observation of a lake undergoing directional change. If a species is going to be excluded because of nutrient competition, it would obtain a smaller and smaller share of the limiting nutrient, which should be reflected in a declining trend of the cell quota or in a declining population density.

Even if long-term studies are not feasible, the "ghost of competition past" should leave some tracks amenable to current observation. If the interannual directional change of a lake is slow relative to the establishment of new populations and the decline of excluded populations, the species pool of a lake should be near competitive equilibrium at the interannual time scale, despite the prevalence of nonequilibrium conditions at the week-to-week time scale of within-season studies. This should be reflected by close relationships between taxonomic composition and resource ratios, if competition is an important factor. The increasing share of blue-green algae with decreasing $N : P$ ratios (Smith, 1983; see Section III,B) is an example at the level of taxonomic composition of biomass. In this section, we are more interested in the level of species numbers or partial diversities. Is there a higher species number or diversity of diatoms, if long-term mean $Si : P$ ratios are high? Is there a higher species number or diversity of blue-green algae, if long-term mean $N : P$ ratios are low? Such predictions can be derived from Figure 36 in Tilman (1982): On the multiannual time scale, within-season changes in resource ratios become undirectional fluctuations around the long-term mean and permit the coexistence of species which are "neighbors" on the resource ratio gradient, i.e., which have similar optimal resource ratios. According to the competition experiments published so far, these tend to be taxonomically related species. It follows that a higher taxon favored by the long-term average of resource ratios should not only be dominant in terms of biomass and abundance, but also in terms of species number.

Thus, competition theory provides a theoretical basis for a reexamination of Nygaard's (1949) plankton indices. Nygaard tried to characterize the trophic status of lakes by quotients of the species numbers of higher taxa (orders or divisions) within a lake. The species number of the supposedly more "eutrophic" taxon was put into the numerator and the species number of the supposedly more "oligotrophic" taxon into the denominator. Thus, the index was expected to increase with increasing eutrophication. The classification of taxa was based on intuition and experience, rather than on experimentation. Nygaard suggested five different indices: The cyanophycean index (number of species of blue-greens divided by number of species of desmids), the chlorococcalean index (number of species of chlorococcales divided by number of species of desmids), the diatom index (number of species of centric diatoms divided by number of species of pennate diatoms), the euglenophycean index (number of species of Euglenophyceae divided by number of species of greens and blue-greens together), and the compound index (number of species of blue-greens, chlorococcales, centric diatoms, and Euglenophyceae together divided by number of species of desmids). The concept of resource ratios is the key to evaluating the relationship of Nygaard's indices to competition theory. Certainly, resource ratios change along the oligotrophic–eutrophic gradient: Si : P, Si : N, light : P, and light : N ratios decline. Among Nygaard's indices, the diatom index is perfectly consistent with the fact that centric diatoms have relatively low optimal Si : P ratios while pennate diatoms have relatively high ones (Tilman *et al.*, 1982). The position of blue-greens, chlorococcales, and centric diatoms in the numerator of the compound index is also justified by the results of competition experiments (Tilman *et al.*, 1986). For further elaboration, however, there is a need to replace the one-dimensional oligotrophic–eutrophic gradient by explicit consideration of which nutrient is limiting. Nevertheless, the fact that a lifelong experience in phytoplankton field observations has led to the plankton indices supports the assumption that past competition has at least partially shaped the phytoplankton species pools of lakes.

If competition for a particular nutrient is important in determining a lake's species pool, the species that are present should, on average, be better competitors for that particular nutrient than phytoplankton species of lakes where other resources limit algal reproduction or than a randomly drawn subsample from all possible planktonic algal species. The data for such a comparison are not yet available. A first step, however, is possible if two assumptions are made. The first is that algal species used for nutrient kinetics studies represent an unbiased subsample of the total. Of course they are not a truly random subsample, but strongly selected by ease of isolation and cultivation. This bias, however, is not crucial as long as there is no systematic relationship between "competitiveness" and ease of isolation and cultivation. The second assumption is that competitiveness for a particular nutrient is inversely related to the minimal cell quota. In fact, Droop's intracellular kinetics describe only one component of competitiveness, for uptake kinetics is an essential counterpart. With these caveats in mind, the Schöhsee data (Sommer, 1988a) can be used for comparison with the literature values of minimal P cell quotas (see Section II,B and references there). The Schöhsee bioassay experiments provided estimates of μ_{rel} and of its weighted community average $(\overline{\mu_{rel}})$. The chemical analyses provided C : P ratios for total plankton and for size fractions. Rearrangement of Eq. (2) gives

$$q_0 = (1 - \mu_{rel})/(C:P) \tag{3}$$

Solving this equation for each size fraction and each sampling date individually gives frequency distributions of q_0 estimates which can be compared with the literature data (Fig. 3). The nonparametric two-sample Kolmogoroff–Smirnoff test showed the difference to the literature data to be significant both for the entire plankton and for all four size fractions. In the worst case (size fraction 10 to 35 μm), the significance level

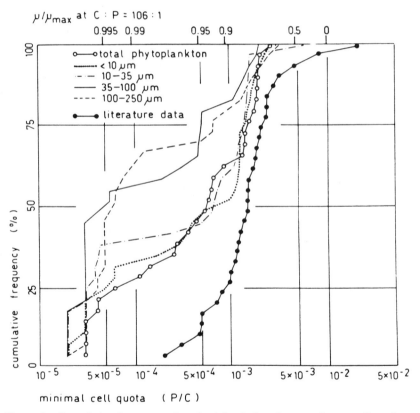

Figure 3 Cumulative frequency plot of minimal phosphorus cell quotas from literature compared to calculated minimal cell quotas for Schöhsee phytoplankton. The top line shows the relative reproductive rates calculated for Redfield stoichiometry.

was 0.0168; in the best case (size fraction 35 to 100 μm), it was 9.8 $\times$ 10^{-7}; for the unfractionated plankton, it was 0.0089. This means that phytoplankton from Schöhsee have, on average, significantly lower minimal cell quotas of P than phytoplankton in general. This is consistent with the assumption that competition for P has played a role in the selection of the phytoplankton species pool of this lake.

V. Concluding Remarks

In this chapter, I have given examples which are consistent with the assumption that phytoplankton communities are partially structured by competition for limiting nutrients. Those examples are located at three different levels: evidence for nutrient limitation, dominance patterns along resource ratio gradients, and tracks of past competition in species lists. These three different levels correspond to different time scales. Nutrient limitation begins as soon as external and intracellular nutrient pools are depleted below saturating levels by organismic uptake and growth. The onset of nutrient limitation is, by definition, also the beginning of the process of competition. Dominance patterns and species lists are the product of competition. This product can only be reached if the process is going on in the same direction for long enough. It usually takes a few weeks until superior competitors become numerically dominant, and it takes a few years until an inferior competitor is totally excluded from a lake. If the boundary conditions for the process of competition change before that, the equilibrium state toward which the plankton community develops also changes. In a changing environment, real community coposition will only imperfectly track the projected equilibrium state. Sometimes, tracking may be so imperfect that no relationships between resource ratios and taxonomic composition become apparent, despite ongoing competition. Testing for nutrient limitation is thus the less restrictive, but sufficient test for nutrient competition.

For a long time, experimental and theoretical competition research has emphasized the equilibrium outcome of competition and given relatively little attention to transient dynamics. This bias was probably caused by the mathematical simplicity of equilibrium solutions. Field ecologists looked for patterns anticipated by equilibrium models and their inferences from patterns to underlying processes were frequently premature. When others showed that those patterns could be explained by alternative mechanisms, including chance, the concept of competition seemed discredited (Simberloff, 1983; Strong *et al.*, 1983). Some ecologists felt a need to construct a "novel ecology" (Price *et al.*, 1984), mainly based on randomness, history, physical factors, and mutualism as opposed to "traditional ecology," which is mainly based on competition and predation. "Traditional ecology" is said to believe in equilibrium, whereas "novel ecology" accepts the nonequilibrium state of nature (for an application to phytoplankton ecology, see Harris, 1986). By extending a sometimes justified critique of premature conclusions to a "revolution" of the ecological world view, the "novel ecologists" have made the same mistake as the naive pattern searchers: confusion of process and product. They inferred the absence of competition from the fact that the observed patterns were not distinct from random ones.

When I started my first competition experiments about a decade ago, I did not expect much use of the experimental results for the prediction of species shifts *in situ*. Initially, I had more theoretical questions in mind (coexistence versus exclusion, Hutchinson's (1961) paradox, maintenance of diversity by perturbations). Rather than taking the species in my chemostats as "real" *Asterionellas* or "real" *Synedras,* I took them as abstract species A and species B, as in a simulation model. Usually, I was forced by the referees of my manuscripts to compare the results of my experiments with species shifts *in situ*. Because of the highly artificial character of the chemostat environment (vigorous mixing, spatial restriction, absence of vertical gradients) I did not expect the *Asterionella* in chemostats to behave like the *Asterionella* in lakes. The examples in this chapter show, however, that much can be extrapolated from culture competition experiments to the *in situ* performance of phytoplankton species. Phytoplankton competition research is on its way from the laboratory to the lake.

VI. Summary

Despite substantial success in experimental research, the concept of exploitative competition as a major determinant of phytoplankton species composition has met opposition from plankton ecologists. Physical perturbation and the rapid and patchy recycling of limiting nutrients have been invoked as mechanisms preventing exploitation of nutrients to limiting levels. A C : N : P stoichiometry of oceanic seston near the "Redfield ratio" (106 : 16 : 1) has been taken as evidence for nutrient-saturated rates of cell division. In this chapter, field tests for nutrient competition and limitation are discussed at three different levels: (1) evidence for nutrient limitation, (2) evidence for dominance shifts within a lake's species pool due to competition, and (3) evidence for the role of nutrient competition in determining a lake's species pool.

1. A review of stoichiometry data shows that N or P limitation in marine and freshwater habitats is quite common. Enrichment bioassays

with phytoplankton from a small North German lake were in good agreement with stoichiometry-based conclusions, which confirms that the Redfield ratio can serve to distinguish between nutrient limitation and saturation. Current evidence suggests that N and P limitation are frequently interrupted and usually of moderate intensity. Silicate limitation of diatoms, however, is more often strong and constant.

2. Three examples showing how algal taxa are distributed along nutrient ratio gradients show qualitative agreement with experimentally derived predictions: Along a horizontal nutrient ratio gradient in Lake Michigan, *Asterionella* : *Cyclotella* ratios correlate with the ratio of dissolved Si to P. During the seasonal succession of phytoplankton in Lake Constance and in Schöhsee (FRG), the contribution of diatoms to total algal biomass follows Si : P ratios with a time lag of 1 to 2 weeks. In many North American lakes, blue-green algae become increasingly important with decreasing total N : P ratios.

3. There is presently only very preliminary evidence for the role of nutrient competition in structuring a lake's species pool over long time scales. Calculated values for the minimal cell quota of P for Schöhsee are on average lower than those found in physiological literature. A reexamination of Nygaard's plankton indices (which characterize lake trophic status by quotients of species numbers in different taxa) shows conceptional compatibility of those indices with competition theory.

References

Ahlgren, G. (1987). Temperature functions in biology and their application to algal growth constants. *Oikos* **49,** 177–190.

Armstrong, R. A., and McGehee, R. (1977). Competitive exclusion. *Am. Nat.* **115,** 151–170.

Berman, T. (1970). Alkaline phosphatases and phosphorus availability in Lake Kinneret. *Limnol. Oceanogr.* **15,** 663–674.

Connell, J. H. (1980). Diversity and coevolution of competitors, or the ghost of competition past. *Oikos* **35,** 131–138.

Droop, M. R. (1973). Some thoughts on nutrient limitation in algae. *J. Phycol.* **9,** 264–272.

Droop, M. R. (1983). 25 years of algal growth kinetics. *Bot. Mar.* **26,** 99–112.

Dugdale, R. C., and Goering, J. J. (1967). Uptake of new and regenerated forms of nitrogen in primary productivity. *Limnol. Oceanogr.* **12,** 196–206.

Elrifi, I. R., and Turpin, D. H. (1985). Steady-state luxury consumption and the concept of optimum nutrient ratios: A study with phosphate and nitrate limited Selenastrum minutum (Chlorophyta). *J. Phycol.* **21,** 592–602.

Elser, J. J., Elser, M. M., MacKay, N. A., and Carpenter, S. R. (1988). Zooplankton-mediated transitions between N- and P-limited algal growth. *Limnol. Oceanogr.* **33,** 1–14.

Gaedeke, A., and Sommer, U. (1986). The influence of the frequency of periodic disturbances on the maintenance of phytoplankton diversity. *Oecologia* **71,** 25–28.

Goldman, J. C. (1979). Physiological processes, nutrient availability, and the concept of relative growth rates in marine phytoplankton ecology. *In* "Primary Productivity in the Sea" (P. G. Falkowski, ed.), pp. 179–194. Plenum, New York.

Goldman, J. C., McCarthy, J. J., and Peavey, D. G. (1979). Growth rate influence on the chemical composition of phytoplankton in oceanic waters. *Nature (London)* **279,** 210–215.

Gotham, I. J., and Rhee, G. Y. (1981). Comparative kinetic studies on phosphate limited growth and phosphate uptake in phytoplankton in continuous culture. *J. Phycol.* **17,** 257–265.

Harris, G. P. (1986). "Phytoplankton Ecology." Chapman and Hall, London.

Heaney, S. I., Smyly, W. J. P., and Talling, J. F. (1987). Interactions of physical, chemical and biological processes in depth and time within a productive English lake during summer stratification. *Int. Rev. Gesamten Hydrobiol.* **71,** 441–494.

Hutchinson, G. E. (1961). The paradox of the plankton. *Am. Nat.* **95,** 137–147.

Ierland, E. T., and Peperzak, L. (1984). Separation of marine seston and density determination by density gradient centrifugation. *J. Plankton Res.* **6,** 29–44.

Kilham, S. S. (1986). Dynamics of Lake Michigan natural phytoplankton communities in continuous cultures along a Si : P loading gradient. *Can. J. Fish. Aquat. Sci.* **43,** 351–356.

Lean, D. J. (1984). Metabolic indicators for phosphorus limitation. *Verh. Int. Ver. Limnol.* **22,** 211–218.

Lehman, J. T. (1985). Cell quotas of nutrients in phytoplankton established by X-ray analysis. *Verh. Int. Ver. Limnol.* **22,** 2861–2865.

Liao, F. F. H., and Lean, D. J. (1978). Seasonal changes in nitrogen compartments of lakes under different loading conditions. *Can. J. Fish. Res. Bd. Can.* **35,** 1095–1101.

Løvstad, Ø. (1986). Biotests with phytoplankton assemblages: Growth limitation along temporal and spatial gradients. *Hydrobiologia* **134,** 141–149.

Olsen, Y., Jensen, A., Reinertsen, H., Børsheim, K. Y., Heldal, M., and Langeland, A. (1986). Dependence of the rate of release of phosphorus by zooplankton on the P : C ratio in the food supply, as calculated by a recycling model. *Limnol. Oceanogr.* **31,** 34–44.

Nygaard, G. (1949). Hydrobiological studies on some Danish ponds and lakes. *Biol. Skr.— K. Dan. Vidensk. Selsk.* **7,** 1–293.

Nyholm, N. (1977). Kinetics of phosphate limited algal growth. *Biotechnol. Bioeng.* **19,** 467–492.

Price, P. W., Slobodchikoff, C. N., and Gaud, W. S. (1984). "A Novel Ecology." Wiley, New York.

Reynolds, C. S., Wiseman, S. W., and Clarke, M. J. O. (1984). Growth- and loss-rate responses of phytoplankton to intermittent artificial mixing and their potential application to the control of planktonic algal biomass. *J. Appl. Ecol.* **21,** 11–39.

Rhee, G. Y., and Gotham, I. J. (1980). Optimum N : P ratios and the coexistence of planktonic algae. *J. Phycol.* **16,** 486–489.

Robinson, J. V., and Sandgren, C. D. (1983). The effect of temporal environmental heterogeneity on community structure: A replicated experimental study. *Oecologia* **57,** 98–112.

Rocha, O., and Duncan, A. (1985). The relationship between cell carbon and cell volume in freshwater algal species used in zooplankton studies. *J. Plankton Res.* **7,** 279–294.

Sakshaug, E., Andresen, K., Myklestad, S., and Olsen, Y. (1983). Nutrient status of phytoplankton communities in Norwegian waters (marine, brackish, and fresh) as revealed by their chemical composition. *J. Plankton Res.* **5,** 175–197.

Sell, D. W., Carney, H. J., and Fahnenstiel, G. L. (1984). Inferring competition between natural phytoplankton populations: The Lake Michigan example reexamined. *Ecology* **65,** 325–328.

Shuter, B. J. (1978). Size dependence of phosphorus and nitrogen subsistence quotas in unicellular microorganisms. *Limnol. Oceanogr.* **23,** 1248–1255.

Simberloff, D. (1983). Competition theory, hypothesis-testing, and other community ecological buzzwords. *Am. Nat.* **122,** 626–635.

Smith, R. E., and Kalff, J. (1982). Size-dependent phosphorus uptake kinetics and cell quota in phytoplankton. *J. Phycol.* **18**, 275–284.

Smith, R. E., and Kalff, J. (1983). Competition for phosphorus among co-occurring freshwater phytoplankton. *Limnol. Oceanogr.* **28**, 448–464.

Smith, V. (1983). Low nitrogen to phosphorus ratios favor dominance by blue-green algae in Lake Phytoplankton. *Science* **221**, 669–671.

Sommer, U. (1983). Nutrient competition between phytoplankton species chemostat experiments. *Arch. Hydrobiol.* **96**, 399–416.

Sommer, U. (1985). Comparison between steady state and non-steady state competition: Experiments with natural phytoplankton. *Limnol. Oceanogr.* **30**, 335–346.

Sommer, U. (1986). Phytoplankton competition along a gradient of dilution rates. *Oecologia* **68**, 503–506.

Sommer, U. (1987). Factors controlling the seasonal variation in phytoplankton species composition—A case study for a deep, nutrient rich lake. *Prog. Phycol. Res.* **5**, 123–178.

Sommer, U. (1988a). Does nutrient competition among phytoplankton occur in situ? *Verh. Int. Ver. Limnol.* **23**, 707–712.

Sommer, U. (1988b). Growth and survival strategies of freshwater diatoms. *In* "Growth and Survival Strategies of Freshwater Phytoplankton" (C. D. Sandgren, ed.), pp. 227–260. Cambridge Univ. Press, Cambridge, England.

Sommer, U. (1988c). Phytoplankton succession in microcosm experiments under simultaneous grazing pressure and resource limitation. *Limnol. Oceanogr.* **33**, 1037–1054.

Sommer, U. (1989). The role of competition for limiting resources in phytoplankton succession. *In* "Plankton Ecology: Succession in Plankton Communities" (U. Sommer, ed.), pp. 57–107. Springer, Heidelberg.

Sterner, R. W. (1986). Herbivores' direct and indirect effects on algal populations. *Science* **231**, 605–607.

Sterner, R. W. (1989). The role of grazers in phytoplankton succession. *In* "Plankton Ecology: Succession in Plankton Communities" (U. Sommer, ed.), pp. 109–172. Springer, Heidelberg.

Strong, D., Simberloff, D., and Abele, L. (1983). "Ecological Communities: Conceptual Issues and Evidence." Princeton Univ. Press, Princeton, New Jersey.

Tett, P., Heaney, S. I., and Droop, M. R. (1985). The Redfield ratio and phytoplankton growth rate. *J. Mar. Biol. Assoc. U.K.* **65**, 487–504.

Tilman, D. (1977). Resource competition between planktonic algae: An experimental and theoretical approach. *Ecology* **58**, 338–348.

Tilman, D. (1981). Test of resource competition theory using four species of Lake Michigan algae. *Ecology* **62**, 802–815.

Tilman, D. (1982). "Resource Competition and Community Structure." Princeton Univ. Press, Princeton, New Jersey.

Tilman, D., Kilham, S. S., and Kilham, P. (1982). Phytoplankton community ecology: The role of limiting nutrients. *Annu. Rev. Ecol. Syst.* **13**, 349–372.

Tilman, D., Kilham, S. S., and Kilham, P. (1984). A reply to Sell, Carney, and Fahnenstiel. *Ecology* **65**, 328–332.

Tilman, D., Kiesling, R., Sterner, R., Kilham, S. S., and Johnson, F. A. (1986). Green, bluegreen and diatom algae: Taxonomic differences in competitive ability for phosphorus, silicon and nitrogen. *Arch. Hydrobiol.* **106**, 473–485.

Uehlinger, U. (1980). Experimentelle Untersuchungen zur Autökologie von *Aphanizomenon flos-aquae*. *Arch. Hydrobiol., Suppl.* **60**, 260–288.

11

Community Theory and Competition in Vegetation

M. P. Austin

I. Introduction

What can vegetation theory tell us about the importance of competition in determining the structure and composition of plant communities?

Many if not most community ecologists have assumed that competition is important but this has been questioned (Strong *et al.*, 1984; Salt, 1983; Simberloff, 1983; Hairston *et al.*, 1960). In contrast to animal ecologists, many plant ecologists have continued to assume competition is the predominant factor determining community structure (e.g., Grime, 1979; Tilman, 1988). This assumption of the primacy of competition needs to be questioned for plant communities, as other factors may be equally or more important.

Vegetation theory cannot provide answers to this question of causal processes. However, current theory can provide a set of descriptive concepts regarding patterns in vegetation (Austin, 1986), and any causal explanation involving competition should be compatible with these concepts and patterns. This chapter examines the extent to which current ideas regarding competition in perennial communities are consistent with the generalized concepts of vegetation pattern as presently conceived by vegetation ecologists.

II. Continuum Concept

Most plant ecologists now accept the continuum as an appropriate description of vegetation (Austin, 1985), and use the term "community" in a very vague sense to refer to vegetation which is sufficiently homogeneous for the purpose of study. Many experimental ecologists studying plant competition adopt this vague useage of the term community. This may have a profound influence on their assessment of competition in natural communities, as indicated below.

A. Experimental Design

Figure 1 compares sampling of vegetation along a one-dimensional environment assuming either the now-outdated community-unit concept (Fig. 1a) or the continuum concept (Fig. 1b). If three replicates are sampled as shown, under the old community assumption the normal statistical assumptions would be met. Under the continuum concept, it is highly questionable whether the replicates can be combined. A particular set of species is not common to all replicates, while the performance of individual species and the degree of competition (direct or diffuse) may vary with position on the environmental gradient.

If we adopt the classic approach of Goodall (1953), that a homogeneous community or association is one in which there is an absence of positive correlations, then if such homogeneous communities exist, the variance in density of member species should be independent of the distance apart of the quadrats used to measure their density. Work on

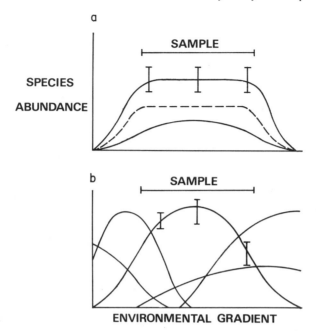

Figure 1   Contrast in sampling variability between replicates depends on which theory of vegetation composition is correct: (a) community-unit theory or (b) continuum concept.

spatial pattern in communities since 1953 has confirmed many times that "apparently homogeneous" vegetation (community?) has a highly non-random structure at numerous scales (Greig-Smith, 1979; see also Watt, 1947a).

Incorporation of spatial autocorrelation at several scales into experimental design has yet to be adequately addressed, but plant communities with morphological, sociological, and environmental scales of pattern (Kershaw and Looney, 1985) will offer difficulties not presently considered by statisticians (Diggle, 1983) or experimentalists (Harper, 1977). Prediction of competitive outcomes in perennial vegetation will require knowledge of the spatial pattern dynamics of species such as those elegantly described by Watt (1947a) for heather (*Calluna vulgaris*) and bracken (*Pteridium aquilinum*) (1947b, 1955), and by Kershaw (1958, 1959) for *Agrostis tenuis* and *Trifolium repens*. Note, however, that these studies suggest that senescence and disturbance may be as important as competition in determining community structure.

The majority of studies of competition have previously avoided this problem by studying annuals, particularly crops and weeds, often in the relatively homogeneous environments of the arable field or glasshouse. Transferring these approaches to natural communities without taking account of the nonrandom structure and the scale of the phenomena is unlikely to be successful. Silander and Pacala (this volume), however, offer some hope that the structure and scale of processes can be simply averaged in some situations, but more work will be needed as their work is also on annuals.

In a recent study of variation in the herbaceous layer of eucalypt forest (Austin and Nicholls, 1988), it was not possible to detect species associations independent of differences in environment. Plots were carefully grouped into relatively homogeneous sets after environmental stratification on topography, seral status of the shrub layer, and multivariate classification of the total species complement. Statistical models of the frequency of individual species were then developed for certain of the homogeneous sets (communities?). For local frequency in quadrats of 75 × 75 cm, no cases were found where individual species could be modeled without the inclusion of environmental variables as predictors. Most species required both environmental variables (often including the seral status of the shrub stratum) and a measure of the "matrix" species for that set of quadrats. The joint measure of abundance of the matrix species can be regarded as a measure of diffuse competition, but it could equally represent a surrogate variable for an unknown environmental variable to which those species respond. Frequently, the modeled response was curvilinear and each species showed individualistic responses. Even at 15 × 15 cm subquadrat scale, associations between species were environmentally determined. The conclusion is that, at the scale at which competition effects are frequently studied, the differential response of species to environment can be equally significant.

The vegetation scientist viewing the experimental design of the majority of recent competition experiments must form the opinion that, while competition ecologists have abandoned Clementsian ideas of the community, most have not yet faced the consequences of adopting either the continuum concept or models of spatial pattern in "communities." Exceptions are provided by Keddy, Grace, and Louda *et al.* (this volume), who recognize that experiments done at a single location and at a particular position along an environmental gradient cannot necessarily be extrapolated to another position in the environmental space. Further conceptual problems regarding the design of field experiments to demonstrate competition are discussed by Tilman (1989), namely, transient dynamics, indirect effects, environmental variability, and the possible existence of multiple stable equilibria. These may be overcome by long-term experiments, but the extent to which conclusions, are location-, design-, and community-specific must remain open unless competition theories and models can be shown to predict general patterns in vegetation.

B. Propositions Regarding the Continuum Concept

The continuum concept has been reviewed elsewhere (Austin, 1985), but it should be recognized that there are several alternative propositions for the patterns of species distribution along a continuum (Fig. 2). Which of these alternative propositions are consistent with competition models? The community-unit theory expressed in terms of a single environmental gradient implies a group of coadapted species, with common limits along the gradient (Fig. 2a; Austin, 1985). The original individualistic continuum concept of Gleason in which species optima and limits are independently distributed is shown in Fig. 2b. Gauch and Whittaker (1972) modified the continuum concept, suggesting "major" species are regularly distributed along the gradient with optima evenly spaced and "minor" species independently distributed (Fig. 2c). A continuum need not consist of individualistically distributed species but could consist of species regularly replacing each other (Goodall, 1963). This would correspond to the usual resource-partitioning approach adopted by animal ecologists (e.g., Pianka, 1981). An extension of this for vegetation would be to suggest that such a continuum exists, but that each stratum (e.g., trees, shrubs, herbs) partitions the gradient independently of other strata (Fig. 2d; Austin, 1985). These alternatives are very difficult to distinguish at present, particularly if the one-dimensional abstract gradient discussed here is replaced with the multidimensional reality. Much

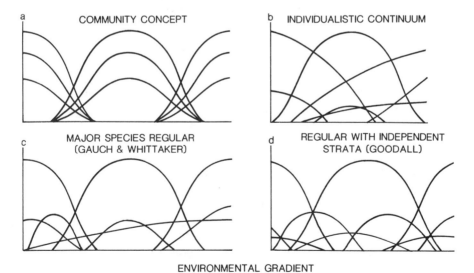

ENVIRONMENTAL GRADIENT

Figure 2 Alternative types of species patterns which may be associated with the continuum concept of variation along an environmental gradient.

more complex patterns based on continuous variation in species composition would then be possible.

How could competitive processes determine these proposed patterns? There is no mechanistic theory linking these theoretical patterns to species physiological and competitive properties. Some researchers (see the review by Austin, 1986) have begun to develop ideas: Grime (1985) with his competitive indices based on morphological attributes and growth rates, and Tilman utilizing ideas on resource competition for light and nitrogen, which he has recently extended to include morphology (Tilman, 1988; see also Keddy, this volume; Grace, this volume). Any attempt to address the link may, however, be premature. The evidence for species response patterns is equivocal at present (Austin, 1987; Shipley and Keddy, 1987) and certainly not sufficient to distinguish between the three alternative continuum concepts (Austin, 1985). Statistical analysis is required to determine the patterns which can be observed in vegetation (see below), while experimentation is necessary to determine mechanism (Shipley and Keddy, 1987). Few attempts have been made to proceed from field observations on patterns of perennial species along gradients to experimental analysis of possible mechanisms. Experimentalists have preferred to presume that competition is important and design experiments accordingly.

C. Species Patterns and Competition: Ellenberg's Contribution

A much-neglected pioneering work which addressed multispecies competition along a gradient was the work of Ellenberg (1953, 1954). Ellenberg published two papers on competition experiments in the early 1950s (1953, 1954; see also Aukland, 1978) which provided the basis of his theoretical ideas, and were subsequently summarized by Mueller-Dombois and Ellenberg (1974; see Fig. 3). In experiments on species response to a watertable gradient from water at the soil surface to a watertable at a depth of 140 cm, he observed that the species biomass optima in monoculture tended to coincide. When the species were grown in multispecies mixtures under the same conditions, their optima were displaced. The ecological optima and shape of response curve differed from the physiological optima and response under monoculture conditions. Ellenberg interpreted this as being due to competition. The shifts in optima corresponded to phytosociological observations on the species relative performance in the field. Examination of the field behavior of a wide variety of species in relation to gradients of light, water, nitrogen, etc. led to the conclusion that competitive ability was distinct from physiological performance and the variety of ecological curves shown in Fig. 3 could occur in nature.

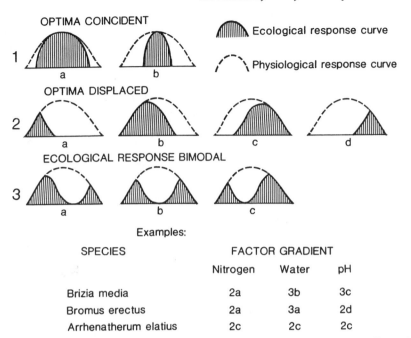

OPTIMA COINCIDENT

Ecological response curve

Physiological response curve

OPTIMA DISPLACED

ECOLOGICAL RESPONSE BIMODAL

Examples:

SPECIES	FACTOR GRADIENT		
	Nitrogen	Water	pH
Brizia media	2a	3b	3c
Bromus erectus	2a	3a	2d
Arrhenatherum elatius	2c	2c	2c

Figure 3 Types of species distribution patterns along an environmental gradient with and without competition following Mueller-Dombois and Ellenberg (1974). Redrawn from Austin (1980) with permission of Klumer Academic.

The species responses shown in Fig. 3 stand in sharp contrast to those in Fig. 2. Anglo-American literature on theory assumes that the ecological response of a species is a bell-shaped curve (see discussions in Austin, 1980, 1985). This seems to be based, at least for plants, on the grounds of mathematical convenience and/or Whittaker's direct gradient analysis results (e.g., Whittaker, 1978). Tabulation of the actual types of curves observed in his and others' data (Austin, 1976) indicates that symmetric bell-shaped curves are uncommon, except in textbooks and theoretical studies. The shape of species response, i.e., the shape of the realized environmental niche, needs to be resolved. Theories of competitive mechanisms should be capable of predicting the shape of the realized niche. Ellenberg (1953, 1954) provides only a descriptive theory of competition. There is no mechanistic theory based on both environmental gradients and species properties which allows one to predict the outcome of competition between species.

There are several technical problems with Ellenberg's watertable experiments, and these have been the subject of comment by Ernst (1978),

Austin (1979, 1982), and Wilson *et al.* (1985), but they remain an elegant design which should be reexamined for competition studies. Their theoretical implications deserve the attention of both experimental and observational community ecologists.

III. Species Response Patterns along Environmental Gradients

Central to any consideration of the relationship between competition theory and vegetation patterns are the patterns of performance of individual species along environmental gradients, under natural conditions (with multispecies competition, herbivory, and pathogens: the realized environmental niche), under competition only, and without competition (monoculture conditions: the fundamental niche or physiological response). Current research results need to be reviewed carefully before attempting to explain patterns in terms of competition.

A. Realized Niche

The response curves in Fig. 3 were based on subjective interpretation of phytosociological field observations. Many recent ordination studies can not provide unequivocal evidence for or against these response curves because of distortion problems associated with the mathematical techniques used (Austin, 1985; Minchin, 1987). Direct gradient techniques (DGA; Whittaker, 1978; Austin *et al.*, 1984) are less ambiguous in terms of technique, but the environmental gradients used are often distal indirect gradients, e.g., altitude, water depth, and aspect, and not resource gradients.

Three broad types of environmental gradient can be distinguished but these are often confused in discussions of continua and competition (Austin, 1980). They are

1. Indirect environmental gradients where the environmental variable does not have a direct physiological influence on plant growth (e.g., altitude or distance from the coast).
2. Direct environmental gradients where the variable does have a direct physiological influence on growth (e.g., temperature or pH) but is not a resource consumed by plants.
3. Resource gradients where the environmental variable (e.g., nitrogen) is an essential resource for plant growth.

Competition experiments along gradients have used various types of gradients. Ellenberg's (1953, 1954) watertable gradient is clearly an indirect gradient, as are the watertable height and exposure gradients used by Wilson and Keddy (1985a,b, 1986a,b; Wilson *et al.*, 1985). The nutrient gradient that I have used (Austin and Austin, 1980; Austin, 1982; Austin *et al.*, 1985) can be considered a direct gradient. The nutrients are resources, but which one is limiting at which point is unknown. The effects of the nutrient ratios on availability of particular nutrients at particular concentrations will have physiological but indirect influence on plant growth. The soil moisture gradient of Pickett and Bazzaz (1978) can also be interpreted as a direct gradient rather than a resource gradient; the highest resource level consisted of a saturated soil which could give rise to indirect aeration effects rather than moisture supply per se. The nearest approach to resource gradients are those where a single resource is varied, e.g., Tilman's (1986) use of soil nitrogen. However, even resource gradients are usually measured in units which are not necessarily those of the direct causal variable. Nitrogen supplied as fertilizer or measured as total soil nitrogen is not equivalent to nitrogen available for uptake at the root surface. Competition experiments along gradients can not be directly compared if the type of gradient is different. Little is known about differences in competitive processes along different gradients.

The results of a DGA with respect to species response shape are therefore subject to unknown environmental effects which may distort the shape. It is surprising therefore to find very clear response curves when suitable and sufficient data are collected (e.g., Fig. 4). The eucalypt forest data for the ecological response curve of species in relation to mean annual temperature, a direct gradient (Fig. 4a,b), were part of a much larger data set and were carefully stratified. The observations are for plots on slopes (>7°) with intermediate exposure (neither north or south facing), no known major disturbance, and a restricted range of mean annual rainfall (Austin, 1987). Keddy and co-workers (Wilson and Keddy, 1985a,b, 1986a,b) have presented a series of observational analyses of aquatic species' responses to water depth and exposure to wave action (Fig. 4c,d). Tilman (1987) provides similar response curves for 15 species in relation to total soil nitrogen from a survey of old fields. Most show skewed curves but some approach the conventional symmetric bell-shaped curve e.g., *Berteroa incana* (Fig. 4e,f). The realized niche or ecological response curves show a variety of shapes but species with clearly defined responses are predominantly skewed (Table 1). Further statistical analysis of the shape of the realized niche has been done simultaneously with three environmental gradients using generalized linear models (Austin *et al.*, in press). The 10 models for the five species studied support the skewed nature of the realized response. The conclusions from these studies are that species have unimodal responses but these

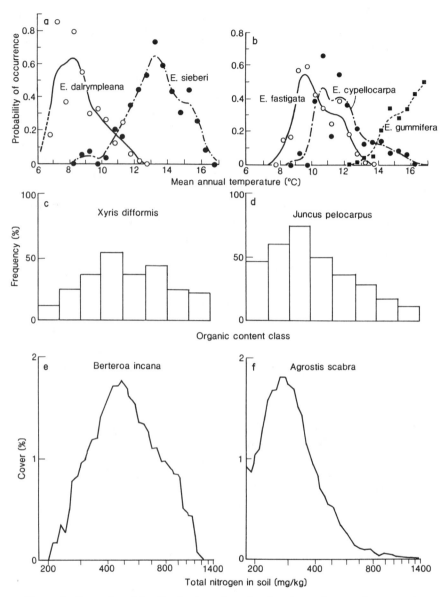

Figure 4   Examples of the distribution of species abundance along environmental gradients. (a,b) Eucalypt species from forests in southeastern Australia. From Austin (1987), with permission of Klumer Academic. (c,d) Aquatic species along an organic sediment gradient from Canadian lakes. From Wilson and Keddy (1985b), with permission of The British Ecological Society. (e,f) Old-field species from Minnesota. From Tilman (1987), with permission of Ecological Society of America.

Table 1 Classification of Species Ecological Response Curves

Gradient (response curve type)	Austin (1987) Mean Annual Temperature	Wilson and Keddy (1985b) Substrate Organic Content	Tilman (1987) Total Soil Nitrogen
Low gradient extreme	2[a]	—	2
Low gradient skewed	7	2	3
Symmetric	7	1	2
High gradient skewed	3	—	3
High extreme	13	3	3
Indeterminate	10	6	2

[a] Types subjectively assessed.

are usually skewed, while no unequivocal bimodal responses have been detected (cf. Austin, 1976).

Various workers have put forward suggestions (hypotheses) about vegetation composition, and their suggestions on species shape are listed in Table 2. Only the hypothesis of Ellenberg is consistent with the observations reported here, those of other more recent workers (Table 2) are not. These observations are based on different types of gradients rang-

Table 2 Shape of Species Ecological Response (Realized Niche)

Theory	Postulated Shape
Individualistic Continuum (Gauch and Whittaker, 1972)	Bell-shaped (Pseudo-Gaussian)
Resource-Partitioned Continuum (Niche Theory; Pianka, 1981)	Bell-shaped (Pseudo-Gaussian)
Ecological Response (Mueller-Dombois and Ellenberg, 1974)	Variable including Bimodal curves
Stress Tolerance/Competition (Grime, 1979)	?
Resource Ratio Competition (Tilman, 1982)	Bell-shaped (Pseudo-Gaussian)

Observation	Predominant Shape
Eucalypt forest (Austin, 1987; Austin *et al.*, in press)	Skewed curve
Canadian lake shores (Wilson and Keddy, 1985, 1986; Wilson *et al.*, 1985)	Skewed curve
American old fields (Tilman, 1987)	Skewed curve

ing from indirect to resource gradients, so the skewness appears reasonably robust to differences in environmental gradient.

B. Fundamental Niche

Any model of competition will first require a model of the physiological response of a species to an environmental or resource gradient. Ellenberg (1953, 1954) put forward the suggestion that the physiological optima of different species tended to coincide (Fig. 3). Austin and Austin (1980) showed that, for grass species along a nutrient gradient, the optima were similar but not coincident. Some of the species concerned were also included in Ellenberg's experiment. Tilman (1986) presents results in his Figure 1 from a field experiment with added nitrogen which show linear regressions for a steady increase in biomass of individual species along a total soil nitrogen concentration gradient. Note, however, that *Agropyron repens* and *Agrostis scabra* could be interpreted as showing optima at 600 and 500 mg nitrogen/kg soil and other species may have curvilinear responses. Vogel (1978), along a gradient of added ammonium nitrate ranging from 0 to 6400 kg/ha/yr, demonstrates clear optima for the several species of grass, sedge, and forb he studied. They are at high levels but again, while similar in shape of response, the optima are not obviously coincident. The main features of these nutrient responses are that the maximum biomass under optimum conditions is very different from species to species and most species have optima far above the conditions usually operating in the field.

Wilson and Keddy (1985b; Wilson *et al.*, 1985) have examined the relationship between physiological optima and ecological optima for two species (*Xyris difformis* and *Juncus pelocarpus*) and found that the absolute biomass yield optimum in monoculture corresponded with the ecological optima in a field experiment. Mitchley (1987) points out that *J. pelocarpus* appears more competitive in a more mesic environment than its optimum position; however, this is in a garden experiment. We do not have enough information to resolve the types of issues regarding physiological tolerance and performance. Most experiments are concerned with shoot biomass. The fundamental niche shape for vegetative shoot growth is not yet well defined, but vegetative growth need not represent fitness when fitness integrates establishment, growth (both root and shoot), and seed production to give a measure of survival. Experimental results on a plant's physiological response provide little definitive information, while physiological theory provides little guidance as to the appropriate measure of, or the expected shape of, a species fundamental niche.

Austin (1982) presented results which indicates that it was possible to predict the vegetative growth performance in multispecies mixture from growth in monoculture at points along a nutrient gradient for grass species. The performance had, however, to be expressed in terms of relative growth (production) *at that point on the gradient.* Relative physiological performance was defined as

$$R_{ij} = Y_{ij}/Y_{mj}$$

where Y_{ij} is the shoot dry weight yield per unit area of species i at factor level j and Y_{mj} is the yield for a monoculture of the most productive species m, at that factor level.

Normalized ecological performance was defined similarly as

$$E_{ij} = Y'_{ij}/Y'_{mj}$$

where Y'_{ij} is the yield per unit area of species i in mixture at factor level j and Y'_{mj} is the yield of the highest yielding species in the mixture at factor level j. This empirical relationship indicates that the shape of the fundamental niche may be of little value in predicting success in mixture. It is the amplitude of the response at any particular point *relative* to that of the other species present at that point which appears critical. Performance in mixture is sensitive to the species present and their relative monoculture performance. Similar results were obtained for thistles (Austin *et al.,* 1985), though the relationship had to be extended to include morphology (root/shoot ratios) to achieve adequate predictions. In addition, effects of plant age have also been demonstrated (Grace, 1988).

The relative physiological optima of species tend to correspond with the optima for the normalized ecological performance, a conclusion contradicting both Ellenberg's and Ernst's proposition about optima (Austin, 1982). Neither does the result correspond to those presented by Wilson and Keddy (1985a,b). They used a specific indirect environmental gradient of wave exposure measured as the amount of organic matter in the substrate. This gradient is clearly a complex-factor gradient, as recognized by the authors. The nutrient gradient used by colleagues and me (Austin and Austin, 1980; Austin, 1982; Austin *et al.,* 1985) is also complex but more clearly related to a direct gradient for plants. It would be of some significance to test these relationships on the species studied by Tilman (1986) in a survey of 22 old fields, particularly as the limitations of short-term experiments for predicting long-term dominance have recently been demonstrated by Tilman (1989).

Competition processes cannot be modeled without reference to the fundamental environmental niche of a species. Vegetation science is unlikely to be able to make useful predictions unless the fundamental niche is known. Experimental results so far are equivocal, but there is some evidence to suggest that it is the relative performance, not the absolute performance of species which is critical. The implication of this hypothesis is that the shape of the fundamental niche is unimportant, but the species context (i.e., which species are present and their relative performance) can be used to predict the outcome of competition in multispecies mixtures.

C. Patterns of Species Optima

Gauch and Whittaker (1972) postulate that major species are regularly distributed along an environmental/resource gradient while minor species are random, i.e., their realized niche optima are evenly spaced along the gradient (major species) or irregular (minor species). Ellenberg's (Mueller-Dombois and Ellenberg, 1974) suggestion that species may have bimodal distributions indicates that he regards the patterns of modes (optima) as variable, depending on the species and the gradient. Other authors (Table 3) are not definitive about this feature of the current descriptive concepts of vegetation composition.

I tested this proposition with eucalypt forest data (Austin, 1987). The position of optima cannot be distinguished from random positioning for either major or minor canopy species (Table 4). The results suggest that the optima may be clumped, but further, more precise analysis is needed to test this speculation (but see Shipley and Keddy, 1987, for an alternative approach).

Resource competition is usually put forward as an explanation for the regular distribution of species (Pianka, 1981). Available empirical evidence does not support this for plants and it is difficult to see how resource partitioning as postulated for animals (e.g., their use of food particles of different size) can apply to resources such as nitrogen or potassium. A test of the pattern of optima for the old-field survey data

Table 3   Pattern of Species Distributions along an
Environmental Gradient[a]

Theory	Postulated Distribution
Individualistic Continuum (Gauch and Whittaker, 1972)	Major species regular Minor species random
Resource-Partitioned Continuum (Niche Theory; Pianka, 1981)	Regular
Ecological Response (Mueller-Dombois and Ellenberg, 1974)	Variable
Stress Tolerance/Competition (Grime, 1979)	?
Resource Ratio Competition (Tilman, 1982)	? Regular

[a] Observed for eucalypt forest (Austin, 1987). The distribution seemed to be random, but was probably clustered.

Table 4   Modal Position of Species in Relation
to Temperature[a,b]

Temperature Class	Maximal Modal Frequency		
	Major Species ≥ 0.3	Minor Species < 0.3	Total
7	1	1	2
7.5	—	1	1
8	—	—	0
8.5	—	—	0
9	2	—	2
9.5	1	1	2
10	—	1	1
10.5	3	—	3
11	—	2	2
11.5	—	—	0
12	—	4	4
12.5	—	—	0
13	3	1	4
13.5	—	1	1
14	—	1	1
14.5	2	1	3
15	2	2	4
15.5	—	—	0

[a] From Austin (1987).
[b] $X^2 = 21.6$; df = 17.

(Tilman, 1988) would provide an opportunity for the observational and experimental analysis to be brought together. This would provide empirical evidence, but the mechanistic theory linking species optima patterns to competition and species physiological and morphological properties remains ill-defined.

D. Species Richness

Gauch and Whittaker (1972) could not determine any consistent relationship of species richness to environmental gradients. Ellenberg (Mueller-Dombois and Ellenberg, 1974) appears silent on the issue, while other authors predict maximum species richness at intermediate conditions, i.e., not extreme or highly productive (Table 5).

Colleagues and I (Margules *et al.*, 1987; Austin, 1987; M. P. Austin, A. O. Nicholls, and C. R. Margules, unpublished observations) have examined this issue using our eucalypt data base. Margules *et al.* (1987) showed that there is a complex dependency of eucalypt species richness on four environmental factors, mean annual rainfall, mean annual temperature, radiation differences due to aspect, and geological substrate.

Table 5   Pattern of Species Richness along an Environmental Gradient[a]

Theory	Postulated Pattern
Individualistic Continuum (Gauch and Whittaker, 1972)	No consistent relationship
Resource-Partitioned Continuum (Niche Theory; Pianka, 1981)	Even
Ecological Response (Mueller-Dombois and Ellenberg, 1974)	?
Stress Tolerance/Competition (Grime, 1979)	Maximum richness at intermediate conditions
Resource Ratio Competition (Tilman, 1982)	Maximum richness at intermediate conditions

[a] Margules *et al.* (1987) showed that there is a complex environmental dependency of species richness in eucalypt forest.

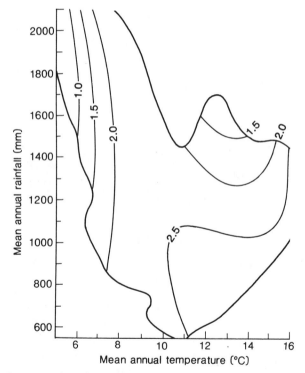

Figure 5   Contours of species richness of *Eucalyptus* species in relation to mean annual rainfall and mean annual temperature for those combinations of rainfall and temperature which occurred in the study area. From Margules *et al.* (1987), with permission of Springer-Verlag.

The pattern in relation to the two major factors of rainfall and temperature is shown in Fig. 5. The precise relationship with any single factor is contingent on the level of the other. Under extreme temperature conditions, i.e., at timberline, a simple relationship exists between tree species number and temperature. Austin (1987) also found that a stratified subset of the eucalypt data base showed a simple relationship of species richness with temperature, but this relationship confounded any attempt to test other propositions of Gauch and Whittaker about the environmental niche breadth of species.

If competition is involved in structuring vegetation, then any mechanistic theory of competition must provide an explanation of the species richness patterns described above.

E. Biomass Patterns

The striking observation of Al-Mufti *et al.* (1977), that species richness shows a very peaked relationship to biomass and litter (Fig. 6), needs to be tested in a variety of vegetation types.

If species richness is significantly related to biomass, then the pattern of biomass production in relation to environmental gradients becomes important as a factor in determining the observed patterns of richness. Austin (1980) presented results from several authors on the patterns of biomass shown in multispecies competition experiments along gradients.

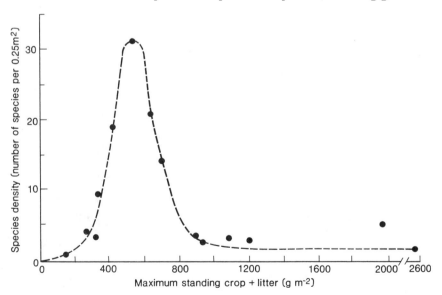

Figure 6 Relationship between species density (richness) of herbs and maximum standing crop plus litter. Redrawn from Al-Mufti *et al.* (1977), with permission of The British Ecological Society.

The relationships between species richness, biomass productivity, competition, and environments should be subjected to experimental investigation. Austin and Austin (1980) presented experimental observations on the patterns of dominance by single species from experimental multispecies mixtures of 5 and 10 species. The relative importance of competition is likely to be contingent on these other properties of vegetation along gradients; competition will contribute to the expression of these vegetation properties along gradients. There is little evidence in the literature to suggest that these various properties are being considered simultaneously with competition in experimental studies.

IV. Discussion

The perspective on competition from the viewpoint of a vegetation scientist is daunting.

The theoretical framework of vegetation science is ill-developed (Austin, 1986), and the relative importance of competition has been assumed rather than investigated in natural or model communities. There has not been a concerted effort to distinguish between the rival hypotheses of experimental community ecologists. Studies of density-dependent competitive control of species interactions and of the population dynamics of individual species have contributed little to the understanding of vegetation patterns as yet. Observational analysis of communities (however they are defined) has yet to provide firm, empirical evidence on species responses and patterns. The current conceptual models of vegetation (Figs. 2 and 3) are very inadequate. The mechanisms (processes) which determine actual observed patterns in vegetation require more empirical investigation. It seems unlikely that competition processes alone will suffice to explain them.

There are grounds for optimism (Austin, 1986), in that experimentalists and observationists are beginning to identify common questions, such as how important are light and nutrients in determining patterns of species performance. Some new experimental designs incorporate gradients, though whether spatial patterns and long-term monitoring of the experiments can be incorporated without enormous cost and effort remains to be determined. The pattern of species richness and the shapes of individual species responses are now often reported in both experimental and observational studies, though few studies contain both observational and experimental analysis of the same assemblage of species (but see Tilman, 1986, 1987, 1988). Tilman (1988) presents a mathematical model which incorporates plant morphology in competition models of vegetation, and Grime (1979) and Mitchley and Grubb (1987) use conceptual models incorporating morphology. These and other models of plant species competition with morphological attributes consider environmental gradients. However, the morphology of plants varies with position on an environmental gradient, both as a community property and for individual species, and the implication of this plasticity for these models requires further testing and extending of the morphological models.

The patterns of response in vegetation composition along environmental gradients are still relatively poorly defined. The physiological response of species is often defined in terms of increasing performance as the resource level increases, and with a limiting factor type of curve (Fig. 7a); the effect of toxicity at higher levels is often ignored. Plants with all other environmental factors at suitable levels will, under conditions of resource sufficiency, form a more or less closed canopy. Light will be the limiting factor under high levels of resource. The tradeoff

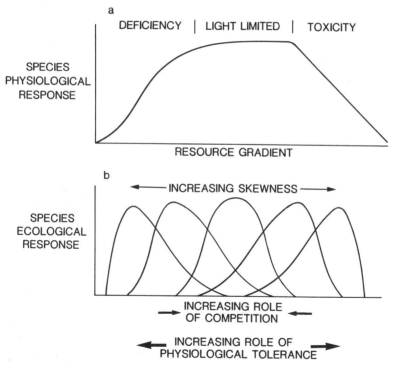

Figure 7   Possible shapes of species response curves along a resource/environmental gradient. (a) Species physiological response without competition, herbivores, or pathogens. (b) Species ecological response curve with competition and other biotic factors. Shape becomes increasingly skewed toward extremes of gradient. Speculations about relative importance of physiological tolerance and competition are also shown.

between resource and light is well documented by Tilman (1988), but there are frequently conditions in nature where resources become toxic. For example, too much water frequently leads to anaerobic conditions for terrestrial plants. Figure 7a represents how little we know about the fundamental environmental niches of plants. The prediction of vegetation patterns and the role of competition may be quite different for the resource-deficient section of the gradient as compared with the resource-toxic section.

The ecological response of species may show a sequence of shapes depending on position on the gradient from highly skewed under deficient/unsuitable conditions, as at low temperatures for eucalypts, to symmetrical bell-shaped curve under mesic conditions (Tables 2 and 3 and Fig. 4). Under supraoptimal or toxic conditions, skewness with the tail in the opposite direction may develop (Fig. 7b). The interplay between physiological tolerance and competition is crucial, but experiments covering the full range of an environmental gradient which might confirm or reject the speculations in Fig. 7 have yet to be done. The postulated existence of bimodal curves (Fig. 3) may be due to greater sensitivity to competition and to a wider physiological tolerance; however, we do not know very much about the ranges of physiological tolerances shown by species. It seems unlikely that they are identical, as suggested by Ellenberg (1953, 1954) (Fig. 3). Many seem to differ more in terms of amount of biomass produced at the species optimum than in the position of the optimum. Biomass production, root/shoot allocation, and height growth may then predict ecological performance through competition for light (Grime, 1979; Tilman, 1988). The relative performance in terms of biomass and height growth at a *particular* position on the gradient would then determine the patterns of ecological response (Austin, 1982; Austin *et al.*, 1985). Under these conditions, the *shape* of the physiological response curve will play a critical role in determining the outcome of competition, as will the morphological plasticity of the species and the potential control of morphology by the environmental gradient itself.

In summary, an opportunity exists for synthesizing competition studies and vegetation science. Three questions need to be considered if competitive processes are to be put into the context of vegetation theory.

1. What determines the relative abundance of species in communities?
2. What determines the shapes of species performance and patterns of community properties (e.g., species richness) along environmental gradients?
3. What determines the relative importance of causal processes, e.g., competition or predation, and how do they vary in importance along environmental gradients?

These questions cannot be answered without contributions from observational analysis, experimentation, and theory. Though vegetation theory is very inadequately developed at the present time, it can contribute two important perspectives on plant competition: the need to consider plant competition experimentally and conceptually in the context of environmental gradients, and second, that the postulated processes of competition must be consistent with the observed patterns of vegetation variation.

V. Summary

Current ideas of competition in perennial plant communities are examined for consistency with the generalized concepts of vegetation pattern as conceived by vegetation ecologists. The design of competition experiments often fails to take account of either the continuum concept or the spatiotemporal structure of plant communities. Failure to consider the vegetation continuum often means that replicates are not sampled from a homogeneous population. Spatial pattern analysis results and the classic pattern and process concept of Watt (1947a) both suggest that spatial autocorrelation analysis is necessary in the experimental design of competition studies in *perennial* communities.

Current mechanistic models of competition are not sufficient to explain the species patterns postulated to occur along environmental gradients by Gauch and Whittaker (1972) and by Ellenberg (Mueller-Dombois and Ellenberg, 1974). The bell-shaped or pseudo-Gaussian curve used to represent the realized niche of a species is shown to be inadequate; it is generally skewed. The fundamental niche or physiological response of species to environmental gradients is not clearly specified by vegetation scientists. Models of competition fail to distinguish the three types of environmental gradient—resource (e.g., nitrogen), direct (e.g., temperature), and indirect (e.g., altitude)—so the possibly different fundamental responses remain undefined. As a consequence, competition models also inadequately define the realized response of species. Some recent experiments suggest that the shape of realized response of a species is context-sensitive, that is, the shape is dependent on which other species are present. The spacing of species optima along gradients is not necessarily regular, as assumed in some competition theories, nor are the patterns of species richness along gradients well predicted or considered by current theory.

In most cases, recent results do not support the hypothesized patterns and concepts of vegetation scientists nor can the competitive processes or models of other plant ecologists provide adequate explanations of them.

Acknowledgments

I thank C. R. Margules and A. O. Nicholls for their collaboration on the eucalypt study and for commenting on the manuscript. I thank J. Stein and E. M. Adomeit for technical assistance.

References

Al-Mufti, M. M., Sydes, C. L., Furness, C. B., Grime, J. P., and Band, S. R. (1977). A quantitative analysis of shoot phenology and dominance in herbaceous vegetation. *J. Ecol.* **65,** 759–791.

Aukland, P. (1978). Translation of two of Heinz Ellenberg's papers. *CSIRO Div. Land Use Res. Tech. Memo.* **78/7.**

Austin, M. P. (1976). On non-linear species response models in ordination. *Vegetatio* **33,** 33–41.

Austin, M. P. (1979). Current approaches to the non-linearity problems in vegetation analysis. *In* "Contemporary Quantitative Ecology and Related Ecometrics" (E. P. Patil and M. Rosenzweig, eds.), Stat. Ecol. Ser. 12, pp. 197–210. International Cooperative Publishing House, Fairland, Maryland.

Austin, M. P. (1980). Searching for a model for use in vegetation analysis. *Vegetatio* **42,** 11–21.

Austin, M. P. (1982). Use of a relative physiological performance value in the prediction of performance in multispecies mixtures from monoculture performance. *J. Ecol.* **70,** 559–570.

Austin, M. P. (1985). Continuum concept, ordination methods, and niche theory. *Annu. Rev. Ecol. Syst.* **16,** 39–61.

Austin, M. P. (1986). The theoretical basis of vegetation science. *Trends Ecol. Evol.* **1,** 161–164.

Austin, M. P. (1987). Models for the analysis of species response to environmental gradients. *Vegetatio* **69,** 35–45.

Austin, M. P., and Austin, B. O. (1980). Behaviour of experimental plant communities along a nutrient gradient. *J. Ecol.* **68,** 891–918.

Austin, M. P. and Nicholls, A. O. (1988). Species associations within herbaceous vegetation in an Australian eucalypt forest. *In* "Diversity and Pattern in Plant Communities (H. J. During, M. J. A. Werger, and J. H. Williams, eds.), pp. 95–114. SPB Academic Publishing, The Hague, The Netherlands.

Austin, M. P., Cunningham, R. B., and Fleming, P. M. (1984). New approaches to direct gradient analysis using environmental scalars and statistical curve-fitting procedures. *Vegetatio* **55,** 11–27.

Austin, M. P., Groves, R. H., Fresco, L. F. M., and Kaye, P. E. (1985). Relative growth of six thistle species along a nutrient gradient with multispecies competition. *J. Ecol.* **73,** 667–684.

Austin, M. P., Nicholls, A. O., and Margules, C. R. (in press). Measurement of the realized niche of plant species: Examples of the environmental niche of five *Eucalyptus* species. *Ecol. Monogr.*

Diggle, P. J. (1983). "Statistical Analysis of Spatial Point Patterns." Academic Press, London.

Ellenberg, H. (1953). Physiologisches und okologisches Verhalten derselben Pflanzenarten. *Ber. Dtsch. Bot. Ges.* **65,** 351–362.

Ellenberg, H. (1954). Uber einige Fortschritte der kausalen Vegetationskunde. *Vegetatio* **5/6,** 199–211.

Ernst, W. (1978). Discrepancy between ecological and physiological optima of plant species: A reinterpretation. *Oecol. Plant.* **13,** 175–188.

Gauch, H. G., and Whittaker, R. H. (1972). Coenocline simulation. *Ecology* **53,** 446–451.

Goodall, D. W. (1953). Objective methods for the classification of vegetation. I. The use of positive interspecific correlation. *Aust. J. Bot.* **1,** 39–63.

Goodall, D. W. (1963). The continuum and the individualistic association. *Vegetatio* **11,** 297–316.

Grace, J. B. (1988). The effects of plant age on the ability to predict mixture performance from monoculture growth. *J. Ecol.* **76,** 152–156.

Greig-Smith, P. (1979). Presidential address 1979: Pattern in vegetation. *J. Ecol.* **67,** 755–779.

Grime, J. P. (1979). "Plant Strategies and Vegetation Processes." Wiley, Chichester, England.

Grime, J. P. (1985). Towards a functional description of vegetation. *In* "Population Structure of Vegetation" (J. White and J. Beeftink, eds.), pp. 503–514. Junk, The Hague, The Netherlands.

Hairston, N. G., Smith, F. E., and Slobodkin, L. G. (1960). Community structure, population control, and competition. *Am. Nat.* **94,** 421–425.

Harper, J. L. (1977). "Population Biology of Plants." Academic Press, London.

Kershaw, K. A. (1958). An investigation of the structure of a grassland community. I. The pattern of *Agrostis tenuis. J. Ecol.* **46,** 571–592.

Kershaw, K. A. (1959). An investigation of the structure of a grassland community. II. The pattern of *Dactylis glomerata, Lolium perenne* and *Trifolium repens.* III. Discussion and conclusions. *J. Ecol.* **47,** 31–53.

Kershaw, K. A., and Looney, J. H. H. (1985). "Quantitative and Dynamic Plant Ecology," 3rd ed. Arnold, London.

Margules, C. R., Nicholls, A. O., and Austin, M. P. (1987). Diversity of *Eucalyptus* species predicted by a multi-variable environmental gradient. *Oecologia* **71,** 229–232.

Minchin, P. R. (1987). An evaluation of the relative robustness of techniques for ecological ordination. *Vegetatio* **69,** 89–107.

Mitchley, J. (1987). Diffuse competition in plant communities. *Trends Ecol. Evol.* **2,** 104–106.

Mitchley, J., and Grubb, P. J. (1987). Control of relative abundance of perennials in chalk grassland in southern England. I. Constancy of rank order and results of pot- and field-experiments on the role of interference. *J. Ecol.* **74,** 1139–1166.

Mueller-Dombois, D., and Ellenberg, H. (1974). "Aims and Methods of Vegetation Ecology." Wiley, New York.

Pianka, E. R. (1981). Competition and niche theory. *In* "Theoretical Ecology: Principles and Applications" (R. M. May, ed.), pp. 167–196. Blackwell, Oxford, England.

Pickett, S. T. A., and Bazzaz, F. A. (1978). Organisation of an assemblage of early successional species on a soil moisture gradient. *Ecology* **59,** 1248–1255.

Salt, G. W. (1983). Roles: Their limits and responsibilities in ecological and evolutionary research. *Am. Nat.* **122,** 697–705.

Shipley, B., and Keddy, P. A. (1987). The individualistic and community-unit concepts as falsifiable hypotheses. *Vegetatio* **69,** 47–55.

Simberloff, D. (1983). Competition theory, hypothesis testing, and other community ecological buzzwords. *Am. Nat.* **122,** 626–635.

Strong, D. R., Simberloff, D., Ahele, L. G., and Thistle, A. B. (eds.) (1984). "Ecological Communities: Conceptual Issues and the Evidence." Princeton Univ. Press, Princeton, New Jersey.

Tilman, D. (1982). "Resource Competition and Community Structure." Princeton Univ. Press, Princeton, New Jersey.

Tilman, D. (1986). Nitrogen-limited growth in plants from different successional stages. *Ecology* **67**, 555–563.

Tilman, D. (1987). Secondary succession and the pattern of plant dominance along experimental nitrogen gradients. *Ecol. Monogr.* **57**, 189–214.

Tilman, D. (1988). "Plant Strategies and the Structure and Dynamics of Plant Communities." Princeton Univ. Press, Princeton, New Jersey.

Tilman, D. (1989). Ecological experimentation: Strengths and conceptual problems. In press.

Vogel, H. H. (1978). "Aufnahme und Speicherung von Stickstoff bei Stickstoffzeigern und Stickstoffmangelzeigern im Grünland," Ph.D. dissertation. Georg-August-Universitat, Göttingen, West Germany.

Watt, A. S. (1947a). Pattern and process in the plant community. *J. Ecol.* **35**, 1–22.

Watt, A. S. (1947b). Contributions to the ecology of bracken. IV. The structure of the community. *New Phytol.* **46**, 97–121.

Watt, A. S. (1955). Bracken versus Heather, a study in plant sociology. *J. Ecol.* **43**, 490–506.

Whittaker, R. H. (1978). Direct gradient analysis. *Handb. Veg. Sci.* **5**, 7–50.

Wilson, S. D., and Keddy, P. A. (1985a). The shoreline distribution of *Juncus pelocarpus* along a gradient of exposure to waves: An experimental study. *Aquat. Bot.* **21**, 277–284.

Wilson, S. D., and Keddy, P. A. (1985b). Plant zonation on a shoreline gradient: Physiological response curves of component species. *J. Ecol.* **73**, 851–860.

Wilson, S. D., and Keddy, P. A. (1986a). Species competitive ability and position along a natural stress/disturbance gradient. *Ecology* **67**, 1236–1242.

Wilson, S. D., and Keddy, P. A. (1986b). Measuring diffuse competition along an environmental gradient: Results from a shoreline plant community. *Am. Nat.* **127**, 862–869.

Wilson, S. D., Keddy, P. A., and Randall, D. L. (1985). The distribution of *Xyris difformis* along a gradient of exposure to waves: An experimental study. *Can. J. Bot.* **63**, 1226–1230.

12

Plant–Plant Interactions in Successional Environments

F. A. Bazzaz

I. Introduction

Except for a few situations, widespread disturbance in a site is followed by colonization and, very soon thereafter, interactions among neighboring plants. The intensity of these interactions varies depending on many factors and may be expected to become greater with the further development of the vegetation. There is now some evidence that plant–plant interactions may actually begin before individuals are in physical contact with each other, or before they begin to share the same resource pools. Apparently, the modification in the quality of the reflected and diffuse light by an individual remotely affects the growth of another individual (Ballaré *et al.*, 1987). Many plant–plant interactions in nature take place in successional environments. It is now widely recognized that, even in stable communities, where species identities change only slowly, small- as well as large-scale disturbances generate successional patches of varying identity of interacting occupants.

It is not my intention in this chapter to evaluate the various definitions of competition (in the broad sense), nor do I wish to discuss in detail the theories of community organization and the role of competition in it. These issues have been extensively treated and hotly, sometimes brutally, debated with no resolution that directly applies to sessile, autotrophic organisms. Rather, my aim is to discuss the role of the different types of negative and positive interactions among plants, within guilds and between guilds, that may influence successional processes in old fields. I rely heavily on data from successional communities in the midwestern United States. I do need, however, to state what I take the terms plant–plant interactions, competition, and interference to mean. This is necessary in order to evaluate available literature and extract what is directly relevant to this chapter. Although Harper (e.g., 1983) has strongly and justifiably argued against the use of the term competition and for replacing it with "interference," the term competition still enjoys a wide usage and is not likely to be abandoned in the near future. Plant–plant interactions occur when individuals are in close proximity such that each may influence the performance of the other in terms of growth, survival, and reproduction. Clearly, interactions among neighboring plants may be negative or positive. "Interference" is used here to mean negative interactions in general. "Mutualism" may be used to describe positive interactions. "Competition" is used to mean the struggle between neighboring plants to obtain a resource (or resources) in short supply. Another form of interference is that of allelopathy, which is the production of chemicals that inhibit plant function (Rice, 1984). Regardless of cause, negative interactions result in changes in morphology, physiology, allocation patterns, etc. and render the target individuals less fecund than they would be without the presence of neighbors. The outcome of these interactions in successional terms would be the local extinction of the target species and the dominance of a superior competitor. The extreme outcome of this process is of course complete competitive exclusion, which is rarely attained in nature on any meaningful scale. Of course, it is always important to keep in mind that plant–plant interactions in successional habitats are further influenced by other biotic interactions. Differential herbivory has been shown to influence competitive outcome greatly (see the review by Edwards and Gillman, 1987). The role of fungi, pathogens, and herbivores in competition is discussed elsewhere in this volume (see chapters by Allen and Allen, Clay, and Louda *et al.*).

Although interactions among species in successional habitats may be classified as competitive, neutral, or facilitative and have been considered to be important in successional change (see Connell and Slatyer, 1977), interactions among individuals are rarely of one type. Interactions among plants may change in intensity and likely become asymmetrical as competing individuals grow larger. Inferior competitors for light become suppressed and may finally become dependent on the winners for survival. For example, if the smaller individuals develop a shade-adapted physiology, the death of the dominants may expose them to stressful, and possibly lethal levels of environmental resources, e.g., high light, low humidity, and high temperature. Suppressed individuals in herbaceous communities may also depend on their bigger neighbors for physical support, which may allow them to allocate more mass to reproduction (Bazzaz, 1984). Thus negative and positive interactions between individuals may occur simultaneously.

II. The Role of Interference in Successional Change

In discussing the role of interference in succession we may find ourselves in a semantic mire. If we consider the entire life of populations in a given location and define competitive superiority as ultimate success relative to other populations, then the whole process of succession will simply be replacement of species by the next relatively more superior competitor! In this view interference is the force that drives succession. On the other hand, if we consider that species individualistically and differentially mature, reproduce, and disperse in space, then successional sequences will be determined largely by propagule arrival time and the time each requires to grow, reproduce, and disperse. Interference in this case only modifies these activities but does not completely govern them. The success of colonizing species in invading and replacing neighboring individuals depends on their life history traits, competitive competence, and complementary behavior in resource acquisition (see the discussion in Bazzaz, 1986).

Most plant ecologists who have considered the role of competition in plant succession have concerned themselves with direct influences of plants on each other, largely through resource depletion (e.g., Werner, 1979). A few have recognized the possible role of chemical inhibition as a factor in species replacement and the resulting successional change (Williamson, this volume). They show that these chemicals may interfere with the availability of nitrogen in soils and its utilization by the plants. Whatever their emphases are, all ecologists agree that competitive interactions play a role in succession. Experimental tests of the role of interference in succession have been most informative when community changes were assessed after the removal from or the addition of species to the community (e.g., Raynal and Bazzaz, 1975; Allen and Forman, 1976; Abul-Fatih and Bazzaz, 1979; Pinder, 1975; Gross, 1980; Armesto and Pickett, 1985, 1986).

III. Who Interacts with Whom in Successional Environments?

Competitive interactions among neighboring plants are almost always of the scramble type (Whittaker, 1975). Each individual preempts as much of a resource as it can. Furthermore, individuals usually interact simultaneously with several other individuals which are in turn interacting with other individuals. Thus, interaction networks are almost always found in field situations. An individual plant may interact with its siblings, with other members of the same species, or with species of completely different taxonomic affinities and growth habits. Except for the clumping of individuals caused by dispersal patterns, the probability that an individual interacts with individuals of different species increases with an increase in species diversity. It is likely that, in early successional habitats and within patches, the probability of interspecies interaction is low, while in highly diverse late successional ecosystems the probability of encountering and therefore interacting with an individual of a different species is usually high.

Irrespective of the identity of interacting individuals early in the life of a target plant, neighborhoods may greatly change over time. In forest ecosystems, especially in the understory, individuals may experience several alternating periods of low competition and high competition for resources. These individuals attain much of their growth during the former and tolerate shortages during the latter. The strength of interactions among neighboring individuals also may change. Slightly asymmetrical interactions may change with time to highly asymmetrical ones as a winner preempts more of the resources by virtue of its size relative to its now suppressed neighbor.

IV. Experimental Investigations of the Role of Plant–Plant Interactions in Successional Change: A Case Study in Illinois Fields

Experimental work to evaluate the role of competition in species replacement began seriously with that of Keever (1950) in the old fields of North Carolina. Her experiments evaluated the causes of replacement of the annuals *Erigeron canadensis* and *Ambrosia artemisiifolia* by *Aster pilosus*, which was in turn replaced by *Andropogon virginicus*. Later investigators examined, in the same sere, the replacement of pines by sweetgum, using physiological attributes to infer competitive replacement (Kozlowski, 1949; Bormann, 1953; Tolley and Strain, 1985). Keever concluded that dispersal patterns, germination time, growth rate, drought resistance, light requirements, competitive superiority, and autotoxicity in *Erigeron* determine the sequence of species in early succession. She also concluded (contrary to the Clementsian view) that the influence of a species on its environment may make it less hospitable for itself but not necessarily more hospitable to the species that succeed it. In Illinois fields, we have studied interactions among neighboring plants to assess their role in community dynamics. In this sere, succession usually proceeds from fields dominated by annuals to perennials, and finally to deciduous trees. Depending on the timing of initial soil disturbance, the winter annuals or the summer annuals dominate first. These are usually followed by *Aster*, then *Solidago*.

A. Interactions between Species Guilds

One of the most obvious aspects of community organization in early succession is that of differential timing of various activities among plants. This can be demonstrated between guilds as well as within guilds (groups of species with similar life forms that occur together). While we can say very little about the origin of these responses (see later), we do find these differences in several successional sequences. They are most obvious, and best studied, in old-field successions on both nutrient-poor and nutrient-rich soils.

1. Winter Annuals versus Summer Annuals   In many temperate successions two groups of annual colonizing species are readily recognized. The so-called winter annuals (usually composites and crucifers) germinate in late summer and fall, grow through the winter, and flower the following spring. The summer annuals emerge in the spring and early summer and flower during the summer. Interactions among the two groups in Illinois fields are now well documented and the mechanisms of these interactions are well understood (reviewed by Bazzaz, 1979). Because of their rosette habit and their ability to adjust their photosynthetic responses to temperature, many winter annuals can fix carbon even during winter (Regehr and Bazzaz, 1976). Therefore some of their rosettes reach considerable size before they bolt in the spring. Thus the winter annuals preempt space and suppress the summer annuals, which germinate in the spring. Resource augmentation and removal experiments with these two groups of plants have revealed much about the nature of interactions among them (e.g., Raynal and Bazzaz, 1975). Suppression of the summer annuals results from depletion of nutrient supply, decreasing light intensity, and altered light quality, all of which are caused by the presence of the winter annuals. It is interesting to note, however, that this suppression results in reduced growth but little or no mortality in the most prominent of the summer annuals, *Ambrosia artemisiifolia*. Plowing, which at once removes the winter annuals and loosens the soil, increases the growth rate and seed production in *Ambrosia* more than only the removal of competing winter annuals (Fig. 1). Thus direct interference between *Ambrosia* and the winter annuals is modified by other environmental factors, a point which I return to later in this chapter. It is also interesting to note that augmentation of the water and nutrient resources did not result in the decline of competitive superiority of the winter annuals. But, while the density of *Ambrosia* in response to fertilizer treatment only slightly changed throughout the season, it declined drastically with addition of water (Fig. 2). Thus, the addition of resources to alleviate competitive interactions resulted in two different responses at the population level.

In fields plowed in the spring, the winter annual rosettes which were recruited in the fall are largely destroyed, and the summer annuals predominate. New seedlings of the winter annuals may emerge under the canopy of the summer annuals. Many of these seedlings die, but those that survive grow in the following fall and winter and dominate the fields during the second year, suppressing the summer annuals. Thus, the predominance and suppression of these two groups of plants does not result from differences in innate competitive abilities but from the timing of disturbance in the fields, which changes competitive hierarchies. Either group of species could suppress the other depending on disturbance time.

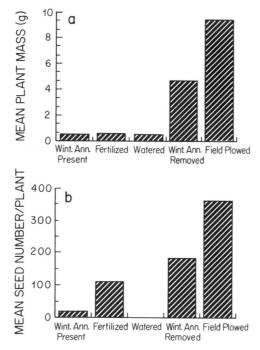

Figure 1 Mean plant mass (a) and mean seed number per plant (b) of *Ambrosia artemisiifolia* in different competitive situations in the field. The treatments are (1) control, (2) plots fertilized, (3) plots watered, (4) winter annuals hand removed, and (5) plots plowed early in the spring. Data from Raynal and Bazzaz (1975).

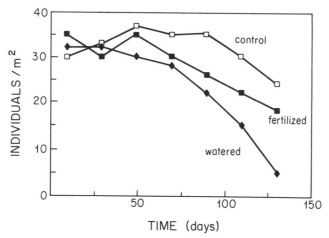

Figure 2 Field density of *A. artemisiifolia* populations in control, fertilized, and watered plots. Data from Raynal and Bazzaz (1975).

2. The Annuals versus Aster   Recruitment of *Aster pilosus* occurs both in the first and the second year after disturbance but rarely later. Seedlings of this species are slow-growing relative to those of the annuals and seed germination may occur in the fall or in the spring. Prominence of *Aster* in successional fields may be attained in the second year or delayed to the third. Therefore, individuals of *Aster* of different sizes may interact with the annuals. The most common response of *Aster* is to be suppressed during the first year because of competition with the much larger annuals, and to bolt and reproduce during the second year. However, when intraspecific and interspecific competition is low, and nutrients are available in large quantities, individuals of *Aster* can grow to maturity during the first year. Suppressed *Aster* seedlings and rosettes found under the canopy of the annuals are quite tolerant of water limitation and have low light saturation requirements for photosynthesis (Peterson and Bazzaz, 1978).

The identity of the competing neighbors is a major determinant of the fate of suppressed *Aster*. For example, equal densities of *Chenopodium album* and *Polygonum pensylvanicum* cause very little *Aster* mortality, while *Amaranthus retroflexus* and especially *Ambrosia artemisiifolia* cause much mortality. Growth reduction of *Aster* is also dependent on the identity of neighbors, with *Ambrosia* being the most effective suppressor and *Amaranthus* the least (Fig. 3). Flowering is completely inhibited by the presence of annual neighbors. End-of-season *Aster* heights were ~24.0 cm in conspecific competition and ranged between 3.0 and 7.5 cm when *Aster* was competing with the annuals. The strategy of *Aster* is therefore: (1) to tolerate resource limitation imposed by the annuals in the first year, (2) to accumulate some resources even in this situation by having certain physiological attributes, (3) to accumulate resources over a long time, in part while its annual competitors are inactive, and (4) to possess the perennial habit which ensures that at the start of the growing season older *Aster* individuals are much larger than the annuals which must start as small seedlings again.

Some of the same physiological attributes, however, may themselves be detrimental to the continued success of *Aster* in the fields. The fact that *Aster* seeds germinate best under high levels of irradiance and moderately high N levels (Peterson and Bazzaz, 1978), conditions associated with some disturbance that removes the vegetation, and the slow growth of young seedlings result in the ultimate elimination of *Aster* seedlings from established *Aster* stands despite the fact that the plants can produce as many as $\frac{1}{2}$ million seeds per m^2.

3. Aster versus Solidago   *Solidago altissima, Solidago canadensis*, or *Andropogon virginicus* usually replace *Aster* in many successions: the first is

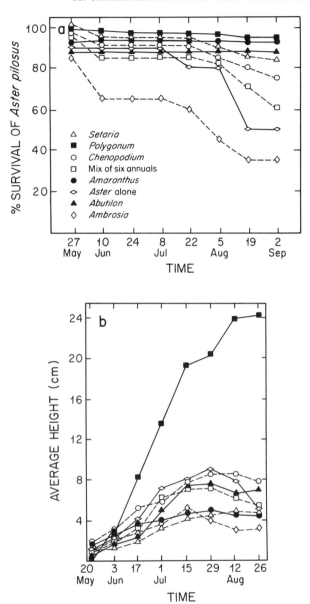

Figure 3 Differences in survivorship (a), and growth (b) of *Aster pilosus* in competition with its potential annual neighbors. From Peterson and Bazzaz (1978).

more likely in the Midwest, the second in the east, and the third in the southeastern part of the eastern deciduous forest in North America. The causes of replacement and the role of competition in the replacement are unknown. Keever (1950) suggested that *Andropogon* deprives *Aster* of water and that *Andropogon* (which she considered very competitive) is delayed in dominance because of its relatively poor dispersal ability.

The replacement of *Aster* by *Solidago* has not been adequately investigated. Goldberg (1987) has shown that *Solidago* transplants survive better in *Solidago* than in *Aster* neighborhoods. We have found that the two genera differ in a number of physiological and morphological traits that influence their interactions. *Aster* has many branches which carry on them a large number of small leaves while *Solidago* ramets are single-stemmed and carry larger leaves. *Aster* genets are not as compact as those of *Solidago* because of differences in rhizome length. *Solidago* has shorter rhizomes producing closely spaced ramets and a compact genet. Thus, above-ground competition for light is usually among sister ramets. In contrast, *Aster* has much longer rhizomes with distantly spaced ramets which may be penetrated by ramets of other individuals or other species. Light penetration is higher in *Aster* canopies than in *Solidago* canopies and *Solidago* ramets drop their lower leaves when they become shaded. The two species do not differ greatly in their photosynthetic rates (B. Schmid and F. A. Bazzaz, unpublished observations). Though no experimental data are available, it is likely that, because of the differences in light penetration, seedlings of *Solidago* can emerge under *Aster* but not the reverse. Furthermore, clonal integration, which is stronger in *Solidago* than in *Aster*, may confer some level of superiority to the former relative to the latter. This high integration may lead to the success of genets even when parts of them are in unfavorable habitats (Bazzaz, 1984; Hartnett and Bazzaz, 1985a). It is puzzling that *Solidago* rosettes appear in old fields 2 to 3 years later than those of *Aster*. There is no reason to suspect that dispersal could account for this. Seeds of both species are well suited for wind dispersal and they do not differ greatly in size. Furthermore, under controlled glasshouse conditions and with ample moisture, light, and nutrients, *Solidago* and *Aster* seeds germinate readily, produce vigorous rosettes, bolt, and flower in what is equivalent to one growing season.

One possible, yet untested, explanation for the persistence of *Solidago* is its ability to penetrate *Aster* stands. After some time the *Aster* genets fragment, while the penetrating *Solidago* individuals grow by forming many connected, physiologically integrated ramets in compact genets (B. Schmid and F. A. Bazzaz, unpublished observations).

4. **Solidago** *versus the Trees*   Life history traits, efficiency of resource capture and use, tolerance of, and, in some cases, preference for low resource levels may all interact to produce the observed changes in community structure from mid to late succession, where, in many geographic locations the herbaceous perennial community is replaced by trees.

A number of differences in biomass allocation patterns may lead to the replacement of *Solidago* by the invading trees. Complete loss of above-ground parts during the dormant season in *Solidago* and rebuilding of that biomass during the following growing season may be the major factor in tree establishment in *Solidago* stands. The trees drop only the leaves but retain their stems and dormant buds and much of their root system. As a result, *Solidago* starts its growing season from rhizomes underground while, in the absence of herbivory and other disturbances, the tree seedlings start successive years' growth from the position of their tip in the previous year. *Solidago* grows largely in two dimensions by clonal spread while the trees grow in three dimensions and, after a few seasons, overtop *Solidago*. *Solidago* invests in much clonal expansion, as the newly developed ramets are very dependent on the underground reserves and their sister ramets for their early support (Hartnett and Bazzaz, 1985b). *Solidago* also invests in sexual reproduction early and, unlike the trees, suffers further loss of biomass by the death of rhizome internodes. There is good evidence that tree seedlings allocate a substantial portion of their yearly accumulated biomass to below-ground growth, developing more extensive and more deeply penetrating roots systems relative to *Solidago*. Partial shading of the young tree seedlings early in their life by *Solidago* may even give them some protection from high irradiance, low humidity, and herbivores. Photosynthetic light saturation curves of some late successional tree species, e.g., *Acer saccharum* show that they may be photoinhibited in full light.

Although there has been much work on the physiological ecology of successional tree species (reviewed by Bazzaz, 1979), competitive relationships have not been adequately studied nor has their role been fully discussed. Inferences about the role of competition have been made but they remain untested. It is likely that different light requirements and shade-tolerance capabilities are major factors in replacements among tree species (e.g., Horn, 1971).

B. Interactions within Species Guilds

Avoidance of competition between species by preferences for certain resource patch types increases the probability of interactions among individuals of the same species and, depending on dispersal patterns, interaction among siblings and with parents. The role of interactions among members of the same species has not been directly implicated in successional change except in cases where autotoxicity may lead to the eventual local extinction, e.g., *Erigeron canadensis* (Keever, 1950). Even if they play only a minor role in species replacement, these within-species interactions may have significant evolutionary consequences for the nature of interactions among species and their outcome. Interactions also occur among members of each group. Again, timing of emergence and growth largely determine the outcome of interactions among individuals. For example, in the winter annual group, the crucifer *Rorippa sessiliflora* emerges in the summer usually after heavy rains, without disturbance, and before the composites. The species forms small monospecific patches consisting of a few to several hundred individuals. Seedlings grow in the fall and overwinter as rosettes. The survivors bolt, flower, and set seed in the spring, before the composites have grown much. At maturity, this species, and several other crucifers, are much smaller than the composites and are undoubtedly inferior to them when started at the same time. Thus *Rorippa* successfully completes its life cycle by avoiding interference, instead of engaging in a losing contest with the composites.

Winter annuals occur in patches of varying densities in the field ranging from a few individuals to more than a thousand individuals per m². The growth of their rosettes during the fall is highly density dependent and the rosettes vary greatly in size at the end of the fall season. Interaction among these plants also varies depending on the nutrient and moisture levels of the sites. Larger rosettes have a higher probability of overwinter survival, and the bolting individuals may produce many more seeds than those that develop from smaller rosettes. For example, individuals of *Erigeron annuus* derived from rosettes with diameter of 1–4 cm produced, on average, 16,000 seeds each, whereas those with diameters above 9 cm produced 72,000 seeds (Regehr and Bazzaz, 1979). One additional consequence of increased density in these plants is the increase in pathogenic infection in some of the species. *Rorippa sessiliflora* mortality, caused mainly by infection with the pathogen *Albugo candida,* was much higher in high-density than in low-density sites (Fig. 4) and, even among the survivors, the degree of infection by the pathogen was still higher in the high-density population (59 versus 11%) (D. L. Regehr and F. A. Bazzaz, unpublished observations).

The Identity and Equivalency of Neighbors   Plant species diversity generally increases with succession and is assumed to reach a maximum before the attainment of a stable community with low levels of disturbance. The rise in diversity with succession is not necessarily monotonic and, in several successions, there are some obvious declines at some stages of succession. The dominance of a single species may be so strong that it drastically reduces the contribution of others. *Solidago altissima* and *Andropogon virginicus* are prominent examples of this situation in old-field successions in the midwestern and the eastern United States. While at first glance one could assume that the identity of neighbors is more

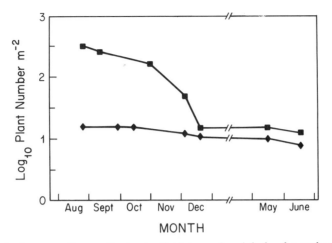

Figure 4 Patterns of mortality in the field (caused mainly by the pathogen *Albugo candida*) in low- and high-density populations of the winter annual *Rorippa sessiliflora* (D. L. Regehr and F. A. Bazzaz, unpublished observations).

predictable in low-diversity than in high-diversity systems, dispersal patterns, environmental patches, and localized seed bank dynamics complicate the predictability. The balance between intraspecific and interspecific competition is very complicated and our knowledge of it is quite limited.

One prediction from the broad responses and the high degree of overlap of early successional plants on environmental gradients is that these species are more or less equivalent in their interactions (Bazzaz, 1987). While plant species within a community could be dissimilar in their response on one or more resource gradients, they are very unlikely to be dissimilar on all gradients, and therefore must experience strong interaction when grown in proximity to each other. Experimental tests with the early successional community have confirmed this prediction. In the diallele competition experiment of Parrish and Bazzaz (1982a), the mean biomass of an individual competing with another individual of the same species was not different between the early successional annuals and the late successional prairie species. But, in the prairie species, the mean biomass of an individual was significantly higher in heterospecific pairs (Table 1). Furthermore, the ratio representing the coefficient of variation of mean weight of an individual of a species in a mixed stand over the mean in pure stands was much lower in the early successional community (0.23) than in the late successional community (0.85). That is, there were fewer differences in performance among species of the

Table 1   Mean Biomass per Individual Competing in Conspecific and in Heterospecific Pairs in Early Successional Annual and in Late Successional Prairie Plants

	Biomass (g)	
	Individual in Conspecific Pair	Individual in Heterospecific Pair
Annuals	1.461	1.092
Prairie species	1.386	1.616
a_u	NS	$p < 0.001$

early successional community. Thus, relative to late successional plants, early successional plants are more equivalent as neighbors. Nevertheless, growth of target species of the annuals differed somewhat in different neighborhoods. For example, *Amaranthus retroflexus* responded to any combination containing C_3 plants in the same way; to it they were more or less equivalent. *Setaria faberii*, another C_4 plant, had somewhat similar response to the presence of the C_3 species. Goldberg and Werner (1983) considered the equivalency of competing plant species having the same growth form and argued that competitive interactions are not usually species specific. Furthermore, Goldberg (1987) investigated competition between *Solidago canadensis* and seven co-occurring herbaceous plants in an old field in Michigan. She showed that the average growth of *Solidago* was reduced by 17–62% depending on the identity of its competitors. However, on a per gram basis, the competitor neighbor identity was unimportant.

Despite the broad responses of many early successional plants and their general equivalency as competitors, there are examples of superior competitive ability. In the summer annual community, *Ambrosia trifida,* when present, dominates the community and decisively outcompetes all its neighbors. Within its guild, it has the largest seeds, earliest emergence, largest cotyledons, highest tolerance to cold nights in the early spring, fastest growth rates, and highest leaf area index. It also fixes a significant amount of carbon in its photosynthetic flowers and seeds (Bazzaz *et al.*, 1979). Removal experiments have shown that its presence lowers community diversity by 90% but increases productivity eightfold (Abul-Fatih and Bazzaz, 1979). In the winter annual community of the southeastern United States originally dominated by *Erigeron canadensis,* *Heterotheca latifolia* is rapidly replacing *Erigeron* as the major pioneer species in the old-field succession. Experimental work by Tremmel and Peterson (1983) showed that competition between the two species re-

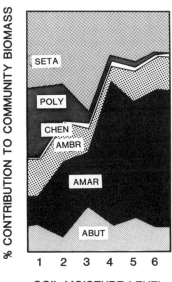

SOIL MOISTURE LEVEL

Figure 5 Differences in competitive hierarchies of six annuals on a controlled soil moisture gradient ranging from wet (level 1) to very dry (level 6). Seta, *Setaria;* Poly, *Polygonium;* Chen, *Chenopodium;* Ambr, *Ambrosia;* Amar, *Amaranthus;* Abut, *Abutilon.* From Pickett and Bazzaz (1978).

sulted in significant reduction in the growth of *Erigeron,* and when seeds of the two species were sown in equal numbers and high density, as might occur in some patches in the field, a pure stand of winter rosettes of *Heterotheca* developed. They concluded that *Heterotheca* outcompetes *Erigeron* in all stages of its life cycle. Field experiments by Miller and Werner (1987) with *Ambrosia artemisiifolia* showed that this species dominates its neighboring annuals.

It must be remembered that competitive performances of species relative to each other may change dramatically depending on the level of resources available to them (Fig. 5). We found experimental evidence for this pattern of competitive response to soil moisture (Pickett and Bazzaz, 1978), nutrient levels (Parrish and Bazzaz, 1982b), and atmospheric CO_2 concentrations (Bazzaz and Garbutt, 1988).

V. Plant–Plant Interactions and the Evolution of Response Breadth

Predictions about the differences in response (niche) breadth of early and late successional plants have been made (e.g., Odum, 1969), based on the $r–K$ continuum of strategies proposed by MacArthur and Wilson

(1967). These predictions are very intimately tied to the strength of competitive interactions among species and their evolutionary outcome. In a series of experiments, we examined the response breadth of a large number of species from early, mid, and late succession, including annual and perennial herbs and trees. We used a number of gradients, including moisture, nutrients, light, temperature, and pollinators in these studies. We assessed response in most cases on one gradient at a time, and in some instances we calculated response on two crossed dimensions. We also considered the separation of response along a time axis and compared the response of species at different times during their ontogeny (reviewed by Bazzaz, 1987).

Of particular interest were the studies on competition for pollinators since biological resources are rarely considered in competition studies. Competition for pollinators among species in both early and late successional plants revealed differences among them (Parrish and Bazzaz, 1978, 1979). In general, seasonal flowering was somewhat clumped. In the winter annual community and the mature grassland there were three flowering assemblages, respectively, in spring–early summer, midsummer, and late summer–early fall. Although the plants in the two communities were pollinated by different vectors, we found evidence for competition for pollinators in all flowering guilds. Analyzing data for daily and seasonal time of visits by pollinators, their identity, and two-dimensional parameters of species of visitor versus daily time revealed differences and similarities among species in the two communities. In general, there was broader response and more overlap among species in the annual community than among species in the grassland communities. The grassland species also showed a higher degree of specialization than did the annuals. In both communities, however, the overlap in response was lower than it would be for a random community, suggesting some degree of niche separation among species. The results of all these studies and others on niche relations have been remarkably consistent with these predictions, despite differences in response among species within guilds and among gradients (Table 2). In every case, whether the comparisons were made between herbs or between trees, early successional species had broader and more overlapping responses than did late successional species.

What do these patterns of response mean in terms of plant–plant interactions in successional environments? This is an area of some confusion and disagreement among investigators. Broad response along environmental gradients results in a higher degree of overlap and stronger interactions among species in the zones of severe overlap. It follows then that, among early successional species, plant–plant interaction ought to be very strong. In contrast, because late successional plants have rela-

Table 2 Response Breadths (Levins' β) and Proportional
Similarities of Plants from Early and Late Succession on
Several Niche Axes

	Early Succession	Late Succession
(a) Mean niche breadth of all species in experimental assemblage		
Herbaceous		
Underground space	.71	.29
Pollinators	.20	.16
Nutrients	.77	.70
Trees		
Nutrients	.91	.82
Moisture	.89	.65
(b) Mean proportional similarity of all species in experimental assemblage		
Herbaceous		
Underground space	.68	.43
Pollinators	.31	.19
Nutrients	.85	.83
Trees		
Nutrients	.94	.83
Moisture	.82	.70

tively narrow responses, their response overlaps are small and competition among them should be less intense. Their coexistence in communities with high diversity is thought to be the result of their differentiation along environmental gradients (higher species packing). It is assumed that over evolutionary time the intensity of interactions among species and their competition for various resources have shaped their responses. Species became generally narrower by directional or stabilizing selection. The result is higher diversity and less competition.

We have experimentally tested the strength of these interactions among plants in early and in late successional communities by comparing the ability of species combinations to obtain resources and to use them to build plant biomass (details in Parrish and Bazzaz, 1982a). We grew plants from these communities singly, in con- and heterospecific pairs, and in within-community pure and mixed stands. Every individual had potential access to an equal quantity of soil. The biomass of individuals in heterospecific pairs relative to the biomass in conspecific pairs was significantly lower for early successional than for late successional plants ($x = 0.72$ and 0.93, respectively). Most of the pairwise combinations of the early successional species had relative biomasses of less than 1. That is, both members of a pair accumulated less biomass in heterospecific than in conspecific competition. In contrast, late successional species had almost equal numbers of combinations with relative biomasses above or below 1. Thus competitive interactions were most prevalent in the early successional community, resulting in decreased ability of both species to obtain resources and to grow. The biomass of mixed species pairs divided by the biomass of individual species grown singly showed that the early successional species experienced significantly more reduction in total biomass than did the late successional species ($x = 0.50$ versus 0.24). Furthermore, the species of the late successional community had a higher relative yield when grown in mixed stands relative to that in pure stands. Total yield in mixed stands was only 7% higher than for all pure stands of the early successional community but was 34% higher in the late successional community. These results strongly suggest that niche separation leads to lower competition within communities. It also suggests that selection to reduce competition could have been more important in the evolution of late successional species than in the early successional ones.

VI. Evidence for Differences in Response among Species of the Same Community: Mechanisms for Reduction of Competition

Although early successional species generally have broad responses on resource gradients, differences among them do exist when grown in competition with other members of the same community. For example, *Polygonum pensylvanicum* shifts strongly toward the wet end on a moisture gradient while *Abutilon theophrasti* shifts slightly toward the dry end of the gradient. *Ambrosia artemisiifolia* and *Setaria faberii* do not shift much. These shifts may contribute to coexistence in this community (Pickett and Bazzaz, 1978).

Differences among early successional plants under field conditions also exist. These differences are best exemplified by location of roots in soil and time of root growth (Parrish and Bazzaz, 1976). *Setaria faberii* has a fibrous root system which tends to be primarily concentrated at the upper parts of soil profile. *Abutilon theophrasti* locates many of its roots in the middle parts, while *Polygonum pensylvanicum* roots extend down to the water table deep in the soil profile. The location of the root system may allow access to different soil water resources and may influence the pattern of their daily water potential in the field and the way they respond during periods of drought and after heavy rains. Photosynthetic response to leaf water potential differs among the three species (Fig. 6) such that all neighboring individuals, irrespective of their identity, could operate at near maximum despite great differences among them in leaf water potential (see Wieland and Bazzaz, 1975, for details).

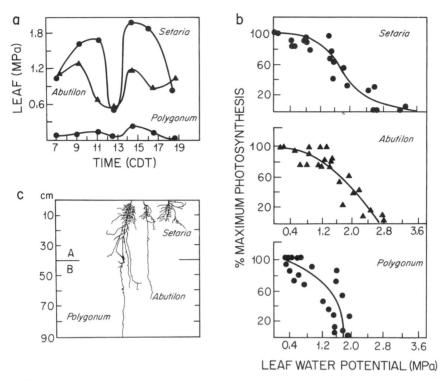

Figure 6   Daily march of leaf water potential in mid-summer on a clear day (a), photo-synthetic response to leaf water potential (b), and root distribution profiles in the field (c) of three cooccurring annuals. Data from Wieland and Bazzaz (1975).

VII. The Opposing Forces of Convergence and Divergence

Theory predicts that intensely competing species populations with a high degree of overlap in resource use may, over evolutionary time, diverge in their response, resulting in reduced competition. If species act in this manner as the selective force on each other, then the result is *niche differentiation* (sensu Bazzaz, 1987). This sort of response is more likely to occur if the importance of the particular resource for which the species are competing overrides all other resources or when other resources covary with the resource in question. Evidence for the existence of such a response is only circumstantial and the action of competition in niche separation along several resource axes has been only inferred. Ecologists have argued about the strength of competition in the shaping of response breadth and in community organization (see Connell, 1980, 1983). While the differences in response among species may contribute to their coexistence (as discussed earlier), the general conclusion is that we still lack strong experimental evidence for the importance of hetero-specific competition leading to niche differentiation.

VIII. Plant–Plant Interactions as Selective Agents on the Genetic Structure of Populations of Early Successional Plants

Although there is strong evidence for the presence of high levels of variation in populations in general (e.g., Hamrick *et al.*, 1979; Ennos, 1983; Loveless and Hamrick, 1984; Gottlieb, 1981) and in successional plant populations in particular (Bazzaz and Sultan, 1987), selection for certain genotypes with successional change has been proposed (see the thorough review by Gray, 1987).

After the first year or two of succession, populations of the annuals usually are represented mainly in the seed bank and by a few individuals in small patches where soil disturbances occur. For a few species, how-ever, populations may persist for a longer time and their individuals are likely to interact strongly with the perennials. What influences might this interaction have on the genetic structure of these persisting annual pop-ulations?

Two populations of *Ambrosia trifida* (a persistent annual in old fields in Illinois), one from an annually disturbed section of a field and another from an adjacent section of the same field plowed 15 years earlier, were compared for a number of characters (Hartnett *et al.*, 1987). The two populations were initially a part of one large population and presumably had a common gene pool. Transplant experiments showed several mor-phological and physiological differences between them. Individuals from the 15-year field emerged earlier, had greater numbers of leaves, greater biomass and seed production, and higher reproductive ratios relative to plants from the young field. Furthermore, individuals from the older field were competitively superior when grown with individuals from the younger field (Fig. 7). But, although the older population showed lower variability for half of the traits examined, there were no statistically sig-nificant differences in overall variation. Thus the differences between the two populations may have resulted from selection imposed by the changing neighborhood over successional time. Biotype selection during succession has been observed in *Erigeron annuus*, a winter annual coloniz-ing species which may persist in some successional fields. Hancock and Wilson (1976) found different genotypes in successional fields of varying ages and concluded that directional selection operates during succession. Therefore, certain genotypes are favored over others with time.

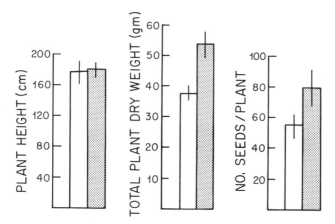

Figure 7   The response in competitive pairs of individuals from populations of *Ambrosia trifida* from two sites of different successional ages. Unshaded bars, individuals from annually plowed part; shaded bars, individuals from a site undisturbed for 15 years. From Hartnett *et al.* (1987).

IX. Conclusions

Plant–plant interactions play an important role in community dynamics, which is influenced by several plant life history features (including propagule arrival time, germination, flowering and dispersal, and the position of the perennating buds) in determining successional outcome.

In common with other plant–plant interaction research, successional studies should recognize that

1. Interaction occurs between individuals and not between species.
2. Plant–plant interactions simultaneously occur between individuals of the same species, individuals of different species, and of different successional guilds.
3. Except in habitats with very spatially patchy environments or for plants with limited dispersal distances, the identity of interacting neighbors is more predictable in low-diversity than in high-diversity communities.
4. The identity, size, distance, developmental stage, and therefore the equivalency of interacting individuals are crucial in understanding and predicting the outcome of interactions.
5. While there may be a high degree of response breadth similarity among species, competition among neighboring individuals may be highly asymmetrical because of size, developmental stages, and other factors mentioned above.

Although the role of plant–plant interactions in successional change was recognized very early in the development of succession theory, there are only a few detailed studies of its importance in nature. Most studies of plant–plant interactions have been concerned with species of the same community, without regard to the successional status of the community or to the outcome of interactions in relation to successional changes. Much of the work aimed at understanding the mechanisms of plant–plant interactions has concentrated on simple herbaceous plants and most of this work considered members of the same population, with only a few having considered more than two species. This emphasis has been promoted largely by the desire to develop simple models to describe competition among plants, e.g., those of de Wit (1960), the various permutations of the logistic equation (van Hulst, 1979), and more recent work by a number of investigators, e.g., Weiner (1982), Firbank and Watkinson (1985), and Pacala and Silander (1985). Predictive models of the consequences of plant–plant interactions to community structure and species replacement need to be developed and very likely will be most informative if they are based on mechanistic understanding of plant behavior. Physiologically based competition models (e.g., Reynolds *et al.*, 1987) which consider position of leaves in the canopy, light distribution carbon gain capacities, and allocation (especially of carbon and nitrogen) hold much promise for understanding competitive interactions in plant communities.

In order to better understand the role of plant–plant interactions in successional change, there is a real need for coordinated research programs that address these interactions in a wide range of successional environments. For example, even within the same geographic or climatic region we need to better understand the influence of differing degrees of resource base changes due to various disturbances (e.g., Tilman, 1986, and this volume) on plant–plant interactions. While we can predict that these events do influence plant–plant interactions, we do not know how they do so, nor can we, at the present, predict their outcome for even a small number of communities.

Our knowledge is very limited for plant–plant interactions in habitats acutely or chronically dominated by shortages, excesses, or great imbalances in resource availability. Furthermore, we have only limited knowledge of how plant–plant interactions are modified in habitats which are governed by interactions between rather than within trophic levels.

There is now some recognition of the possible impacts on plant–plant interactions of infrequent but severe, large-scale climatic departures from normal. The 1983 El-Nino, for example, has had detectable effects on plant performance over a wide area, especially in the tropics. Such events could greatly influence the interaction among species within and especially between trophic levels, and, consequently, influence successional trends. Furthermore, the rapid change in the geosphere–biosphere–atmospheric interactions, such as the rising CO_2 concentrations, may also have strong influences on plant–plant interactions, including those that have direct relevance to successional change. Evidence is beginning to accumulate that these factors may interact with each other (e.g., Bazzaz and Carlson, 1984; Bazzaz and Garbutt, 1988) and with the soil environment to produce enormous changes in the outcome of plant–plant interaction in successional environments.

References

Abul-Fatih, H. A., and Bazzaz, F. A. (1979). The biology of *Ambrosia trifida* L. I. Influence of species removal on the organization of the plant community. *New Phytol.* **83**, 813–816.

Allen, E. B., and Forman, R. T. T. (1976). Plant species removals and oldfield community structure. *Ecology* **57**, 1233.

Armesto, J. J., and Pickett, S. T. A. (1985). Experiments on disturbance in oldfield communities: Impact on species richness and abundance. *Ecology* **66**, 230–240.

Armesto, J. J., and Pickett, S. T. A. (1986). Removal experiments to test mechanisms of plant succession in old fields. *Vegetatio* **66**, 85–93.

Ballaré, C. L., Sánchez, R. A., Scopel, A. L., Casal, J. J., and Ghersa, C. M. (1987). Early detection of neighbor plants by phytochrome perception of spectral changes in reflected sunlight. *Plant Cell Environ.* **10**, 551–557.

Bazzaz, F. A. (1979). Physiological ecology of plant succession. *Annu. Rev. Ecol. Syst.* **10**, 351–371.

Bazzaz, F. A. (1984). Demographic consequences of plant physiological traits: Some case studies. *In* "Perspectives in Plant Population Ecology" (R. Dirzo and J. Sarukhan, eds.), pp. 324–346. Sinauer, Sunderland, Massachusetts.

Bazzaz, F. A. (1986). Life history of colonizing plants: Some demographic, genetic and physiological features. *In* "Ecology of Biological Invasions" (H. A. Mooney and J. A. Drake, eds.), pp. 96–110. Springer-Verlag, New York.

Bazzaz, F. A. (1987). Experimental studies on the evolution of niche in successional plant populations. *In* "Colonization, Succession and Stability" (A. J. Gray, M. J. Crawley, and P. J. Edwards, eds.), pp. 245–272. Blackwell, Oxford, England.

Bazzaz, F. A., and Carlson, R. W. (1979). Photosynthetic contribution of flowers and seeds to reproductive effort of an annual colonizer. *New Phytol.* **82**, 223–232.

Bazzaz, F. A., and Carlson, R. W. (1984). The response of plants to elevated CO_2. I. Competition among an assemblage of annuals at two levels of soil moisture. *Oecologia* **62**, 196–198.

Bazzaz, F. A., and Garbutt, K. (1988). The response of annuals in competitive neighborhoods: Effects of elevated CO_2. *Ecology* **69**, 937–946.

Bazzaz, F. A., and Sultan, S. E. (1987). Ecological variation and the maintenance of plant diversity. *In* "Differentiation Patterns in Higher Plants" (K. M. Urbanska, ed.), Chap. 4, pp. 69–93. Academic Press, Orlando, Florida.

Bormann, F. H. (1953). Factors determining the role of loblolly pine and sweetgum in early old-field succession in the Piedmont of North Carolina. *Ecol. Monogr.* **23**, 339–358.

Connell, J. H. (1980). Diversity and the coevolution of competitors, or the ghost of competition past. *Oikos* **35**, 131–138.

Connell, J. H. (1983). On the prevalence and relative importance of interspecific competition: Evidence from field experiments. *Am. Nat.* **122**, 661–696.

Connell, J. H., and Slatyer, R. O. (1977). Mechanisms of succession in natural communities and their role in community stability and organization. *Am. Nat.* **111**, 1119–1144.

de Wit, C. T. (1960). On competition. *Versl. Landbouwkd. Onderz.* **660**, 1–82.

Edwards, P. J., and Gillman, M. P. (1987). Herbivores and plant succession. *In* "Colonization, Succession and Stability" (A. J. Gray, M. J. Crawley, and P. J. Edwards, eds.), pp. 295–314. Blackwell, Oxford, England.

Ennos, R. A. (1983). Maintenance of genetic variation in plant populations. *Evol. Biol.* **16**, 129–155.

Firbank, L., and Watkinson, A. R. (1985). On the analysis of competition within two-species mixtures of plants. *J. Appl. Ecol.* **22**, 503–517.

Goldberg, D. E. (1987). Neighborhood competition in an old-field plant community. *Ecology* **68**, 1211–1223.

Goldberg, D. E., and Werner, P. A. (1983). Equivalence of competitors in plant communities: A null hypothesis and a field experimental approach. *Am. J. Bot.* **70**, 1098–1104.

Gottlieb, L. D. (1981). Electrophoretic evidence and plant populations. *In* "Progress in Phytochemistry" (L. Reinhold, J. B. Harborn, and T. Swain, eds.), pp. 1–46. Pergamon, Oxford, England.

Gray, A. J. (1987). Genetic change during succession in plants. *In* "Colonization, Succession and Stability" (A. J. Gray, M. J. Crawley, and P. J. Edwards, eds.), pp. 273–293. Blackwell, Oxford, England.

Gross, K. L. (1980). Colonization by Verbascum thapsus (Mullein) of an old-field in Michigan: Experiments on the effects of vegetation. *J. Ecol.* **68**, 919–927.

Hamrick, J. L., Linhart, Y. B., and Mitton, J. B. (1979). Relationship between life-history characteristics and electrophoretically detectable genetic variation in plants. *Annu. Rev. Ecol. Syst.* **10**, 173–200.

Hancock, J. F., and Wilson, R. E. (1976). Biotype selection in *Erigeron annuus* during old field succession. *Bull. Torrey Bot. Club* **103**, 122–125.

Harper, J. L. (1983). A Darwinian plant ecology. *In* "Evolution from Molecules to Man," pp. 323–345. Cambridge Univ. Press, Cambridge, England.

Hartnett, D. C., and Bazzaz, F. A. (1983). Physiological integration among intraclonal ramets of *Solidago canadensis*. *Ecology* **64**, 779–788.

Hartnett, D. C., and Bazzaz, F. A. (1985a). The integration of neighborhood effects by clonal genets of *Solidago canadensis*. *J. Ecol.* **73**, 415–427.

Hartnett, D. C., and Bazzaz, F. A. (1985b). The regulation of leaf, ramet and genet densities in experimental populations of the rhizomatous perennial *Solidago canadensis*. *J. Ecol.* **73**, 429–443.

Hartnett, D. C., Hartnett, B. B., and Bazzaz, F. A. (1987). Persistence of *Ambrosia trifida* populations in old fields and responses to successional changes. *Am. J. Bot.* **74**, 1239–1248.

Horn, H. S. (1971). "The Adaptive Geometry of Trees." Princeton Univ. Press, Princeton, New Jersey.

Keever, C. (1950). Causes of succession on old fields of the Piedmont, North Carolina. *Ecol. Monogr.* **20**, 230–250.

Kozlowski, T. T. (1949). Light and water in relation to growth and competition of Piedmont forest trees species. *Ecol. Monogr.* **19**, 207–231.

Loveless, M. D., and Hamrick, J. L. (1984). Ecological determinants of genetic structure in plant populations. *Annu. Rev. Ecol. Syst.* **15**, 65–95.

MacArthur, R. H., and Wilson, E. O. (1967). "The Theory of Island Biogeography." Princeton Univ. Press, Princeton, New Jersey.

Miller, T. E., and Werner, P. A. (1987). Competitive effects and responses between plant species in a first-year old-field community. *Ecology* **68**, 1201–1210.

Odum, E. P. (1969). The strategy of ecosystem development. *Science* **164**, 262–270.

Pacala, S. W., and Silander, J. A., Jr. (1985). Neighborhood models of plant population dynamics. I. Single-species models of annuals. *Am. Nat.* **125**, 385–411.

Parrish, J. A. D., and Bazzaz, F. A. (1976). Underground niche separation in successional plants. *Ecology* **57**, 1281–1288.

Parrish, J. A. D., and Bazzaz, F. A. (1978). Pollination niche separation in a winter annual community. *Oecologia* **35**, 133–140.

Parrish, J. A. D., and Bazzaz, F. A. (1979). Difference in pollination niche relationships in early and late successional plant communities. *Ecology* **60**, 597–610.

Parrish, J. A. D., and Bazzaz, F. A. (1982a). Competitive interactions in plant communities of different successional ages. *Ecology* **63**, 314–320.

Parrish, J. A. D., and Bazzaz, F. A. (1982b). Responses of plants from three successional communities to a nutrient gradient. *J. Ecol.* **70**, 233–248.

Peterson, D. L., and Bazzaz, F. A. (1978). Life cycle characteristics of *Aster pilosus* in early successional habitats. *Ecology* **59**, 1005–1013.

Pickett, S. T. A., and Bazzaz, F. A. (1978). Organization of an assemblage of early successional species on a soil moisture gradient. *Ecology* **59**, 1248–1255.

Pinder, J. E. (1975). Effects of species removal on an old field plant community. *Ecology* **56**, 747.

Raynal, D. J., and Bazzaz, F. A. (1975). Interference of winter annuals with *Ambrosia artemisiifolia* in early successional fields. *Ecology* **56**, 35–49.

Regehr, D. L., and Bazzaz, F. A. (1976). Low temperature photosynthesis in successional winter annuals. *Ecology* **57**, 1297–1303.

Regehr, D. L., and Bazzaz, F. A. (1979). The population dynamics of *Erigeron canadensis,* a successional winter annual. *J. Ecol.* **67**, 923–933.

Reynolds, J. F., Skiles, J. W., and Moorhead, D. (1987). "SERECO: Simulation of Ecosystem Response to Elevated CO_2. Parts I–III," Response of Vegetation to Carbon Dioxide Ser., Rep. 041. Carbon Dioxide Res. Div. U.S. Dep. Energy, Washington, D.C.

Schmid, B., and Bazzaz, F. A. (1987). Clonal integration and population structure in perennials: Effects of severing rhizome connections. *Ecology* **68**, 2016–2022.

Tilman, D. (1986). Evolution and differentiation in terrestrial plant communities: The importance of the soil resource : light gradients. *In* "Community Ecology" (J. Diamond and T. J. Case, eds.), pp. 359–380. Harper & Row, New York.

Tolley, L. C. and Strain, B. R. (1985). Effects of CO_2 enrichment and water-stress on gas-exchange of *Liquidambar styraciflua* and *Pinus taeda* seedlings grown under different irradiation levels. *Oecologia* **65**, 166–172.

Tremmel, D. C., and Peterson, K. M. (1983). Competitive subordination of a piedmont old field successional dominant by an introduced species. *Am. J. Bot.* **70**, 1125–1132.

van Hulst, R. (1979). On the dynamics of vegetation: Markov chains as models of succession. *Vegetatio* **40**, 3–14.

Weiner, J. (1982). A neighborhood model of annual plant interference. *Ecology* **63**, 1237–1241.

Werner, P. A. (1979). Competition and coexistence of similar species. *In* "Topics in Population Biology" (O. T. Solbrig, S. Jain, G. B. Johnson, and P. H. Raven, eds.), Chap. 12, pp. 287–310. Macmillan, London.

Whittaker, R. H. (1975). "Communities and Ecosystems," 2nd ed. Macmillan, New York.

Wieland, N. K., and Bazzaz, F. A. (1975). Physiological ecology of three codominant successional annuals. *Ecology* **56**, 681–688.

13

Competitive Hierarchies and Centrifugal Organization in Plant Communities

Paul A. Keddy

I. Introduction

A. Choosing Research Goals

If we are going to study plant competition, we need to have some long-term research objectives. Otherwise, with at least a quarter of a million species of angiosperms on this planet, and a vastly larger number of possible interactions among them, our discipline could turn into simply a collection of special cases. Choosing long-term goals and the right questions is simultaneously the most important and most subjective part of scientific research programs (Keddy, 1989). Perhaps this is the reason why long-term goals are often neither explicitly stated nor extensively discussed. The long-term objective I propose is as follows: To be able to predict measurable aspects of competition and their effects on populations and communities from a knowledge of environmental conditions and the traits of the species involved.

I propose this goal for three main reasons:

1. It indicates the measurable state variables which will comprise a body of theory (e.g., Lewontin, 1974; Rigler, 1982; Keddy, 1987). These include state variables describing both mechanism (e.g., competition intensity, degree of asymmetry) and pattern (e.g., biomass, species richness, life form).
2. It emphasizes prediction as an essential element of understanding (Peters, 1980a,b; Rigler, 1982).
3. Such objectives, if met, would allow plant ecologists to make useful contributions to the three goals of the World Conservation Strategy (International Union for the Conservation of Nature and Natural Resources, 1980).

Other goals, are of course, possible. This particular goal has two basic assumptions about motivation for research. First, I assume that we have an interest in building theory (Austin, 1986). Second, I assume that the research and theory should apply in some way to real problems of living organisms, which is why I included reference to the World Conservation Strategy. Whether these assumptions are justified is open to discussion. One could argue that other motivations are both possible and reasonable. They could include visiting exotic locales, appreciating natural beauty, building one's reputation, keeping one's job, providing entertainment, filling in time, and so on.

One way of approaching the above goal would be to divide it into two components: The first would be "assembly rules" for plant communities. Although some of Diamond's (1975) methods have been justifiably criticized (e.g., Connor and Simberloff, 1979; Weins, 1983), the objective of producing rigorous rules for community assembly is still useful. A second related goal would be to develop "response rules" to predict accurately changes in plant communities after specified perturbations (e.g., Nobel and Slatyer, 1980; van der Valk, 1981). I have discussed these objectives in more detail elsewhere (Keddy, 1989). In this chapter, I cover three topics: (1) The obstacles to general theory, and some possible antidotes to these obstacles. (2) Some general rules about plant competition which we already possess. (3) The implications of these general rules for the way in which wetland plant communities (and perhaps other plant communities) are organized.

B. Obstacles to the Development of Theory

I assume that the objective of science is the detection, testing, and refinement of general principles. This is not a new idea in plant ecology. Tansley (1914) stated in his Presidential address to the British Ecological Society that "Quantitative results are of no use . . . unless they have some kind of general validity." The large number of species and environments makes the search for generality essential. Suppose we make two simplifying assumptions: (1) that communities can be reconstructed from pairwise interactions, and (2) that these interactions do not change with changing environments (including herbivores and mycorrhizae). The number of interactions we need to study is then simply $\binom{S}{2}$ where S is the number of species in the pool. In wetlands near Ottawa we can easily locate 160 species, which would require $\binom{160}{2} = 12,720$ comparisons. Wilson and Keddy (1986a) explored $\binom{7}{2} = 21$ comparisons in 1 year. At this rate, it would take approximately 600 years to run the necessary experiments to characterize interactions in one comparatively small region of the planet, and Colinvaux (1986) estimates there are more than 250,000 angiosperms on the planet. Rigler (1982) and Wimsatt (1982) provide two similar illustrations. There are, of course, many other obstacles to the development of rigorous competition theory, as I have discussed elsewhere (Keddy, 1989). The obsession with collecting special cases, whether multivariate descriptions of site x or autecological studies of species y, is, however, probably the greatest obstacle faced by plant ecology.

C. A Path to General Theory (Antidotes to the Obstacles)

How do we search for general principles? We could still try to justify collecting observations of special cases. We could optimistically assume that if we patiently record enough such observations, generalizations will eventually emerge through induction. This is, however, more a statement of faith than a demonstrable fact. While we may count on later scientists to carry out reviews such as those of Schoener (1983) and Connell (1983) and put together the pieces, the least damning criticism of this approach is its inefficiency. Clement's pioneering work (e.g., Clements *et al.*, 1929) showed countless examples of competition; in 1933 he attempted syntheses of his own experiments noting that, in general, "the taller grasses enjoyed a decisive advantage over the shorter." Yet 50 years later, we are still adding up the examples.

We could take a more visionary approach by designing our work to detect trends and test principles right now. Instead of emphasizing differences, we could pay more attention to similarities. Once we find similarities and make predictions using them, we can concentrate on improving our predictions by exploring the deviations from the general principles. The context of theory would then guide the selection of critical case studies. Peters (1980a,b) has elaborated on these arguments and argued in favor of "predictive ecology."

There are several tools which now exist to allow us to pose important general questions here and now (Keddy, 1989). I briefly consider four here.

1. Using Gradients in Comparative Studies   Instead of studying allegedly homogeneous patches of vegetation, we could seek out natural environmental gradients. These gradients provide opportunities for "comparative studies" (sensu Keddy, 1989) or "natural experiments" (sensu Diamond, 1983) where vegetation characteristics, plant traits, and species composition all covary. By comparing such variables across a range of vegetation types, we can test for general patterns and detect exceptions from them.

2. Empiricism   There is still little agreement about the choice and measurement of state variables for describing plants and plant communities. As a consequence, plant ecologists have pursued sterile debates such as the community unit–continuum controversy. Such controversies are unresolvable precisely because they are stated and debated in unfalsifiable form with nonoperational state variables (Shipley and Keddy, 1987). Proper choice of measurable state variables is one step toward rigorous theory (Lewontin, 1974; Peters, 1980a; Rigler, 1982; Keddy, 1987, 1989), as is illustrated by the clear relationships between α diversity and biomass (e.g., Grime, 1973, 1979; Al-Mufti *et al.*, 1977; Silvertown, 1980; Tilman, 1982; Morre *et al.*, 1989; Wisheu and Keddy, 1989a).

3. Plant Traits   Instead of using species nomenclature, we could build our theories on plant traits. One may not be able to generalize from the ecology of *Sabatia kennedyana* in the Tusket River valley, Nova Scotia (Keddy, 1985) to other situations. One may, however, be able to draw generalizations about the distribution of evergreen rosette species on shorelines (e.g., Keddy, 1983; Boston and Adams, 1987; Day *et al.*, 1988;

Wisheu and Keddy, 1989a) or, even more importantly, evergreen species in general (Grime, 1977; Chapin, 1980). The importance of theory built on plant traits is increasingly accepted (e.g., Grime, 1974, 1979; Box, 1981; Rorison *et al.*, 1987; Gaudet and Keddy, 1988; Tilman, 1988; Keddy, 1989). Rereading Clements *et al.* (1929) or du Rietz (1931) shows that this is hardly a new idea, yet theories built on plant traits have been slow to arise.

4. Nested Hierarchies of Models We can recognize that general models need more specific models nested within them, and explore our data using a continuum of models from the site-specific to the most general. For example, Grime's (1973, 1974, 1979) theories about plant strategies were criticized by Grubb (1985) as being incorrect, but a rereading of Grubb's paper will show that, instead of a criticism of Grime, it could have been presented positively as the need for and an enumeration of some system-specific submodels. Table 1 illustrates this concept by presenting one possible hierarchical organization of ecological models describing plant communities in wetlands.

Table 1 One Possible Nested Hierarchy of Models for Plant Communities, Ranging from General (Top) to Most Specific (Bottom)[a]

Level of Organization	Gradients			Plants
State variables		Biomass Species richness		Traits (e.g., growth rate, height)
General process and pattern	Fertility (stress, adversity)		Disturbance	Functional groups, strategies (e.g., ruderals, stress tolerators)
Vegetation type (riverine wetlands)	Loss on ignition Soil nutrients		Ice damage Wave damage	Wetland functional groups (e.g., annuals, reeds, isoetids)
Region (Ottawa River Valley)		DCA axes 1 and 2		Species nomenclature (e.g., *Phalaris arundinacea, Eriocaulon septangulare*)
Site (Westmeath)		Points in DCA space		Species nomenclature (e.g., *Phalaris arundinacea, Eriocaulon septangulare*)

[a] A majority of work occurs at the very lowest levels. The concrete examples come from riverine wetlands (Day *et al.*, 1988; Shipley *et al.*, 1989).

II. Evidence for Predictable Patterns in Plant Competition

Are there general principles about plant competition which would take us in the direction of the goal stated above? In this section, I review some of the experimental data showing that competition varies in a predictable manner in plant communities. I then explore two sets of consequences for the organization of plant communities. The first set of consequences are some questions these data raise about the generality and significance of competitive hierarchies. The second consequence is the implications of such data for larger scale patterns of plant community organization along competition gradients. This yields a model of "centrifugal organization" which predicts changes in life form, α diversity, and species pools along biomass gradients.

A. Competition Intensity

Competition intensity can be defined as the combined (negative) effects of all neighbors on the performance of an individual or population (Keddy, 1989). The reason for measuring competition intensity is that an individual plant experiences the negative effects of all neighbors simultaneously; pairwise designs measuring each interspecific interaction separately measure something very different. Competition intensity can be measured by comparing the performance of plants in cleared plots with the performance of those surrounded by neighbors. One could either introduce phytometers (sensu Clements, 1935) into cleared and uncleared plots or else remove all plants but one from the cleared plots.

Wilson and Keddy (1986b) used this design to test whether competition intensity varied along a natural environmental gradient. Given the many species in the vegetation, they assumed, but did not test, that they had measured diffuse competition, the effects of many species combined. Individuals of three different plant species ("phytometers") were exposed to both above- and below-ground competition along a standing crop gradient. In some plots, phytometers were planted in pots containing intact cores of soil and plants which were inserted into established vegetation; in other plots, both pots and surrounding vegetation were weeded regularly so that above- and below-ground competition was minimized. Three general results emerged (Fig. 1). (1) Competition varied from one site to the next. (2) The intensity of competition increased with standing crop. (3) The intensity of competition increased with soil organic content. Therefore, not only did competition intensity change from site to site, but it was predictable from both biotic and abiotic factors. This competition bioassay is probably the most direct approach to asking questions about the predictible variation in competition intensity in nature.

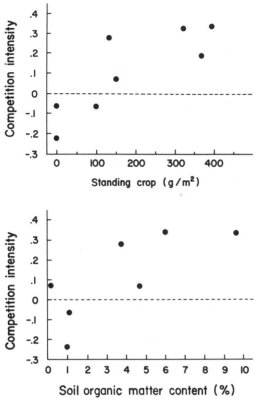

Figure 1   Competition intensity plotted against standing crop (top) and soil organic matter content (bottom) on lake shores. Competition intensity was measured as the total effects of neighbors on three species of phytometer. From Keddy (1989); data from Wilson and Keddy (1986b).

B. Asymmetric Competition

There is a tendency among ecologists to compare competition within pairs of species that are very similar (e.g., Harper and Chancellor, 1959; Harper *et al.*, 1961; Harper and McNaughton, 1962; Harper and Clatworthy, 1963; Werner, 1979). In fact, similarity in size is an important assumption of the replacement series design (de Wit, 1960; Harper, 1977) to ensure that density and biomass are not confounded. There is consequently a tendency to assume that competitive interactions between species in general are symmetric (Keddy, 1989), that is, that each species is more or less capable of suppressing the growth of the other. In nature, however, plant communities may be made up of species having very different morphologies and life histories. Are interactions symmetric in such cases?

To explore this, we need a measure of asymmetry. Let Y_{ii} be the yield of an average individual of species i grown in monoculture, and Y_{ij} be the yield of an average individual of species i when grown in mixture with species j. Further, let $RYP_{ij} = Y_{ij}/Y_{ii}$. Pairwise asymmetric competition can be said to exist when $RYP_{ij} > 1$ and $RYP_{ji} < 1$. That is, species i grows better in mixture with species j but species j grows worse in mixture with species i. Therefore species i should eventually exclude species j.

Keddy and Shipley (1989) used this definition to examine asymmetry in eight published competition studies including species from sea cliffs, lake shores, and chalk grassland. In seven of the eight examples, the matrices were dominated by asymmetric interactions, that is, in pairwise interactions there was consistently a winner and loser. The sole exception was the study by Harper (1965) of competition among different varieties of the same species. This exception is entirely consistent with the prediction that the degree of asymmetry in competitive interactions is lowest when species are nearly identical. But few real plant communities consist of nearly identical species, suggesting that asymmetric interactions are the exception rather than the rule. I have agreed elsewhere that this has important implications for studies of coexistence and the competitive exclusion principle (Keddy, 1989).

C. Competitive Hierarchies

The usual way of approaching the organization of plant communities is to explore pairwise interactions by growing (or removing) component species in all possible pairwise combinations. While the replacement series (de Wit, 1960; Harper, 1977) is the most widely used design, it is not the only one possible. However, this is the design which has been used in many recent studies of natural communities, including sea-cliff vegetation (Goldsmith, 1978), lakeshore vegetation (Wilson and Keddy, 1986a), and chalk grassland (Mitchley and Grubb, 1986). What evidence of community structure or effects of competition can be extracted from such matrices? Keddy and Shipley (1989) have proposed a quantitative measure of transitivity for exploring such matrices, and have found a highly significant pattern of transitive community structure in seven of eight published matrices. That is, in most cases, there was a significant tendency for plant communities to be organized in a competitive hierarchies. Again, the single exception was not a plant community, but intraspecific interactions among genotypes of a single species (Harper, 1965; Keddy and Shipley, 1989).

Of course, all pot experiments are subject to the criticism of any laboratory experiment: that the results cannot be extrapolated to real plant communities. One way out of this difficulty is to test whether predictions from the laboratory are consistent with patterns in the field. Wilson and

Keddy (1986a) found that position in the competitive hierarchy was correlated with field distributions along lakeshore exposure gradients, and that the field distributions were consistent across many study sites. Similarly, Mitchley and Grubb (1986) found that the position in the competitive hierarchy was positively correlated with abundance of species in chalk grassland, and that these patterns of abundance were consistent across many study sites. Goldsmith (1978) did not perform any statistical tests, but similarly observed that the competitive dominants in his pot experiments tended to occupy least exposed sites with lower salinities.

D. Traits Conferring Competitive Ability

If plant communities tend to be organized in competitive hierarchies, and if there are competition intensity gradients, we may well ask about plant traits associated with positions higher in the competitive hierarchy, or equally, about plant traits associated with habitats having higher competition intensity. There are two ways to explore this: to review the literature, or to design experiments specifically to test for empirical relationships. Consider these approaches in turn. Clements (1933) summarized the results of hundreds of transplant and removal experiments in prairie vegetation (e.g., Clements *et al.*, 1929) and concluded that, in general, "the taller grasses enjoyed a decisive advantage over the shorter." Goldsmith (1978) studied sea-cliff plants and showed that the larger species suppressed the smaller (Fig. 2). Wilson and Keddy (1986b) experimentally derived a competitive hierarchy for seven shoreline species. The dominant was a tall species whereas the subordinate was a small rosette species. Keddy and Shipley (1989) reanalyzed the Wilson and Keddy data and showed that more than one-third of the competitive ability of these species in mixture could be predicted from knowledge of their heights ($r^2 = 0.37$). Similarly, in the chalk grassland study, Mitchley and Grubb (1986) derived a dominance hierarchy for six plant species and found a significant correlation between position in the hierarchy and mean turf height in monocultures; Mitchley and Grubb noted that "the plants with the tallest leaves were the most effective in interference." Mitchley (1988) has since shown that there is a positive correlation between the height of grassland species and their relative abundance. Givnish (1982) has presented a general model for the evolution of leaf height in herbaceous plants that are competing for access to light.

Since diallele designs increase in size by the square of the number of species examined, there are obvious upper limits on the number of species which can be studied to relate traits to competitive ability. To overcome this problem, Gaudet and Keddy (1988) used a modified additive design to measure competitive ability of 44 wetland plant species. Each

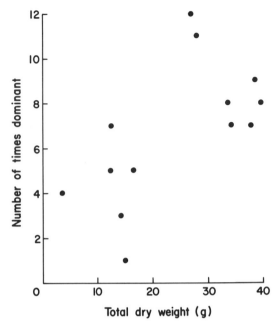

Figure 2   Position in competitive hierarchy (measured as number of times dominant in mixture) plotted against plant size in pure stand for sea-cliff plants. From Keddy (1989); data from Goldsmith (1978); $r = 0.61$, $p < 0.05$.

species was grown with a phytometer (*Lythrum salicaria*), and competitive ability measured as the ability to suppress this phytometer (competitive effect sensu Goldberg and Fleetwood, 1987; Goldberg, this volume). We showed that simple traits such as biomass, height, and canopy diameter could account for 74% of the measured competitive ability. Above-ground biomass was the best predictor ($r^2 = 0.62$). Height also was significant ($r^2 = 0.43$). A subset of the species was tested against a different phytometer and similar results were obtained. Figure 3 plots percent reduction in the phytometer biomass plotted against the above-ground biomass for 44 species.

E. A Possible Mechanism for the Above Patterns

The above evidence comes from patterns detected in experimental studies. Are such patterns consistent with our understanding of mechanisms by which pairs of plants interact? If two plants are growing close enough to one another to interact, the taller plant will not only have access to incoming light for its own growth, but it will simultaneously reduce the growth of the smaller plant by setting up a positive feedback loop where the larger plant continually improves its access to light and the smaller

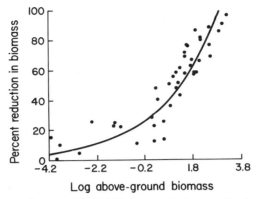

Figure 3 Screening for competitive ability across 44 plant species. Percent reduction in the biomass of a phytometer (*Lythrum salicaria*) when grown with each species of neighbor is plotted against mean above-ground biomass of the neighbors. [44 species, each point is the mean of $n = 5$ replicates, $y = \exp(3.34 + 0.44\alpha)$, $r^2 = 0.69$.] Points on the left are small evergreen species while points on the right are from one large leafy species. The remaining 31% of variation includes both experimental error and the contribution of other plant traits. From Keddy (1989); data from Gaudet and Keddy (1988).

plant is increasingly denied access to it (Givnish, 1982; Keddy and Shipley, 1989).

This will in turn influence ability to forage for nutrients. Reduced access to light will reduce carbohydrates available for root growth, thereby reducing rates of nutrient uptake. It is therefore a mistake to treat roots and shoots as separate entities, since they are part of one physiologically integrated unit. Caldwell *et al.* (1987) illustrated this by showing that shading or defoliation can reduce root growth and mineral uptake within 24 hr.

The taller plant will therefore not only deny light to its shorter competitor, but will also reduce the ability of its neighbor to forage for nutrients. Simultaneously, the increased energy reserves of the larger plant should increase resources for root construction, which will increase the depletion zone of nutrients around the larger plants. The result is that the smaller individual will likely be trapped in an asymmetric interaction in which positive feedback loops operating above and below ground are confining it within light and nutrient depletion zones created by the larger plant. A situation like this has been studied using the vine *Ipomea tricolor*, where above- and below-ground competition can be neatly separated. Weiner (1986) has shown that when above- and below-ground effects are separated, above-ground interactions (competition for light) are asymmetric, whereas below-ground effects (competition for nutrients) are symmetric. When above- and below-ground effects co-occur, interactions are again asymmetric.

An alternative approach which arrives at the same conclusion begins with a consideration of the characteristics of two classes of resources: light and nutrients. There is a fundamental distinction between them. Nutrient gradients can occur in the absence of neighbors since soil fertility varies along natural environmental gradients. In contrast, light gradients are almost invariably produced by the presence of neighbors. Species responses to gradients in soil fertility might therefore be considered part of their fundamental niches, whereas species responses to light gradients may be a part of their realized niches. This difference would generate a situation where species' field distributions along fertility gradients could be determined at one end by their fundamental niche (physiological tolerance limits) and at the other end by their realized niche (ecological tolerance limits determined by ability to compete for light). That is, ability to tolerate low fertility may produce a special "physiological response curve," whereas ability to forage for light or tolerate low light may produce a species' "ecological response curve." Inclusive fundamental niche structure (sensu Miller, 1967; Colwell and Fuentes, 1975) for nutrients combined with a hierarchy of abilities to compete for light could produce the dominance hierarchies described above, as well as the zonation patterns observed along natural environmental gradients. I have called this the "competitive hierarchy model," and have contrasted it with traditional resource partitioning models elsewhere (Keddy, 1989).

F. New Questions

The above examples illustrate the insights into plant communities which can be gained by looking for experimental designs that avoid traditional pairwise approaches, by explicitly considering plant traits, and by testing for patterns using operationally defined state variables. The potential for such approaches has barely been explored, and many new questions can be raised. These include the following.

1. Are competition intensity gradients found in most vegetation types? If so, what environmental variables and plant traits are correlated with competition intensity gradients?
2. Are competitive hierarchies a general feature of plant communities?
3. For a given plant community, are hierarchies invariant across the range of environmental conditions occupied by that community?
4. What plant traits predict positions in competitive hierarchies?
5. Do the traits conferring competitive ability vary among vegetation types?

The approaches for answering some of these questions are straightforward. Question 3 poses some more difficult issues, so I will briefly expand on it below. Two kinds of experiments are possible to answer this question. Let us consider them in turn.

G. Constraints on Competitive Hierarchies

1. Pairs of Species along Gradients   One approach would be to take a pair of species and allow them to interact along an environmental gradient. If we measured the performance of each species in mixture relative to pure stand, we could ask whether or not competitive outcome varied with environment. There are some difficulties with this approach. The design incorporates only two species. Communities are made up of large numbers of interacting species, and there is no reason why a pair of species selected ad hoc will necessarily provide inferences about the community as a whole.

There are also significant problems in the interpretation of such experiments. One good example illustrating the difficulty in testing for reversals in competitive ability comes from Goldsmith (1973). The experiment involved a small rosette species found in saline habitats (*Armeria maritima*) and a rapidly spreading, taller perennial grass (*Festuca rubra*) found in less saline habitats. Goldsmith grew them in a replacement series design, with two replicates receiving fresh water and two receiving salt water (Fig. 4). Reference to the pure stands shows that both species

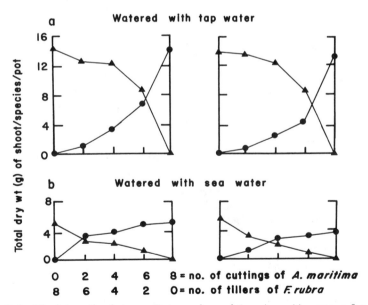

Figure 4   The interaction between *Festuca rubra* and *Armeria maritima* grown for 1 year in a replacement series watered with tap water (a) and sea water (b). Both replicates are shown. After Goldsmith (1973).

showed dramatic declines in performance at high salinity. Comparison of the degree to which the lines deviated from linearity reveals that with fresh water (Fig. 4a) interspecific competition was more intense than with salt water (Fig. 4b). *Festuca rubra* was clearly the competitive dominant with fresh water. The key question is whether *Armeria maritima* became a better competitor under saline conditions, or was simply better able to tolerate the saline conditions. It appears from inspection that *Armeria* shows little ability to suppress *Festuca* at high salinities (as judged by the weak skewing of the *Festuca* line at high salinity).

Let us assume that in fact *Festuca* was significantly suppressed by *Armeria* at high salinities. Two facts remain. *Festuca* shows more than a 50% reduction in performance in this environment in the absence of competition (pure stand); the added reduction (if any) due to competition is very small by comparison. The salinity reduced *Festuca*'s performance far more than competition. Therefore, on one hand we might argue that a reversal took place. On the other hand, the pure stands show that these species have inclusive niches, and that interspecific competition was more intense in the more productive environment.

2. Diallele Experiments   An alternative approach to choosing a pair of species would be to examine the structure of community matrices in large diallele experiments. Two questions could be asked: (1) Do transitivity and asymmetry vary among experiments? (2) Does the rank order of species in the hierarchy vary among experiments? The measures proposed by Keddy and Shipley (1989) provide the means to test these hypotheses, but practical constraints are the main problem.

Experiments examining many pairwise interactions increase in size as the square of the number of species considered. Thus, practical problems of research budgets and space may discourage people from creating diallele designs in different environments. Presumably, this is why studies such as those of Goldsmith (1978), Wilson and Keddy (1986a), and Mitchley and Grubb (1986) have all looked at only one environment.

The largest test to date was done by Fowler (1982), who looked at hierarchies among four weedy species in a North Carolina grassland. Let us look at her results in some detail. First, consider the constraints. Only four species were used. Second, the species were all small herbaceous species (weights in pure stand: *Anthoxanthum odoratum* 5.98, *Plantago lanceolata* 5.84, *Poa pratensis* 3.73, *Rumex acetosella* 5.22 g/pot in watered and fertilized treatments). Fowler created dry, unfertilized, and dry and unfertilized treatments in addition to moist and fertilized treatments, and explored variation in hierarchical organization. Given the initial similarity of the species, this is a strong test for invariance because these are precisely the conditions where asymmetry would be expected to be small, and therefore variance in the hierarchy might be expected. Nonetheless, Fowler found remarkably invariant ordering: in all environments *Anthoxanthum* was competitively dominant to *Poa* and *Rumex*. *Rumex* and *Poa* were the two species lowest in the hierarchy (as predicted from pure stand biomass) and for these species outcome was dependent on environment. Fowler's data support the contention that hierarchies are relatively invariant.

3. Relationship to Real Communities Two aspects of experiments need to be considered here: the environmental conditions used, and the similarity of the species, since the outcome of a particular experiment and its applicability to real communities will likely depend on both.

First, consider the environmental conditions. Experiments enable us to create environments which may be rare or absent in natural systems. If we observe changes in competitive hierarchies in experiments, this may or may not be relevant to real communities. The question is therefore not whether (or where) hierarchies can reverse in theory, but whether reversals occur within the range of conditions actually found in real communities. If, following the Goldsmith (1973) example, reversals occurred only at salinity levels higher than those normally found in the field, the pot experiment would not necessarily tell us anything about real sea-cliff vegetation. If, in general, reversals occur only at very low fertilities where plants have very low growth rates, normal levels of natural disturbance could eliminate both species from real communities. This could have major consequences for field distributions, but is something we would not detect in experiments where biomass removal (loss rates) is kept to a minimum. In the studies by both Goldsmith (1973) and by Fowler (1982), we could ask how much the results would have differed if different treatments had been used.

Second, natural plant communities contain a wide array of plant morphologies, but given the current paradigms (sensu Kuhn, 1970) of plant ecology, the choice of species is not likely to represent this reality. There is an overwhelming tendency to select pairs of "similar" species, often congenors, for competition experiments. Such similar pairs of species are exactly those where asymmetry and hierarchies may be least important, and therefore where switches in competitive ability are most likely (Keddy, 1989). Switches in competitive ability among very similar species may tell us little about hierarchical organization among very different species. Such studies will therefore need to include objective means for testing how similar the pairs of species are (Green, 1980; Legendre and Legendre, 1983; Keddy, 1989).

4. A Potential Reconciliation Dichotomies are part of our Western intellectual heritage, but undoubtedly there are cases where the search for a synthesis is more appropriate than struggling for a yes/no answer. As Dayton (1979) observed, "Many of my own hypotheses were carefully designed to force yes or no answers from nature when, in fact, nature may have been crying out *mu*. . . ." "Mu" is the only response possible when both yes and no are wrong (Watts, 1958).

Competitive hierarchies may be the rule along gradients of resource quantity, or where one environmental factor alone is of overwhelming importance. In wetlands, for example, major soil nutrients are often correlated along gradients from sand beaches to organic bays (Wilson and Keddy, 1985, 1986a,b); this would be a gradient of nutrient availability (see Chapin *et al.*, 1986). On sea cliffs, salinity is the major factor controlling plant community structure (Goldsmith, 1973, 1978). In contrast, resource partitioning may occur along gradients of resource quality. If soil nutrients levels are relatively constant, ratios of two resources may be opportunities for species to partition resources along a ratio gradient (Tilman, 1982). The critical question may turn out to be, How many communities are organized along each kind of gradient? Critical tests will be necessary to resolve such points.

III. Large-Scale Patterns and Long-Term Goals

A. Patterns in Wetland Vegetation

At this point it is important to return to our long-term objectives. If we lose sight of clear goals, it is easy to be side-tracked into collecting natural historical detail (Peters, 1980a) or distracted by unresolvable arguments about untestable concepts (Shipley and Keddy, 1987). As stated in the Introduction, I think our long-term goals should be nested models which allow us to make quantitative predictions about the structure of plant communities and their responses to perturbations. As well as doing intellectually satisfying science we should be able to say something useful about global environmental problems. Unfortunately, our intellectual training is usually to dissect our problems into smaller and smaller subproblems. The regrettable result can be a loss of perspective—with negative consequences for development of general theory and for our own biosphere. I therefore deliberately end this chapter not by listing a series of more precise questions which need to be addressed (see the list in Section II,F), but instead by considering how the empirical relationships reviewed here suggest a general community model with both theoretical and applied relevance.

Let us return to the search for general principles. There is good evidence that plant communities are structured by competitive hierarchies.

There is also evidence that competition is most intense in high-biomass sites and that such sites are occupied by plants occurring at the top of hierarchies. These general patterns are consistent with natural patterns in wetland vegetation. Fertile undisturbed sites are usually dominated by *Typha* spp.; *Phragmites communis, Phalaris arundinacea, Calamagrostis canadensis,* or *Carex* spp. can also form dense, nearly monospecific stands. These species all share certain traits: height and clonal spread. These traits are recognized in other published studies as traits which allow plant species to dominate communities (Grime, 1973, 1979; van der Valk, 1981; Givnish, 1982; Day *et al.,* 1988; Gaudet and Keddy, 1988; Shipley *et al.,* 1989).

The remarkable morphological convergence suggests only one type of morphology is appropriate to fertile undisturbed sites with high biomass. This occurs at ecological time scales (competitive exclusion) and evolutionary time scales (selection for size). In contrast, a vast array of life forms and morphologies is found in disturbed and/or infertile sites: annuals (van der Valk, 1981; Keddy and Reznicek, 1986), isoetids (Boston, 1986; Boston and Adams, 1987; Wisheu and Keddy, 1989b; Day *et al.,* 1988; Moore *et al.,* 1989), and carnivorous plants (Keddy, 1983; Wisheu and Keddy, 1989a,b). Thus, once the constraints of intense competition are released, a wide array of life forms apparently becomes possible. Table 2 lists, for a few representative sites, the morphological and ecological diversity of plants in wetlands that are infertile or exposed to damage from waves and flowing water.

Figures 5 and 6 attempt both to portray these patterns in wetlands accurately and express them in a testable form. The variation in life form types has already been discussed, but Fig. 6 emphasizes that increasing biomass may produce changes in other state variables such as the total species pool, α diversity, and number of vegetation types. Although the species pool may decrease with increasing biomass, α diversity has different patterns. Grime (1973) proposed that species density (α diversity) reaches a maximum at intermediate levels of community biomass. It is important to recognize the distinction between the total species pool and α diversity. The latter has received most emphasis in conservation management (e.g., Grime, 1973), but when planning nature reserve *systems* to maintain biodiversity, it is the species pool which is most important. Since the number of kinds of vegetation and species increases with decreasing biomass, any reserve system representing wetlands must include this array of low-biomass environments. Moore *et al.* (1989) have recently explored one part of this model by testing whether there was a negative relationship between the number of nationally rare plant species on wetlands and standing crop. They found that nationally rare species occurred only in peripheral habitats (less than 500 gm^{-2}).

Table 2 Variation in Life Form and Life History Type in Five Wetlands with Low Standing Crop[a]

Low Standing Crop Areas	Annual Species	Reeds[b]	Isoetids[c]	Insectivorous Species[d]
Nova Scotia Gillfillan Lake[e] (n = 114)		Eleocharis smallii Eleocharis tenuis Equisetum arvense Equisetum fluviatile Juncus filiformis	Eleocharis acicularis Eriocaulon septangulare Gratiola aurea Isoetes acadiensis Juncus pelocarpus Lobelia dortmanna Lycopodium inundatum* Ranunculus reptans Sabatia kennedyana Xyris difformis*	Drosera intermedia Drosera rotundifolia Sarracenia purpurea Utricularia cornuta Utricularia geminiscapa Utricularia subulata Utricularia vulgaris
Wilsons Lake[f] (n = 67)	Elatine minima	Eleocharis robbinsii Eleocharis smallii Eleocharis tenuis Equisetum fluviatile Juncus filiformis Scirpus validus	Eleocharis acicularis Eriocaulon septangulare Isoetes tuckermani Juncus pelocarpus Lobelia dortmanna Lycopodium inundatum* Myriophyllum tenellum Ranunculus reptans Sabatia kennedyana Xyris difformis*	Drosera intermedia Drosera rotundifolia* Utricularia cornuta Utricularia purpurea* Utricularia resupinata Utricularia vulgaris*
Ontario Axe Lake[g] (n = 65)	Bidens sp.	Eleocharis smallii Juncus filiformis Scirpus torreyi	Eriocaulon septangulare Juncus pelocarpus Lobelia dortmanna Lycopodium inundatum* Myriophyllum tenellum Xyris difformis*	Drosera intermedia Drosera rotundifolia* Utricularia cornuta Utricularia gibba Utricularia intermedia Utricularia purpurea Utricularia resupinata Utricularia vulgaris

Site				
Ottawa River[h] (n = 55)		Eleocharis calva Eleocharis smallii Equisetum fluviatile Scirpus acutus Scirpus americanus	Eleocharis acicularis Eriocaulon septangulare Isoetes echinospora Juncus pelocarpus Myriophyllum tenellum Potamogeton gramineus Ranunculus reptans Sagittaria graminea	Drosera intermedia** Utricularia cornuta** Utricularia vulgaris**
Westmeath[i] (n = 75)	Bidens cernua Bidens frondosa Fimbristylis autumnalis Gratiola neglecta Impatiens capensis Ludwigia palustris Polygonum lapathifolium Polygonum neglectum Polygonum persicaria Sporobolis vaginiflorus	Eleocharis elliptica Eleocharis smallii Equisetum fluviatile Scirpus acutus Scirpus americanus	Eleocharis acicularis Eriocaulon septangulare Juncus pelocarpus Ranunculus reptans	

[a] All have uncommon vegetation types. Wetlands with high biomass are dominated by one life form, large leafy rhizomatous perennials (e.g., *Typha angustifolia*, *Phalaris arundinacea*, *Calamagrostis canadensis*).
[b] As defined by Day *et al.* (1988); species with a single "leafless" aerial shoot.
[c] From Table 1 in Boston and Adams (1987); * indicates two similar species I have added.
[d] Species not in cited sources but * added from personal observations or ** added from unpublished report.
[e] Unpublished data analyzed by Keddy (1984a,b).
[f] From Wisheu (1987); see also Keddy (1985).
[g] From Keddy (1981); see also Keddy (1983).
[h] From Day *et al.* (1988).
[i] Unpublished data of C. Gaudet, D. Moore, and P. Keddy.

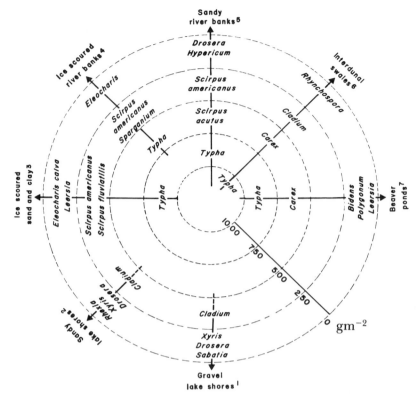

Figure 5 Community organization along biomass gradients in some wetlands. 1, Wilsons Lake (Wisheu and Keddy, 1989a,b); 2, Axe Lake (Keddy, 1981); 3, Luskville, Ottawa River (Moore *et al.,* 1989); 4, Ottawa River at four locations (Day *et al.,* 1988); 5, Westmeath, Ottawa River (Moore *et al.,* 1989); 6, Presqu'ile Park, Lake Ontario (Moore *et al.,* 1989); 7, Beaver ponds, Lanark County, Ontario (P. A. Keddy, field observations). From Moore *et al.,* 1989, by permission of Elsevier, Applied Science Publishers, Ltd.

This model therefore integrates a number of state variables of interest to plant ecologists. It summarizes apparent general patterns in wetland vegetation. It has patterns of plant traits consistent with our current empirical understanding of competitive hierarchies and plant competitive ability. It makes testable predictions about how species pools and vegetation types should change along biomass gradients. It also predicts the negative effects for biodiversity which would result from eutrophication. And finally, it guides managers of natural areas regarding the priorities for selection of sites for a comprehensive system of protected wetlands. It is therefore one preliminary attempt to provide some "assembly rules" and "response rules" for a plant community.

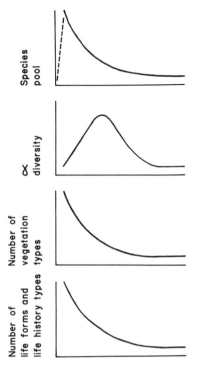

Figure 6 Predicted relationships between state variables of communities and landscapes as a function of biomass.

B. Centrifugal Community Organization in Wetlands

The graphical model in Figs. 5 and 6 appears consistent with the "centrifugal community organization" model proposed for desert rodents by Rosenzweig and Abramsky (1986). The essential element of their centrifugal organization model is that there is a central or core habitat which is preferred by all species, and a series of peripheral habitats with different species specialized upon each. Interspecific competition is therefore intense in this central habitat, but each species has a refugium from interspecific competition in its preferred peripheral habitat. The peripheral habitats therefore permit coexistence. The underlying mechanism is (1) that the species have shared preference (inclusive niche structure, sensu Colwell and Fuentes, 1975) and (2) that secondarily they have distinct preference for peripheral habitats.

This model appears to describe some aspects of the situation in wetlands. There is a central preferred habitat (high fertility, low disturbance) with species radiating out from it, each responding to different kinds of stress or disturbance. There is experimental evidence that these species have shared preference for the central habitat (Wilson and Keddy, 1985) with competitive hierarchies excluding species different distances along radial axes (Wilson and Keddy, 1986a).

The application of this model to species-rich plant communities simultaneously suggests a number of possible extensions of and refinements to the centrifugal organization model.

1. In plant communities, it appears that a few species can dominate the central habitat without having refugia along another axis.
2. Rather than individual species radiating outward from a shared central habitat, entire niche axes radiate outward. Each axis appears to represent a different set of increasingly severe environmental constraints along which an entire group of species can be arrayed.
3. Rather than simple inclusive niche structure where all peripheral species prefer the central habitat, it may be that the end of the axis near the center has species with inclusive niche structure, but that species at the periphery tend toward distinct niche structure. In other words, as the environmental constraints become more severe at the margins, there is selection for distinct preference. In lake shores, there is evidence that stress tolerators, such as the evergreen rosette species *Lobelia dortmanna,* which occupy extremely infertile sandy shorelines, do not grow better in the central habitat (Wilson and Keddy, 1985, 1988).
4. There is the possibility of a niche structure that is neither shared nor distinct, but one sided. In such a case, species would have fundamental niches which overlap several adjacent species occurring toward the central habitat, but not toward the periphery. Competition would then displace such species down the axis toward the periphery; removal experiments would be predicted to show the possibility of competitive release in one direction but not the other ("competitive hierarchy model"; Keddy, 1989). The amount of competitive release toward the central habitat would then be predicted to be highest near the center (shared preference) and decrease through the middle of the axis (one-sided preference) to the margins (distinct preference).

Centrifugal organization is one possible model to provide assembly rules and response rules for communities. The degree to which one can generalize from the patterns and processes observed in wetlands is unclear, and can only be determined by testing for similar patterns and processes in other vegetation types.

IV. Summary

To illustrate the possibilities for general theory based on measurable state variables, I present evidence that some plant communities are organized along competition gradients with distribution and abundance controlled by asymmetric competition producing competitive hierarchies. The evidence comes from three sources. (1) A field experiment has tested for variation in competition intensity along an environmental gradient. (2) Published diallele experiments including sea cliff, grassland, and lakeshore vegetation have been analyzed to test for asymmetric competition and competitive hierarchies. (3) A comparative study using phytometers as a bioassay of competitive ability of 44 plant species has provided evidence of the direct relationship between plant size and competitive ability. These empirical relationships are probably a direct consequence of the fact that tall plants shade short plants but short plants cannot shade tall ones.

Two sets of consequences are explored. First, the presence of new state variables (e.g., competition intensity, transitivity) raises basic questions about general patterns of organization in plant communities. Some critical questions are suggested. Second, competition gradients and competitive hierarchies can produce large-scale patterns. An example is "centrifugal organization" in wetlands. In fertile undisturbed sites, a few large rhizomatous perennials dominate the vegetation. As fertility declines or disturbance increases, a much wider array of vegetation types radiates outward from the region dominated by clonal perennials. As biomass decreases, life form variation, the number of vegetation types, α diversity, and the total species pool change in a predictable manner.

Acknowledgments

I thank A. Payne and I. Wisheu for assistance with preparation of the manuscript. Many of the ideas expressed here are undoubtedly the result of discussions with C. Gaudet, D. Moore, S. Wilson, B. Shipley, and I. Wisheu. I thank L. Aarssen, N. Fowler, J. Grace, S. McCanny, D. Moore, B. Shipley, and R. Turkington for helpful comments on early drafts. M. Rosenzweig contributed helpful thoughts on centrifugal organization. The research presented here was supported by the Natural Sciences and Engineering Research Council of Canada and World Wildlife Fund (Canada).

References

Al-Mufti, M. M., Sydes, C. L., Furness, S. B., Grime, J. P., and Band, S. R. (1977). A quantitative analysis of shoot phenology and dominance in herbaceous vegetation. *J. Ecol.* **65,** 759–791.

Austin, M. P. (1986). The theoretical basis of vegetation science. *Trends Ecol. Evol.* **1,** 161–164.

Boston, H. L. (1986). A discussions of the adaptations for carbon acquisition in relation to the growth strategy of aquatic isoetids. *Aquat. Bot.* **26,** 259–270.

Boston, H. L., and Adams, M. S. (1987). Productivity, growth and photosynthesis of two small "isoetid" plants, *Littorella uniflora* and *Isoetes macrospora. J. Ecol.* **75,** 333–350.

Box, E. O. (1981). "Macroclimate and plant forms: An introduction to predictive modelling in phytogeography." Junk, The Hague.

Caldwell, M. M., Richards, J. H., Manwaring, J. H., and Eissenstat, D. M. (1987). Rapid shifts in phosphate acquisition show direct competition between neighbouring plants. *Nature (London)* **327,** 615–616.

Chapin, F. S., III (1980). The mineral nutrition of wild plants. *Annu. Rev. Ecol. Syst.* **11,** 233–260.

Chapin, F. S., III, Vitousek, P. M., and Van Cleve, K. (1986). The nature of nutrient limitation in plant communities. *Am. Nat.* **127,** 48–58.

Clements, F. E. (1933). Competition in plant societies. *News Serv. Bull., Carnegie Inst. Washington* **April 2.**

Clements, F. E. (1935). Ecology in the public service. *Ecology* **16,** 246.

Clements, F. E., Weaver, J. E. and Hanson, H. C. (1929). "Plant Competition: Analysis of Community Functions." Carnegie Inst. Washington, Washington, D.C.

Colinvaux, P. (1986). "Ecology." Wiley, New York.

Colwell, R. K., and Fuentes, E. R. (1975). Experimental studies of the niche. *Annu. Rev. Ecol. Syst.* **6,** 281–309.

Connell, J. H. (1983). On the prevalence and relative importance of interspecific competition: Evidence from field experiments. *Am. Nat.* **122,** 661–696.

Connor, E. F., and Simberloff, D. (1979). The assembly of species communities: Chance or competition? *Ecology* **60,** 1132–1140.

Day, R., Keddy, P. A., McNeill, J., and Carleton, T. (1988). Fertility and disturbance gradients: A summary model for riverine marsh vegetation. *Ecology* **69,** 1044–1054.

Dayton, P. K. (1979). Ecology: A science and a religion. *In* "Ecological Processes in Coastal and Marine Systems" (R. J. Livingston, ed.), pp. 3–18. Plenum, New York.

de Wit, C. T. (1960). On competition. *Versl. Landbouwkd. Onderz.* **66,** 1–82.

Diamond, J. M. (1975). Assembly of species communities. *In* "Ecology and Evolution of Communities" (M. L. Cody and J. M. Diamond, eds.), pp. 342–444. Belknap, Cambridge, Massachusetts.

Diamond, J. M. (1983). Laboratory, field and natural experiments. *Nature (London)* **304,** 586–587.

du Rietz, G. E. (1931). Life-forms of terrestrial flowering plants. *In* "Acta Phytogeographica Seucica. III." Almqvist & Wiksell, Uppsala, Sweden.

Fowler, N. (1982). Competition and coexistence in a North Carolina grassland. III. Mixtures of component species. *J. Ecol.* **70,** 77–92.

Gaudet, C. L., and Keddy, P. A. (1988). Predicting competitive ability from plant traits: A comparative approach. *Nature (London)* **334,** 242–243.

Givnish, T. J. (1982). On the adaptive significance of leaf height in forest herbs. *Am. Nat.* **120,** 353–381.

Goldberg, D. E., and Fleetwood, L. (1987). Competitive effect and response in four annual plants. *J. Ecol.* **75,** 1131–1143.

Goldsmith, F. B. (1973). The vegetation of exposed sea cliffs at South Stack, Anglesey. II. Experimental studies. *J. Ecol.* **61,** 819–829.

Goldsmith, F. B. (1978). Interaction (competition) studies as a step towards the synthesis of sea cliff vegetation. *J. Ecol.* **66,** 921–931.

Green, R. H. (1980). Multivariate approaches in ecology: the assessment of ecologic similarity. *Annu. Rev. Ecol. Syst.* **11,** 1–14.

Grime, J. P. (1973). Competitive exclusion in herbaceous vegetation. *Nature (London)* **242,** 344–347.

Grime, J. P. (1974). Vegetation classification by reference to strategies. *Nature (London)* **250,** 26–31.

Grime, J. P. (1977). Evidence for the existence of three primary strategies in plants and its relevance to ecological and evolutionary theory. *Am. Nat.* **111,** 1169–1194.

Grime, J. P. (1979). "Plant Strategies and Vegetation Processes." Wiley, Chichester, England.

Grubb, P. J. (1985). Plant populations and vegetation in relation to habitat, disturbance and competition: Problems of generalization. *In* "The Population Structure of Vegetation" (J. White, ed.), pp. 595–621. Junk, Dordrecht, The Netherlands.

Harper, J. L. (1965). The nature and consequence of interference amongst plants. *Genet. Today* **2**, 465–482.

Harper, J. L. (1977). "Population Biology of Plants." Academic Press, London.

Harper, J. L., and Chancellor, A. P. (1959). The comparative biology of closely related species living in the same area. IV. *Rumex:* Interference between individuals in populations of one and two species. *J. Ecol.* **47**, 679–695.

Harper, J. L., and Clatworthy, J. N. (1963). The comparative biology of closely related species. VI. Analysis of the growth of *Trifolium repens* and *T. fragiferum* in pure and mixed populations. *J. Exp. Bot.* **14**, 172–190.

Harper, J. L., and McNaughton, I. H. (1962). The comparative biology of closely related species living in the same area. VII. Interference between individuals in pure and mixed populations of *Papaver* species. *New Phytol.* **61**, 175–188.

Harper, J. L., Clatworthy, J. N., McNaughton, I. H., and Sagar, G. R. (1961). The evolution and ecology of closely related species living in the same area. *Evolution* **15**, 209–227.

International Union for the Conservation of Nature and Natural Resources (IUCN) (1980). "World Conservation Strategy." IUCN, Gland, Switzerland.

Keddy, P. A. (1981). Vegetation with Atlantic coastal plain affinities in Axe Lake, near Georgian Bay, Ontario. *Can. Field-Nat.* **95**, 241–248.

Keddy, P. A. (1983). Shoreline vegetation in Axe Lake, Ontario: Effects of exposure on zonation patterns. *Ecology* **64**, 331–344.

Keddy, P. A. (1985). Lakeshores in the Tusket River Valley, Nova Scotia: The distribution and status of some rare species, including *Coreopsis rosea* Nutt, and *Sabatia kenneduyana* Fen. *Rhodora* **87**, 309–320.

Keddy, P. A. (1987). Beyond reductionism and scholasticism in plant community ecology. *Vegetatio* **69**, 209–211.

Keddy, P. A. (1989). "Competition." Chapman and Hall, London.

Keddy, P. A., and Reznicek, A. A. (1986). Great lakes vegetation dynamics: The role of fluctuating water levels and buried seeds. *J. Great Lakes Res.* **12**, 25–36.

Keddy, P. A., and Shipley, B. (1989). Competitive hierarchies in herbaceous plant communities. *Oikos* **54**, 234–241.

Kuhn, T. S. (1970). "The Structure of Scientific Revolutions." Univ. of Chicago Press, Chicago.

Legendre, L., and Legendre, P. (1983). "Numerical Ecology." Elsevier, Amsterdam.

Lewontin, R. C. (1974). "The Genetic Basis of Evolutionary Change." Columbia Univ. Press, New York.

Miller, R. S. (1967). Pattern and process in competition. *Adv. Ecol. Res.* **4**, 1–74.

Mitchley, J. (1988). Control of relative abundance of perennials in chalk grassland in southern England. II. Vertical canopy structure. *J. Ecol.* **76**, 341–350.

Mitchley, J., and Grubb, P. J. (1986). The control of relative abundance of perennials in chalk grassland in southern England. I. Constancy of rank order and results of pot and field experiments on the role of interference. *J. Ecol.* **74**, 1139–1166.

Moore, D. R. J., Keddy, P. A., Gaudet, C. L., and Wisheu, I. C. (1989). Conservation of wetlands: Do infertile wetlands deserve a higher priority? *Biol. Conserv.* **47**, 203–217.

Nobel, I. R., and Slatyer, R. O. (1980). The use of vital attributes to predict successional changes in plant communities subject to recurrent disturbances. *Vegetatio* **43**, 5–21.

Peters, R. H. (1980a). From natural history to ecology. *Perspect. Biol. Med.* **23**, 191–203.

Peters, R. H. (1980b). Useful concepts for predictive ecology. *Synthese* **43**, 257–269.

Rigler, F. H. (1982). Recognition of the possible: An advantage of empiricism in ecology. *Can. J. Fish. Aquat. Sci.* **39**, 1323–1331.

Rorison, I. H., Grime, J. P., Hunt, R., Hendry, G. A. F., and Lewis, D. H. (1987). Frontiers of Comparative Plant Ecology. *New Phytol.* **106** (suppl.), 1–317.

Rosenzweig, M. L., and Abramsky, Z. (1986). Centrifugal community organization. *Oikos* **46**, 339–348.

Schoener, T. W. (1983). Field experiments on interspecific competition. *Am. Nat.* **122**, 240–285.

Shipley, B., Keddy, P. A., Moore, D. R. J., and Lemky, K. (1989). Regeneration and establishment strategies of emergent macrophytes. *J. Ecol.* **77**.

Shipley, B., and Keddy, P. A. (1987). The individualistic and community-unit concepts as falsifiable hypotheses. *Vegetatio* **69**, 47–55.

Silvertown, J. (1980). The dynamics of a grassland ecosystem: Botanical equilibrium in the park grass experiment. *J. Appl. Ecol.* **17**, 491–504.

Tansley, A. G. (1914). Presidential address. *J. Ecol.* **2**, 194–203.

Tilman, D. (1982). "Resource Competition and Community Structure." Princeton Univ. Press, Princeton, New Jersey.

Tilman, D. (1988). "Plant Strategies and the Structure and Dynamics of Plant Communities." Princeton Univ. Press, Princeton, New Jersey.

van der Valk, A. G. (1981). Succession in wetlands: A Gleasonian approach. *Ecology* **62**, 688–696.

Watts, A. W. (1958). "The Spirit of Zen." Random House, New York.

Weiner, J. (1986). How competition for light and nutrients affects size variability in *Ipomea tricolor* populations. *Ecology* **67**, 1425–1427.

Weins, J. A. (1983). Avian community ecology: An iconoclastic view. *In* "Perspectives in Ornithology" (A. H. Brush and G. A. Clark, Jr., eds.)., pp. 355–403. Cambridge Univ. Press, Cambridge, England.

Werner, P. A. (1979). Competition and coexistence of similar species. *In* "Topics in Plant Population Biology" (O. T. Solbrig, S. Jain, G. G. Johnson, and P. H. Raven, eds.), pp. 287–310. Columbia Univ. Press, New York.

Wilson, S. D., and Keddy, P. A. (1985). Plant zonation on a shoreline gradients: Physiological response curves of component species. *J. Ecol.* **73**, 851–860.

Wilson, S. D., and Keddy, P. A. (1986a). Species competitive ability and position along a natural stress/disturbance gradient. *Ecology* **67**, 1236–1242.

Wilson, S. D., and Keddy, P. A. (1986b). Measuring diffuse competition along an environmental gradient: Results from a shoreline plant community. *Am. Nat.* **127**, 862–869.

Wilson, S. D., and Keddy, P. A. (1988). Species richness, survivorship, and biomass accumulation along an environmental gradient. *Oikos* **53**, 375–380.

Wimsatt, W. C. (1982). Reductionistic research strategies and their biases in the units of selection controversy. *In* "Conceptual Issues in Ecology" (E. Saarine, ed.), pp. 155–201. Reidel, Dordrecht, The Netherlands.

Wisheu, I. C. and Keddy, P. A. (1989a). Species richness—standing crop relationships along four lakeshore gradients: constraints on the general model. *Canad. J. Bot.* **67**, 1609–1617.

Wisheu, I. C., and Keddy, P. A. (1989b). The conservation and management of a threatened coastal plain plant community in eastern North America (Nova Scotia, Canada). *Biol. Conserv.*, **48**, 229–238.

14

Disorderliness in Plant Communities: Comparisons, Causes, and Consequences

Norma L. Fowler

I. Introduction

The study of competition and other processes that affect the dynamics and structure of plant communities is a search for order and organization in complex systems. Sometimes, despite well-designed and well-executed experiments, this search fails to uncover order: the experiments "fail," i.e., they produce negative results. The usual response is to seek larger samples sizes and more elaborate experimental designs and analyses. In many cases this response is entirely appropriate. In some cases, however, no adjustment of sample sizes or experimental design and analysis will reveal much order in the particular interaction or process under consideration; it is not there to be found.

It is this absence or weakness of order in plant communities that I address here. I begin by defining the sense in which I use the terms "order" and "orderliness." I then briefly outline some of the causes of disorderliness in the interactions between individuals and between species and of the consequent disorderliness in the behavior of plant populations and communities. I suggest some ways in which the degree of disorderliness could be measured. Finally, I call attention to the potential consequences of disorderliness in plant communities, which are as yet largely unclear.

II. Definitions

The operation and outcome of competition between individuals of two or more plant species may be consistent and predictable, that is, *orderly*. For example, there may be a consistent "winner" and "loser," as occurs when shade-intolerant tree species (e.g., *Pinus taeda*) compete with the shade-tolerant tree species that replace them during succession (Spurr and Barnes, 1980). Alternatively, the outcome of competition between two or more plant species may be quite inconsistent and hence unpredictable, that is, *disorderly*. The more factors upon which the outcome depends, and the larger the stochastic component of the interaction, the more inconsistent and unpredictable will be the outcome of competition. For example, the outcome of competition between plants quite often depends on the sizes of the individuals as well as on the species involved (Harper, 1977). Hence the outcome depends on germination date, which in turn implies that the timing of the arrival of seeds in a site and annual variations in weather, among other factors, will affect the outcome of competition between two species.

Orderly and disorderly are useful adjectives, but orderliness is clearly a matter of degree. A continuum exists from the most orderly to the most disorderly competitive relationships. At the latter extreme lie those competitive interactions whose outcome is effectively random.

Other ecological processes can also be characterized by their degree of disorderliness, including the interactions between particular plants and their herbivores, pathogens, and mutualists. For example, some pathogens always kill their host, a simple, orderly, interaction. In most cases, however, the interaction between an individual plant and a particular pathogen is much less orderly, since the effect of the pathogen depends on the genotype, age, and physiological status of the individual plant and on the microenvironment in which it is growing (Burdon, 1987).

The regulation of the size of a population and the regulation of the relative abundances of species in a community can also be characterized by their relative degrees of orderliness. I use the word "regulate" in a strict sense, to indicate the action of density-dependent processes that tend to produce an increase in population size when it is relatively low, and a decrease when it is high. I also include the processes that tend to produce an increase in the relative abundance of a species within a community when it is sparse, and a decrease when it is abundant. The regulation of species' abundances in this strict sense is by definition frequency dependent, even if the population of each species is regulated completely independently of those of other species.

If population regulation is orderly, population size is effectively and consistently regulated by density-dependent processes (e.g., as during self-thinning in some monocultures; Westoby, 1984). At the other extreme, population size may be essentially unregulated (in the strict sense just defined), that is, the relationship between population size and changes in population size is essentially random. This situation could occur if, for example, density-independent factors keep the population at very low densities, as might happen in a population becoming extinct in an unsuitable site. Between these two extremes, population regulation may be disorderly: regulation may be weak, or sporadic, or occur only when populations are very large or very small, or occur at a spatial scale larger than a single population or study site or over long periods of time.

The regulation of the relative abundances of species in a community can also be characterized by its relative degree of orderliness. The regulation of relative abundances is orderly if it is effectively and consistently regulated by frequency-dependent processes; disorderly if the regulation is weak, sporadic, or at another scale than that being measured; random if relative abundances are not regulated. At least two different situations can theoretically produce the latter situation: (1) the population of each species is unregulated, and the different species do not compete with each other, or (2) the overall density (total number of individuals of all species) is regulated, but the species are ecological equivalents (the species are competitive equals, are favored by the same environments, are not regulated by different herbivores, etc.). In either of these situations the relative abundances of species would be free to drift randomly, in a fashion analogous to the genetic drift of neutral alleles, to extinction or "fixation."

In general, orderliness in the regulation of the relative species' abundances in a community should be correlated with the effectiveness and consistency (orderliness) of the mechanisms that promote the persistence of species. These mechanisms may involve differential resource use, the use of different microenvironments, or regulation by different herbivores or pathogens (Grubb, 1977; Fowler, 1988).

Note that I use the term "persistence," not "coexistence." "Persistence" encompasses a variety of mechanisms that tend to maintain species in a community without producing a stable equilibrium in the mathematical sense. [Chesson (1986) provides an excellent discussion of this distinction.] A large number of models have recently been constructed which model communities that are disorderly in the sense used here. Most, but not all, of these are nonequilibrium models, although they all contain, explicitly or implicitly, the differences between species and the resulting frequency dependence that tends to promote the persistence of species in a community (e.g., Chesson and Warner, 1981; Chesson, 1985; see reviews by Shmida and Ellner, 1984, and DeAngelis and Waterhouse, 1987). A few models of truly random interactions among species (no frequency-dependent regulation) have also been constructed (e.g., Hubbell, 1979; Hubbell and Foster, 1986).

Disorderliness is not a synonym for stochasticity. Deterministic, as well as stochastic, factors can increase disorderliness (see below). Both appear as sources of variation in our experiments, either as "unexplained variation" in the error term or as variation associated with block effects and covariates. Both can decrease the strength of the relationship between two variables of interest (e.g., between population size and plant performance; between the number of neighboring individuals and the magnitude of their competitive effect). Disorderliness is more, therefore, than just the amount of "noise" (stochastic variation) in a system: a community could theoretically have little noise in its structure or dynamics but be highly disorderly.

III. Causes of Disorderliness

Disorderliness, in the sense of the word used here, may arise from a very large number of different sources.

1. *Complexity*, that is, a large number of species and of interactions among species in a community and many different factors affecting the outcome of any given interaction, will produce disorderliness in species interactions, because the strength of the relationship between any two variables is, in practice, usually reduced as the number of additional variables increases. For example, if the population dynamics of a species are affected by the densities of a large number of competing species, the

effect of any given competitor species is likely to be weak. Likewise, density-dependent regulation is likely to be particularly weak and sporadic for sparse species (Grubb, 1986).

2. *Weak stabilizing biotic interactions* produce disorderliness, by definition. For example, if niche overlap is high, species may be close to competitive equivalence, and, if so, in the absence of other regulating factors their relative abundances will be strongly affected by chance events. Weak or sporadic regulation of a population by a herbivore or pathogen is another example of a weak stabilizing biotic interaction.

3. *Spatial and temporal variation in the environment* is undoubtedly the source of much of the disorderliness of natural communities. I include disturbances such as treefalls in this category, because they can be considered a source of environmental heterogeneity in both space and time. I have discussed the roles of environmental variation in the regulation of populations and communities at some length elsewhere (Fowler, 1988). It can be a source of variation in demographic parameters (e.g., Watkinson and Harper, 1978; Keddy, 1980, 1981; Gartner *et al.,* 1986), in the outcomes of competition (e.g., Raynal and Bazzaz, 1975; Berendse, 1983) and herbivory (e.g., Cox and McEvoy, 1983; Bryant, 1987; Rice, 1987), and in the effects of disturbances (e.g., Platt, 1975).

4. *Strong destabilizing biotic interactions* include the many forms of complex dynamics discovered by the analysis of theoretical models: limit cycles, overshoots, "chaos," and so on (reviewed by DeAngelis and Waterhouse, 1987, as "biotic feedback instability").

5. If a community or a population is *regulated only at a larger scale* than the one in which we are interested, its dynamics at the scale at which we are working may be extremely disorderly. A familiar example of this situation is a fugitive species that can never persist for long in any given patch. The number of individuals of such a species is determined by the frequency and sizes of new open patches; within-patch population sizes may be unregulated (in the strict sense used here). However, the inverse relationship between competitive ability and dispersal ability that characterizes fugitive species is not always necessary for regional persistence despite unstable cell (i.e., patch) dynamics (Slatkin, 1974). [See DeAngelis and Waterhouse (1987) for a review of the numerous models of this situation.]

6. *Regulation may occur only at extreme population sizes or relative abundances,* when a species is very abundant or very sparse. It will therefore occur only sporadically. At intermediate densities and abundances, regulation may be absent, and population sizes, competitive outcomes, etc. will be affected by whatever other factors may be operating. Strong (1986a) has called this phenomenon "density-vague" population regulation.

7. *Time lags* will produce disorderliness. For example, if population growth lags behind changes in resources, there may be quite long periods in which a population is so far below carrying capacity that competition for resources, and hence density-dependent population regulation, is absent or very weak (Wiens, 1977; Fowler, 1988). A population may be small because the species has only recently arrived in the site; a species may be absent because it has not yet arrived in the site, although it is part of the flora of the area and could persist in the community.

8. *Historical effects* are a special case of time lags. The effects of earlier events that affected germination or early growth can persist years later, with the result that community composition reflects events many years in the past. Size hierarchies, once formed, tend to be maintained (Harper, 1977) and together with the dependence of the outcome of competition on plant size, may make plant populations and communities particularly susceptible to the effects of time lags. The preemption of space by colonists who happened by chance to arrive there first would also fall into this category.

9. *The dispersal of plant propagules is highly stochastic* and introduces an important source of disorderliness into the dynamics and structure of plant communities. In this category, we can include the presence of species that persist in a site only because of repeated colonization from another, more favorable site (Shmida and Ellner, 1984).

10. *Genetic variation within a population* can affect the outcome of interactions between different species (Turkington and Harper, 1979; Burdon, 1980; Gouyon *et al.*, 1983; Aarssen and Turkington, 1985), as well as interactions between members of the same population (Shaw, 1986).

This list includes a mixture of stochastic and deterministic factors. In fact, most of the factors contain both stochastic and deterministic elements. For example, winter is a predictable form of temporal heterogeneity, but the date of the first frost is highly variable and unpredictable. As a second example, consider the situation in which population regulation occurs only when a population is very large. If the size of the population essentially drifts at other times, the time at which it will drift to a large enough size for regulation to occur may be stochastic, but at this size, regulation may be a highly deterministic process.

IV. A Shift in Our Views of Plant Communities

To some extent, orderliness represents a view of plant communities that was once more popular than it is now. As more detailed demographic data are collected and more experimental studies are made of natural plant communities, interactions among species and the regulation of populations and species' abundances are being found to be less orderly, in the sense just defined, than many of us perhaps once expected. This shift in our thinking is a response to accumulating data, not to new theory. It is a quantitative shift, not a qualitative change, in our way of thinking about plant communities. Similar shifts have occurred in the way ecologists think about succession (Drury and Nisbet, 1973; Connell and Slatyer, 1977) and animal communities (e.g., Wiens, 1977, 1984; den Boer, 1981; Strong, 1986a,b).

Probably no group of ecologists would agree on the extent of the disorderliness they expect to find in a natural plant community, or on the extent to which their own views have shifted, or on the timing of that shift. Casual discussion of this topic with a variety of plant ecologists has suggested that each of us holds views that are highly influenced by the particular plant communities with which we have worked, the type of studies we have done, and by the period, place, and nature of our training. That an expectation of an orderly world still exists is demonstrated by the numerous talks given each year at the annual meeting of the Ecological Society of America in which the speaker tells us how especially variable in space or time his or her particular system is. However, random species interactions (i.e., interactions whose outcomes are independent of the particular species involved) seem recently to have become a more popular hypothesis.

V. Measuring the Degree of Disorderliness

Since disorderliness is a relative, not an absolute, property, we can ask where on the continuum from most to least orderly a particular competitive interaction falls, in comparison with an interaction between the same species in a different site or year, or in comparison with a different pair of species. We could also compare the orderliness of other sorts of interactions, for example, asking which of two herbivores has the most predictable, consistent effects on a particular plant population. We could compare the orderliness of the regulation of two or more populations, of the same or of different species, or of the same population at different times. Likewise, we could compare the regulation of the relative abundances of different pairs or groups of species, or the same pair or group of species in a different place or time.

These comparisons would tell us a great deal about how natural plant communities function. How frequent in time and space is population regulation (in the strict sense used here), and how intensely does it operate under different conditions? That is, how close is the relationship between density and plant performance? How often does the outcome of a particular interaction between two species accord with the regulating processes, and how often do other factors prevent this outcome? Are there groups of species within a community that are not effectively regulated, but function as random assemblages of ecological equivalents, and, if so, how common are such groups? All of these questions, and many others, are aspects of the more general question of the extent to which the dynamics of a community are the result of orderly processes and the extent to which they are not.

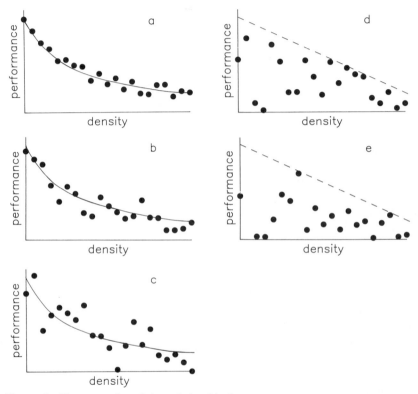

Figure 1 Five examples of the relationship between a measure of individual plant performance (size, fecundity) and density (the number of conspecific individuals present). Plant performance has the same underlying relationship to density in a, b, and c (the same hyperbolic function, indicated by the solid lines, was used to generate each data set; Weiner, 1982), but the variation around this underlying relationship increases from a to c. Density-dependent population regulation therefore occurs in all three examples, but is a less orderly process in b than in a and is least orderly in c. Graphs d and e represent a different kind of relationship between plant performance and density (Firbank and Watkinson, 1987; Goldberg and Fleetwood, 1987). In these two examples, density sets the same upper limit (indicated by the dashed lines) to plant performance; d can be considered more orderly than e because the upper limit is better-defined in d than in e, that is, more plants approach the upper limit in d than in e.

To answer these questions and test these hypotheses, we need ways to quantify the degree of disorderliness. Many different approaches are possible; I outline a few. The frequency with which competition can be detected is a measure of the orderliness of competition (Connell, 1983). If a competitive "winner" and "loser" have been identified, we could use a measure of the extent to which the outcome varies (the "loser" wins 5% of the time? 10%?) as measure of (dis)orderliness. More probably, we are interested in the magnitude of competitive effect or competitive response (Miller and Werner, 1987). In this case, some measure of the magnitude of the variation around the mean effect or response would be appropriate.

If population regulation is sporadic, we could use the proportion of years in which regulation occurs as a measure of disorderliness. With sufficient data, a curve can be fitted to the relationship between density and per-capita performance (ignoring, for the sake of simplicity, complex life histories); our measure of disorderliness could then be a measure of the deviations of the points from the fitted curve (Fig. 1). Several studies of natural plant populations have fitted such curves. A comparison of their results gives us a comparison of the relative degree of orderliness in the regulation of these populations (if we are willing to ignore, for the sake of this example, differences between these studies in their measures of plant performance and of density, and in the equations used to fit the curves). The relationship between density and individual performance was stronger in *Polygonum* spp. (83–86% of the variance explained; Weiner, 1982) than in *Pinus rigida* (53% of the variance explained; Weiner, 1984), and still weaker in most populations of *Viola* spp. (5–59% of the variance explained; Waller, 1981). Thus, population regulation appears to be a more orderly process in the annual *Polygonum* species than in the pine, and more orderly in the pine than in the violet species.

VI. Disorderliness at the Community Level

In view of the large number of factors that affect the degree of disorderliness, there is no reason to expect that different interactions among the species of a particular community should be characterized by the same degree of disorderliness. Likewise, the populations of some of the component species of a community may be regulated in a very orderly, i.e., consistent and predictable, fashion, while the populations of other species may be regulated weakly, sporadically, or not at all. Thus, assigning to a community a single position along the continuum of orderliness is problematical. Some sort of average or other composite function of the orderliness of each interaction between species, and/or of the regulation of each population, would seem best.

Such a composite function would necessarily be crude. However, it could be used to cast some light on a variety of interesting questions. Are communities of fewer species characterized by more orderly dynamics and interactions? All else being equal, this should be so, because an important source of complexity is reduced as the number of species is reduced. By the same reasoning, are communities occupying sites with less temporal and spatial variation in the environment, on average, more orderly? Other hypotheses will no doubt occur to the reader.

Another situation in which the use of a measure of the disorderliness of an entire community may be appropriate is the comparison of different models. Here, measures such as the rate of loss of species from a community, or the consistency of the composition of a community (relative abundances of species) between runs, patches, or times, may prove useful. Models can provide useful insights into interpreting patterns in natural communities. For example, how strong and how frequent need regulation be to maintain species in a community? How constant would the composition of communities of different degrees of disorderliness be from site to site and year to year?

Disorderliness may be an intrinsic property of a particular interaction or process. In some cases, however, some of the disorderliness may disappear if we alter the scale of our study. For example, if the density-dependent regulation of a population operates via a specialist insect, the population may appear to be unregulated if short-term responses to the removal of neighboring plants are examined. To detect the regulation of this population, experimental manipulations of plant density must be at the spatial scale of insect movement, and sufficient time must be allowed for the insect (or insect population) to respond to changes in plant density. In this example, the apparent disorderliness in the regulation of this population was a result of the use of an inappropriate scale of manipulation and measurement.

One process or interaction may be disorderly and another not. Consider, for example, a hypothetical community in which the abundances of the dominant plant species are regulated by herbivory and in which, moreover, differential herbivory is the primary factor determining the fate of individual plants. A study restricted to competitive interactions between neighboring plants in this hypothetical community will therefore reveal that these interactions are extremely disorderly, and no change of scale will alter this conclusion. If our real question is simply, What is going on in this community? then the appropriate response to this finding is to shift the focus of our study, in this case from competition to herbivory, since in our hypothetical community, the effects of herbivory are not only important but also orderly (i.e., consistent and predictable). If, on the other hand, our real question concerns relationships between neighboring plants, not herbivory, then we correctly conclude that the relationships between neighbors in this community are intrinsically disorderly.

In many cases, our ultimate goal is indeed simply to "figure out what is going on" or, more formally, to describe the factors and processes determining the abundances, spatial distributions, and so on of the component species of a community. When this is the case, our most appropriate response to the discovery of disorder in one process is to look for order in other processes. Sometimes our interests are more specific and circumscribed. We may be interested in competition per se (or herbivory per se), whether or not it is important in the particular community. When our interests are circumscribed in this fashion, and the process or phenomenon of interest is disorderly, the disorderliness itself is of interest.

In some communities there may be no orderly process that structures the community, and hence no change of focus that can reveal it. This possibility suggests a different definition of the orderliness of a community: the degree to which the important processes that structure the community are orderly. This definition is intuitively appealing, but presents some formidable problems of application. To be confident that there really is no orderly process structuring a community would require eliminating all possibilities: not just competition and predation by the more obvious herbivores, but pathogens, mycorrhizae, herbivorous nematodes, and so on. Second, this definition of community orderliness requires us to compare apples and oranges. How are we to compare the orderliness of competitive interactions with the orderliness of the effects of herbivores on plants? Despite these problems, however, this definition of community orderliness may prove conceptually useful.

VII. Consequences for Communities

All sources of disorderliness will tend to weaken the effectiveness of the processes that promote species coexistence. They will also tend to weaken the process of competitive exclusion. The outcome is a community in which random events become more important, but it is not clear whether the net effect of weakening both coexistence and competitive exclusion will be to increase or to decrease diversity. No doubt the level of disorderliness of a community, or of specific processes and interactions within a community, will affect many other aspects of community structure and function, but it is not yet clear what these effects will be.

This would therefore seem to be a productive area for theoretical modeling.

The immediate evolutionary consequence of a greater degree of disorderliness is probably most often a weakening of selective pressures. If, for example, the outcome of a competitive interaction is unpredictable, being influenced by many extraneous factors, then the selective pressure for enhanced competitive ability in both species may be fairly weak.

In the longer term, however, the situation is complicated by the fact that many of the sources of disorder also provide opportunities for ecological specialization and thus for additional species to persist in the communities. This will produce a new set of selective pressures on these species and on the species with which they interact. Many of the sources of disorder can also act as factors that maintain genetic variation within populations, if different genotypes produce phenotypes with different ecological specializations.

The most obvious opportunities for ecological specialization are provided by environmental heterogeneity. If species can specialize on patch types, different seasons, wet or dry years, different ages of disturbances, and so on, in such a way that they are negatively affected more by conspecific individuals than by individuals of other species, then environmental variation can be the basis for their persistence in the community (Fowler, 1988). Such environmental variation will therefore no longer be a direct source of disorder in the community, although the resulting increase in the number of species present will increase the complexity of the system, which in turn may increase its disorderliness. Other sources of disorder can also provide opportunities for additional species: additional species (an increase in complexity) can in turn create niches for yet more species; the time lags involved in processes like succession create opportunities for fugitive species.

VIII. Future Directions

A number of potentially fruitful avenues of investigation arise from consideration of the disorderliness that characterizes most natural plant communities. First and simplest, it suggests that "negative results" are not necessarily just frustrating failures. If the experimental design is a sufficiently powerful one (that is, if the probability of a type II error is sufficiently low), negative results may be telling us something important about the process or interaction being studied, and about the structure and dynamics of the community. Disorderliness is itself an interesting and important property of population dynamics, of interactions among species, and of plant communities.

Second, it suggests that, if our goal is simply to understand the processes by which a particular community is structured, an appropriate response to a "failed" experiment (i.e., one that produces negative results despite a reasonably powerful, appropriate design and analysis) may be to shift the focus or scale of the study, looking for order in a different place.

Third, the degree of orderliness that characterizes any given interaction between species or individuals is itself interesting and useful information. Comparisons among sites, among years, among species, and among levels of specific environmental factors are all possible and potentially revealing. Comparisons could also be made between sites, years, or species as to the causes of disorderliness in the interaction under consideration. I have listed a number of causes of disorderliness; it would be informative to know their relative contributions.

Finally, and perhaps most important, what are the consequences for a community of a given kind and degree of disorderliness? I would argue that the discovery that plant communities may be less orderly than some of us perhaps expected should not be discouraging. Although disorderliness makes it more difficult to detect the operation of competition and other processes, it also challenges us to reexamine the ways in which we think about plant communities. Do we need to modify our ideas about how natural plant communities are structured in light of accumulating evidence of a high degree of disorderliness in their functioning? Are very disorderly communities different in any fundamental way from more orderly ones? These questions are far from being answered.

IX. Summary

Instances of processes such as competition and predation may be characterized by their degree of orderliness, that is, the degree to which the outcome is consistent and hence predictable. Similarly, the regulation of a particular population can also be characterized by the degree to which it is orderly, that is, the degree to which it is effectively and consistently regulated by density-dependent processes. The regulation of the relative abundances of different species in a community can also be so characterized.

Disorderliness can have many different sources, including complexity (e.g., a large number of species in the community), spatial and temporal variation in the environment, regulation only at extreme population sizes or relative abundances, time lags, and historical effects. Both stochastic and deterministic factors can increase disorderliness.

The comparison of the relative degree of orderliness in, for example, competitive interactions involving different species, or in the regulation of different populations, would be informative. Several quantitative measures of the degree of orderliness are suggested.

Since different processes within a single community will have different degrees of orderliness, the comparison of the orderliness of different communities presents problems. Two possible approaches are identified.

When a study finds disorderliness, a change of scale or a change of focus (e.g., from competition to herbivory) may sometimes reveal orderliness. In other cases, the system is intrinsically disorderly; this itself may be an interesting and important finding.

Acknowledgments

I thank P. Chesson, J. Connell, S. Hermann, D. Goldberg, P. Keddy, P. Morin, W. Platt, A. Thistle, D. Strong, J. Travis, and two anonymous reviewers for helpful comments on various versions of the manuscript. The participants in the Cedar Creek symposium, especially S. Pacala, J. Silander, J. Connell, and D. Goldberg, offered comments and discussion of the initial presentation of the ideas discussed in this manuscript. Tall Timbers Research Station and Florida State University provided a stimulating environment in which to work. Funding was provided by NSF grant BSR 8600068 and by a Faculty Research Assignment funded by the University Research Institute of the University of Texas.

References

Aarssen, L. W., and Turkington, R. (1985). Biotic specialization between neighbouring genotypes in *Lolium perenne* and *Trifolium repens* from a permanent pasture. *J. Ecol.* **73,** 605–614.

Berendse, F. (1983). Interspecific competition and niche differentiation between *Plantago lanceolata* and *Anthoxanthum odoratum* in a natural hayfield. *J. Ecol.* **71,** 379–390.

Bryant, J. P. (1987). Feltleaf willow–snowshoe hare interactions: Plant carbon/nitrogen balance and floodplain succession. *Ecology* **68,** 1319–1327.

Burdon, J. J. (1980). Variation in disease-resistance within a population of *Trifolium repens. J. Ecol.* **68,** 737–744.

Burdon, J. J. (1987). "Diseases and Plant Population Biology." Cambridge Univ. Press, Cambridge, England.

Chesson, P. L. (1985). Coexistence of competitors in spatially and temporally varying environments: A look at the combined effects of different sorts of variability. *Theor. Pop. Biol.* **28,** 263–287.

Chesson, P. L. (1986). Environmental variation and the coexistence of species. *In* "Community Ecology" (J. Diamond and T. J. Case, eds.), pp. 240–256. Harper & Row, New York.

Chesson, P. L., and Warner, R. R. (1981). Environmental variability promotes coexistence in lottery competitive systems. *Am. Nat.* **117,** 923–943.

Connell, J. H. (1983). On the prevalence and relative importance of interspecific competition: Evidence from field experiments. *Am. Nat.* **122,** 661–696.

Connell, J. H., and Slatyer, R. O. (1977). Mechanisms of succession in natural communities and their role in community stability and organization. *Am. Nat.* **111,** 1119–1144.

Cox, C. S., and McEvoy, P. B. (1983). Effect of summer moisture stress on the capacity of tansy ragwort (*Senecio jacobaea*) to compensate for defoliation by cinnabar moth (*Tyria jacobaeae*). *J. Appl. Ecol.* **20,** 225–234.

DeAngelis, D. L., and Waterhouse, J. C. (1987). Equilibrium and nonequilibrium concepts in ecological models. *Ecol. Monogr.* **57,** 1–21.

den Boer, P. J. (1981). On the survival of populations in a heterogeneous and variable environment. *Oecologia* **50,** 39–53.

Drury, W. H., and Nisbet, I. C. T. (1973). Succession. *J. Arnold Arbor. Harvard Univ.* **54,** 331–368.

Firbank, L. G. and Watkinson, A. R. (1987). On the analysis of competition at the level of the individual plant. *Oecologia* **71,** 308–317.

Fowler, N. L. (1988). The effects of environmental heterogeneity in space and time on the regulation of populations and communities. *Symp. Br. Ecol. Soc.* **28,** 249–269.

Gartner, B. L., Chapin, F. S., and Shaver, G. R. (1986). Reproduction of *Eriophorum vaginatum* by seed in Alaskan tundra. *J. Ecol.* **74,** 1–18.

Goldberg, D. E., and Fleetwood, L. (1987). Competitive effect and response in four annual plants. *J. Ecol.* **75,** 1131–1143.

Gouyon, P. H., Fort, P., and Caraux, G. (1983). Selection of seedlings of *Thymus vulgaris* by grazing slugs. *J. Ecol.* **71,** 299–306.

Grubb, P. J. (1977). The maintenance of species-richness in plant communities: The importance of the regeneration niche. *Biol. Rev.* **52,** 107–145.

Grubb, P. J. (1986). Problems posed by sparse and patchily distributed species in species-rich plant communities. *In* "Community Ecology" (J. Diamond and T. J. Case, eds.), pp. 207–225. Harper & Row, New York.

Harper, J. L. (1977). "Population Biology of Plants." Academic Press, New York.

Hubbell, S. P. (1979). Tree dispersion, abundance, and diversity in a tropical forest. *Science* **203,** 1299–1309.

Hubbell, S. P., and Foster, R. B. (1986). Biology, chance, and history and the structure of tropical rain forest tree communities. *In* "Community Ecology" (J. Diamond and T. J. Case, eds.), pp. 314–329. Harper & Row, New York.

Keddy, P. A. (1980). Population ecology in an environmental mosaic: *Cakile edentula* on a gravel bar. *Can. J. Bot.* **58,** 1095–1100.

Keddy, P. A. (1981). Experimental demography of the sand-dune annual, *Cakile edentula,* growing along an environmental gradient in Nova Scotia. *J. Ecol.* **69,** 615–630.

Miller, T. E., and Werner, P. A. (1987). Competitive effects and responses between plant species in a first-year old-field community. *Ecology* **68,** 1201–1210.

Platt, W. J. (1975). The colonization and formation of equilibrium plant species associations on badger disturbances in a tall-grass prairie. *Ecol. Monogr.* **45,** 285–305.

Raynal, D. J., and Bazzaz, F. A. (1975). The contrasting life-cycle strategies of three summer annuals found in abandoned fields in Illinois. *J. Ecol.* **63,** 587–596.

Rice, K. J. (1987). Interaction of disturbance patch size and herbivory in *Erodium* colonization. *Ecology* **68,** 1113–1115.

Shaw, R. G. (1986). Response to density in a wild population of the perennial herb *Salvia lyrata:* Variation among families. *Evolution* **40,** 492–505.

Shmida, A., and Ellner, S. (1984). Coexistence of plant species with similar niches. *Vegetatio* **58,** 29–55.

Slatkin, M. (1974). Competition and regional coexistence. *Ecology* **55,** 128–134.

Spurr, S. H., and Barnes, B. V. (1980). "Forest Ecology," 3rd ed. Wiley, New York.

Strong, D. R. (1986a). Density-vagueness—Abiding the variance in the demography of real populations. *In* "Community Ecology" (J. Diamond and T. J. Case, eds.), pp. 257–268. Harper & Row, New York.

Strong, D. R. (1986b). Population theory and understanding pest outbreaks. *In* "Ecological Theory and Integrated Pest Management" (M. Kogan, ed.), pp. 37–57. Wiley, New York.

Turkington, R., and Harper, J. L. (1979). The growth, distribution and neighbour relationships of *Trifolium repens* in a permanent pasture. IV. Fine scale biotic differentiation. *J. Ecol.* **67**, 245–254.

Waller, D. M. (1981). Neighborhood competition in several violet populations. *Oecologia* **51**, 116–122.

Watkinson, A. R., and Harper, J. L. (1978). The demography of a sand dune annual: *Vulpia fasciculata*. I. The natural regulation of populations. *J. Ecol.* **66**, 15–33.

Weiner, J. (1982). A neighborhood model of annual-plant interference. *Ecology* **63**, 1237–1241.

Weiner, J. (1984). Neighbourhood interference amongst *Pinus rigida* individuals. *J. Ecol.* **72**, 183–195.

Westoby, M. (1984). The self-thinning rule. *Adv. Ecol. Res.* **14**, 167–225.

Wiens, J. A. (1977). On competition and variable environments. *Am. Sci.* **65**, 590–597.

Wiens, J. A. (1984). On understanding a non-equilibrium world: Myth and reality in community patterns and processes. *In* "Ecological Communities: Conceptual Issues and the Evidence" (D. R. Strong, D. Simberloff, L. G. Abele, and A. B. Thistle, eds.), pp. 439–457. Princeton Univ. Press, Princeton, New Jersey.

15

The Role of Competition in Structuring Pasture Communities

Roy Turkington   **Loyal A. Mehrhoff**

I. Introduction

There are many interacting factors that influence the structure of plant communities. In attempting to understand community structure, ecologists have pursued two major avenues of research: the initial description of patterns in community structure, and an investigation of the processes and mechanisms that generate the patterns. Processes and mechanisms particularly are not easy to examine in a complex ecological world and include competition (Schoener, 1983) and predation (Connell, 1975) as major actors on a stage defined by habitat heterogeneity and environmental fluctuations. In this chapter, we try to unravel some of this complexity and particularly focus on the role of competition.

The world's grasslands have been classified in many different ways, related chiefly to climate (see Moore, 1964; Wilson, 1978; French, 1979). To address the questions of relevance to this chapter we focus on one grassland type, the managed permanent pasture, although in some cases, examples will be drawn from other grassland types.

These pastures typically contain a set of cooccurring herbaceous perennials and are maintained by grazing or mowing. This pasture community has a wide distribution and occurs throughout northern Europe, most of Britain, New Zealand, and in temperate North America. Pastures and pasture species have been a subject of intensive research effort by ecologists and agronomists (Donald, 1963; Lowe, 1970; Wilson, 1978; Haynes, 1980; Frame, 1986; Turkington *et al.*, 1988). Pastures are subject to intensive grazing (Watkin and Clements, 1978) and are exposed to various types and intensities of disturbance (Platt, 1975; Parish, 1987). The communities found in long-grazed permanent pastures have three important floristic attributes: a large number of coexisting herbaceous species (up to about 50; Snaydon, 1978, 1985; Turkington and Harper, 1979a; Fowler and Antonovics, 1981), the presence of nonrandom species arrangement (Aarssen *et al.*, 1979; Turkington and Harper, 1979b; Thorhallsdottir, 1983; Aarssen and Turkington, 1985a), and high genetic diversity within species (Burdon, 1980a,b; Burdon and Harper, 1980; Gliddon and Trathan, 1985). These patterns are not static, but ever changing with time. Typically, the abundance of pasture and grassland species changes considerably from year to year (Rabotnov, 1966) and there is rapid population flux (Charles, 1961; Sarukhan and Harper, 1973; Williams and Roe, 1975; Fowler, 1984, 1986). An exception is the relatively infertile chalk grassland, which is apparently much more stable (Grubb *et al.*, 1982). Establishment from seed has been reported for nearly all grassland types but is often of irregular occurrence (Sarukhan and Harper, 1973; Williams and Roe, 1975; Parish, 1987). These factors combine to produce a complex system in which competition is likely to have important consequences on community structure.

There have been at least three interrelated problems that have impeded our progress in understanding the role of competition. First, it has been notoriously difficult to demonstrate unequivocally that competition actually occurs in nature and that it is important; this has often been due to poor experimental design (Connell, 1983; Underwood, 1986). In fact, many researchers argue that it is only infrequently important (den Boer, 1986; Silvertown and Law, 1987) and that other factors such as predation (Connell, 1975) and disturbance (Wiens, 1977; Pickett, 1980) are much more important as structuring forces in communities. Often, the evidence accepted as demonstrating competition can be produced by other types of interactions, e.g., apparent competition (Connell, 1983, and this volume). A second problem has been the attempt to infer the process of competition by studying patterns (Thompson, 1980; Colwell and Winkler, 1984); a typical example of this is, when apparent niche differences are detected in field studies, past competition is usually invoked as having generated the observed patterns. This interpretation is based on the assumption that observed differences arose during prolonged periods when the species were coexisting (i.e., past competition generated the pattern, but present competition maintains it). This relationship, or lack of relationship, between the consequences of competition in the past versus that occurring in the present is a critical distinction. Grant (1975) and Connell (1980) make this distinction and argue that such studies do not exclude the possibility that the species may have evolved separately so that, when they later came together, they coexisted because they were already different, and present competition keeps them apart (i.e., past competition was unimportant in generating pattern, but present competition maintains it). Another aspect of this problem has been the neglect in distinguishing between actual genetic change and long-term plastic responses (multigenerational carryover effects). Appropriate breeding experiments are required to determine genetic changes and these have seldom been done.

Much attention has been focused on the evolutionary consequences of competition, both in terms of inter- and intraspecific interactions (MacArthur and Levins, 1967; Gill, 1974; Slatkin, 1980; Arthur, 1982; Pianka, 1983). Intraspecific competition is generally predicted to have a disruptive effect on population distribution and result in broadened resource use (Pianka, 1983). In contrast, interspecific competition for limiting resources (exploitation competition) is predicted to select for divergence in the type, timing, or rate of resource use (MacArthur and Levins, 1967; Lawlor and Maynard Smith, 1976; Pianka, 1983).

Alternative outcomes of competition have been proposed, such as (1) the evolution of interference mechanisms, rather than changes in resource use (Gill, 1974; Aarssen, 1983; Boyce, 1984; Hairston *et al.*, 1987), (2) convergence in resource use (MacArthur and Levins, 1967;

Ghilarov, 1984; but see Lawlor and Maynard Smith, 1976), and (3) that competition between similar species need not necessarily lead to competitive exclusion (Tilman, 1981; Aarssen, 1983; Abrams, 1983; Agren and Fagerstrom, 1984; Ghilarov, 1984).

In an essay on plant competition in pastures, we need first of all to demonstrate that competition is, or has been, occurring. Second, we need to demonstrate that it is having an effect on community structure. In this chapter, we examine the floristic patterns found in one type of plant community, permanent pastures, and try to determine the role that competition has played in establishing or maintaining these patterns. We do this by first looking at large-scale (between species) patterns and then focusing on smaller scale patterns found within species and at the individual plant level. We conclude by integrating these patterns into a picture of competition's role in the pasture community.

II. Between-Species Patterns

Ideally, one would prefer to follow the development of a pasture system from inception (initial sowing) until all patterns have become established. Because of time considerations, this is seldom practical and alternative strategies must be devised. A potentially useful shortcut is to compare and contrast patterns in communities at different stages of succession (Hancock and Wilson, 1976; Roos and Quinn, 1977; Parrish and Bazzaz, 1982; McNeilly and Roose, 1984). While one gains convenience from this shortcut, one also loses resolution, since each pasture community will probably be unique in some aspect of its development due to environmental conditions or to historical differences in management, initial gene pool, and immigration.

This shortcut method has been applied to several pastures in British Columbia ranging in age from 0 to 46 years. Aarssen and Turkington (1985a) recorded the relative abundance of species in three of these pastures from June 1979 to March 1982 when the pastures were initially 2, 21, and 40 years old. Parish (1987) continued these surveys to October 1984. Species composition was not appreciably different between pastures. However, the relative frequencies and stability of the frequencies of these species were different between pastures. Overall floristic characteristics of the three pastures were summarized in an ordination of all surveys based on total percentage cover (Aarssen and Turkington, 1985a) and rooted frequency (Parish, 1987) for each species. Time series lines linking successive sampling events show that the three pastures are readily distinguished, with the youngest pasture showing the most seasonal and yearly variability in botanical composition, while the oldest

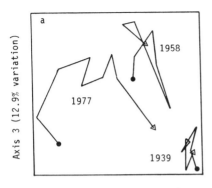
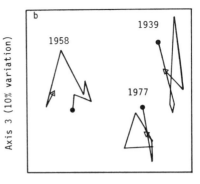

Figure 1   (a) Ordination of percentage cover values for nine surveys from June 1979 to March 1982 (redrawn from Aarssen and Turkington, 1985a), and (b) rooted frequency for eight surveys from July 1982 to October 1984 (redrawn from Parish, 1987) in three pastures in British Columbia. The pastures were sown in 1939, 1958, and 1977. Lines join the points for each successive survey in each pasture, progressing from the first (solid circle) to the last (open arrow) survey. Points which are closer together have a more similar botanical composition than points which are further apart.

pasture is much more stable (Fig. 1a). However, after three additional growing seasons the composition of the youngest pasture has become much less variable (Fig. 1b).

Concurrent with the previous study, Aarssen and Turkington (1985a) recorded species associations as being temporary, seasonal, or stable for the duration of the study. A similar recording of species associations over a 2-year period is reported by Aarssen *et al.* (1979) from pastures in southern Ontario. Both studies show that older pastures have fewer temporary associations and slightly more persistent or stable associations (Table 1). The general picture is that of a transition from an essentially unorganized assemblage of species to a more organized community where the dominant species are involved in more or less permanent associations.

A. Ecological Significance

In order to establish the ecological significance of competition and other factors in structuring pastures, it is necessary to manipulate the system experimentally. Perturbation experiments have been increasingly used to analyze the structure of communities (Allen and Foreman, 1976; Fowler, 1981; Hils and Vankat, 1982; Moore, 1982) and to determine relationships between species (Berendse, 1983). This involves the removal of existing populations and then monitoring the response of remaining community members. Selective removal experiments are useful

Table 1   Number of Significant[a] Temporary, Seasonal, and Stable Associations Detected between Plant Species in Three Different-Aged Pastures in British Columbia and Southern Ontario

Age of Pasture at Initial Survey (yr)	Between-Species Associations		
	Temporary	Seasonal	Stable
British Columbia[b]			
2	52	2	0
21	57	0	4
40	35	1	6
Southern Ontario[c]			
13	37	2	2
18	27	1	3
31	26	3	4

[a] $p < 0.05$.
[b] From Aarssen and Turkington (1985a).
[c] From Aarssen et al. (1979).

as general indicators of changing patterns in community structure and in demonstrating that competition is occurring. However, the results are often difficult to interpret and the magnitude of a species response cannot be used as an estimate of the magnitude of competition (Goldberg and Werner, 1983; Bender *et al.*, 1984).

Fowler (1981) selectively removed species singly, and in groups, from a 30-year-old grassland in North Carolina. After 2 years, 14 (19%) of the 72 pairwise effects showed significant response to removal. Parish (1987) conducted a selective removal study in a 6-year-old and a 44-year-old pasture in British Columbia to compare time-related responses. From four replicate 1 m² plots in each pasture, she removed *Trifolium repens, Holcus lanatus,* or *Lolium perenne* in single-species treatments, and in all possible two- and three-species combinations, a total of seven treatments plus a control. The experiment ran for 2 years after initial removals, with regrowth of the removed species cleared after 1 year. Because different amounts of material were removed in different treatments, all responses were standardized and expressed as an expansion rate calculated by using the abundance of the removed species and the final abundance of the respondent species. In the young pasture, 4 of the 11 most abundant species showed significant response to single-species removals; this represented 17% of the maximum number of responses to single-species removals. In contrast, in the older pasture, there were no responses to single-species removals, but two species responded to multiple removals (3% of maximum responses).

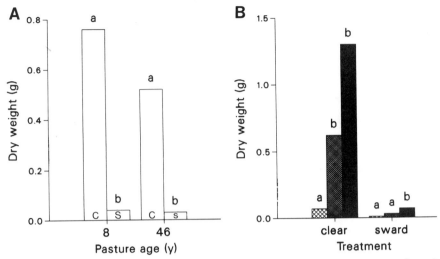

Figure 2   Growth of *Trifolium repens* and *Lolium perenne* from three pastures of age 0 (seed), 8, and 46 years planted into 8- and 46-year-old pastures. (A) Average performance of genets planted into cleared plots (c) and into undisturbed sward (s). (B) Average performance of genets of age 0 (⊠), 8 (▨), and 46 years (■) planted in cleared plots and undisturbed sward. Bars sharing the same letter are not significantly different ($p > 0.05$, median test).

Another approach to the study of competition is the controlled addition of transplants into both competitive and noncompetitive situations. Use of this technique by Turkington and Harper (1979c) and ourselves (Mehrhoff and Turkington, unpublished observations) demonstrates that the negative effects of neighbors are a very real phenomenon in pasture situations (Fig. 2A). However, the experimental design does not allow us to determine whether these results are due to actual competitive interactions or to apparent competition.

B. Evolutionary Consequences

In at least some instances, competition is apparently an important organizing force and it is tempting to invoke a competition argument to explain the stabilizing of these pasture communities. However, it is not clear that the ecological effects of competition are having an evolutionary consequence.

Parrish and Bazzaz (1982) investigated changes in traits of species characteristic of abandoned agricultural fields of different successional age. The response of a species along an environmental gradient, an estimate of its niche breadth, was determined for species from early (annual), mid (early perennial), and late (prairie) successional communities. It was generally found that early successional species have broader and more overlapping response breadths (niches) than do later successional species. They also found that later successional species showed less difference between growth in mixture versus pure stand than did early successional species, suggesting that niche differentiation had occurred in late successional species. However, it cannot be determined from these data if competition (or historical niche differences) were actually responsible for the patterns observed in the later successional stages.

Martin and Harding (1981) collected seed of *Erodium cicutarium* and *Erodium obtusiplicatum*, common plant species in Californian annual grasslands, from sites where the two species coexisted (= sympatric sites) and from sites where they were not growing together (= allopatric). Competition experiments were done in a greenhouse using a de Wit replacement design. The question they addressed was whether plants from sympatric situations would exhibit evidence of coevolution (overyielding) whereas those from allopatric situations would not. Specifically, the hypothesis was that the presence or absence of coexistence in the history of a population would influence its fitness in interspecific competition. They showed that total seed output and total reproductive rates of sympatric mixtures were higher than those of allopatric mixtures, suggesting that there has been evolution (= coevolution?) in response to competition by the two species from the sympatric sites.

Evans *et al.* (1985) showed a similar pattern. Coexisting individuals of *T. repens* and *L. perenne* were sampled from five long-established pastures in Switzerland, France, and Italy, and a 7-year-old pasture in southern England. *T. repens* was grown in two-species mixtures either with its coexisting *L. perenne* or with one of two other *L. perenne* companions, both cultivars from the Welsh Plant Breeding Station, Aberystwyth. Despite the varied origins of the plants, by the end of the second harvest year, those mixtures based on coexisting populations yielded over 20% more on average than the other mixtures. The yield of all five *T. repens* populations was highest when grown with their coexisting grass companion (Table 2). Joy and Laitinen (1980) demonstrated a related phenomenon of overyielding in sympatric *Phleum pratense* and *Trifolium pratense*.

The results of these last three studies are consistent with the prediction of interspecific competition leading to resource partitioning (niche divergence). It would, however, be extremely interesting to follow these studies with controlled field manipulations.

Aarssen and Turkington (1985d) examined the competitive relationships between five pasture species. All possible pairs of *Dactylis glomerata, H. lanatus, L. perenne, Poa compressa,* and *T. repens* were collected where the species pairs occurred in close proximity in pastures that were 2, 21, and 40 years old. Each clone was grown with its natural neighbor-

Table 2 Dry Matter Yield of *Trifolium repens* and *Lolium perenne* When Grown in Mixtures[a]

Origin of T. repens	L. perenne Companion	Dry Matter Yield (g m⁻²)		
		T. repens	*L. perenne*	Mixture
Switzerland	1	121	211	332
	2	344	275	619
	Coexisting popn	464	353	817
Switzerland	1	596	247	843
	2	675	266	941
	Coexisting popn	749	320	1069
Italy	1	636	218	854
	2	793	317	1110
	Coexisting popn	840	307	1147
France	1	550	268	818
	2	644	339	983
	Coexisting popn	704	345	1049
England	1	645	227	872
	2	755	285	1040
	Coexisting popn	787	277	1064
Overall means	1	510	234	744
	2	642	296	938
	Coexisting popn	709	320	1029

[a] Mixtures were based on *T. repens* growing with its coexisting *L. perenne* and with two other *L. perenne* companions. Data from Evans *et al.* (1985).

ing clone in clipped, but ungrazed, experimental field plots. Differences in competitive relations between two species from pastures of different ages were assessed by comparing the total mixture yield, and by comparing the yield quotients of the highest and lowest yielding components in each mixture. For 5 of the 10 species pairs, there was no significant change in total mixture yield from youngest to oldest pasture (no increased overyielding). There was, however, a significant trend in pairs from the oldest pasture toward a more equal contribution to the mixture total by the two components. From these results they concluded that competition was an important force, but that the consequence was frequently a balancing of competitive abilities without niche differentiation. However, additional data are required to substantiate these conclusions.

We (Mehrhoff and Turkington, unpublished observations) have attempted to resolve some of these apparent discrepancies in evolutionary outcome by examining differentiation in pasture plants. Reciprocal transplants of *T. repens* and *L. perenne* were conducted in three different-aged pastures (0 years, near Vancouver, British Columbia, and 8 and 46 years, both 80 km away). The experiments were planted in spring, clipped or grazed during summer and fall, allowed to grow until May, then harvested. *T. repens* showed a significant age-related increase in growth when planted into permanent pastures (Fig. 2B), indicating that microevolution can apparently occur quite rapidly. Both cleared and uncleared (sward) treatments produced similar patterns. Further comparison of grazed and ungrazed experiments shows that the rank order of plant weight (Fig. 3) was more influenced by grazing or environmental conditions than by the presence of an interspecific competitor. *L. perenne* rank orders (Fig. 3C,D) showed the effects of both grazing/environment

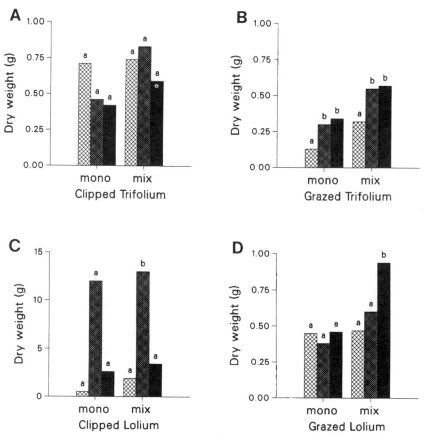

Figure 3 Comparison of mean dry weight production of *Trifolium* (A,B) and *Lolium* (C,D) from pastures of age 0, 8, and 46 years. Plants were grown in monoculture and in mixture under (A,C) ungrazed, but clipped, conditions at Vancouver, British Columbia and (B,D) under grazed conditions in pastures near Aldergrove, British Columbia. Within a graph, those bars sharing a common letter are not significantly different ($p < 0.05$, median test). Bars as in Fig. 2.

and interspecific neighbors. Reasons for site-related differences in rank order are not readily apparent and could be due to a multitude of inter-related factors such as grazing, climatic conditions, soil fertility, and microbial interactions. However, grazing may be the most important factor, since it is the process which arrests the pasture from developing into shrub and forest, presumably by suppressing competitive dominants.

The evolutionary consequences of competitive interactions were tested in another series of experiments using some of the same species as in the Aarssen and Turkington (1985d) studies: *D. glomerata, L. perenne, H. lanatus,* and *T. repens.* Genets of each species were collected from different pastures ranging in age from 0 (freshly sown) to 46 years. Monocultures and two-species mixtures were planted in common garden and field manipulative studies. The results showed that, for each of the three species combinations, total ramet production was significantly greater in mixtures from older pastures than from younger pastures (Fig. 4). Relative yield (RY) (sensu de Wit and van den Bergh, 1965) was used to decouple genotypic increases in yield with pasture age from changes in competitive performance or resource utilization. Since RY compares performance in mixture with performance in pure stand, it can discriminate between noncompetitive and competitive changes and can also be used to estimate the relative importance of intra- versus interspecific interactions. The summation of the RYs for both species is the relative yield total (RYT). Aarssen (1983) and Berendse (1983) use

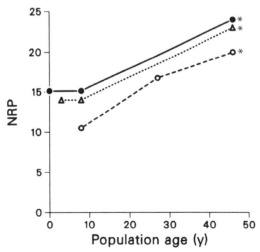

Figure 4 Total ramet production (NRP) of two-species mixtures of *Lolium* with *Dactylis* (△), *Holcus* (○), and *Trifolium* (●) from various aged pastures. Significant ($p < 0.05$, t-test) changes in ramet production with age of pasture of origin are denoted by an asterisk (*).

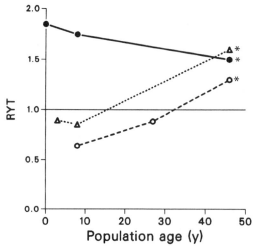

Figure 5 Relative yield total (RYT) of two-species mixtures of *Lolium* with *Dactylis* (△), *Holcus* (○), and *Trifolium* (●) from various aged populations. Significant ($p < 0.05$, t-test) changes in RYT with age of pasture of origin are denoted by an asterisk (*).

RYT as a measure of overlap in resource use by the two species and, thus, as an estimate of niche overlap.

Combinations of *D. glomerata* with *L. perenne*, and *H. lanatus* with *L. perenne* from younger pastures were more negatively affected by interspecific competition and showed significant increases in RYT with pasture age (Fig. 5); a reduction in niche overlap is implied. Both *T. repens* and *L. perenne* were more affected by intraspecific competition when mixed together and showed a significant decrease in RYT, which is interpreted as a slight increase in niche overlap. Reexamination of competitive relationships by Aarssen (1988) also provides evidence of niche divergence in older pastures. These results are consistent with the prediction that interspecific competition (Slatkin, 1980; Pianka, 1983) will promote niche divergence while intraspecific competition results in niche expansion. They also show that competition plays an important evolutionary role in community development.

In the case of *T. repens* and *L. perenne,* grazing, or some other noncompetitive factor, appears to be a major factor generating age-related patterns. If competition were the dominant interaction, then one should see (but does not see) an age-related increase in size and survivorship in *both* the grazed and ungrazed mixtures. This does not mean that competition is unimportant or that it does not generate community structure, only that it does so within the parameters established by other factors. Two results indicate that plant interactions are clearly important: (1) *T. repens* planted into cleared, but grazed, plots had significantly greater survivorship and growth (Fig. 2) than did the same genets planted into grazed grass patches, and (2) age-related increases in *L. perenne* growth were evident only when grown in mixture with *T. repens,* not in monoculture (Fig. 3).

III. Within-Species Patterns

Trifolium repens is the only pasture species which has received much attention at this scale of pattern. From a sheep pasture in north Wales, Turkington and Harper (1979c) collected ramets of *T. repens* from within patches dominated by each of four different grasses. The *T. repens* populations were grown in all possible combinations with the four species of grass in greenhouse flats containing a standard potting compost. Each clover tended to grow best with the grass from which it had originally been sampled, i.e., a principal diagonal effect. It is evident that the different grasses impose different constraints on the growth of *T. repens* and that the *T. repens* population has differentiated into subpopulations defined by the identity of the grass neighbor.

A. Ecological Significance

In addition to the greenhouse studies, Turkington and Harper (1979c) also replanted the *T. repens* from the four grass patches back into patches of each of the four grasses in the pasture. Again, each clover grew best in its natural grass patch; the "native" clovers outperformed alien clovers. While competition is the most likely explanation for the observed patterns, apparent competition (Connell, this volume) due to the presence of a common grazer or soil microorganism has not been ruled out. Given other studies of *T. repens* in this same pasture (Cahn and Harper, 1976; Burdon, 1980a,b; Turkington and Burdon, 1983; Gliddon and Trathan, 1985), it seems likely that the observed patterns are due to competition and probably have a genetic basis, but again, this is equivocal.

From an old pasture in British Columbia, Evans (1986) collected 100 ramets of *T. repens* from each of four neighborhoods dominated by *D. glomerata, H. lanatus, L. perenne,* and *P. compressa.* These were transplanted into a common garden, without competitors, and later scored for 12 morphological characters. For 10 of the characters, a significant proportion of the variation between sampled ramets was accounted for by the identity of the neighboring grass species with which the *T. repens* ramet had been growing in the pasture (Table 3). These same clovers were grown for 2 more years in the common garden and were then

Table 3 Summary of Analyses of Variance for Measured Morphological Characters of *Trifolium repens* from Neighborhoods Dominated by *Dactylis glomerata* (D), *Holcus lanatus* (H), *Lolium perenne* (L), and *Poa compressa* (P)[a,b]

	1982			1984	
	Percent Variation Accounted for	Significance[c]	Multiple Range Test	Percent Variation Accounted for	Significance[c]
Root weight	4.8	**	LD HP	0.0	NS
Shoot weight	4.8	**	LD HP	0.0	NS
Total weight	5.3	**	LD HP	0.0	NS
Primary stolon number	5.7	**	DL HP	Not measured	
Total stolon number	2.7	*	DL HP	0.0	NS
Internode number	0.7	NS	Not measured		
Primary stolon length	0.7	NS		0.17	NS
Secondary stolon length	2.0	*	LD PH	Not measured	
Total stolon length	Not measured		—	0.0	NS
Internode length	2.4	*	LD HP	0.0	NS
Petiole length	20.2	**	LD HP	0.0	NS
Leaf weight	11.8	**	—	0.0	NS
Leaf length	19.6	**	—	0.0	NS

[a] Data from Evans (1986) and Evans and Turkington (1988).

[b] Values represent the percentage variation in measured characters accounted for by the variable "neighbor." The 1982 measurements were made shortly after the original collections, and the 1984 measurements were made on the same material four generations (2 years) later. The multiple range tests (Duncan's) are on the means for measured characters classified by neighborhood type from which the ramets of *T. repens* were collected; means have been ranked from smallest to largest, and underlined sets of means are not significantly different ($p > 0.05$).

[c] *, $p < 0.05$; **, $p < 0.01$; NS, not significant.

scored for morphological characters as before. In no case was a significant proportion of the variation in characters now found to be due to the previous grass neighbors (Table 3). The clear implication of this study is that divergence patterns in *T. repens* morphology have a plastic rather than a genetic basis. It seems probable that these patterns may only be expressed under specific conditions such as the presence of grazing, competition, or certain microorganisms.

To examine these effects further, three additional sets of studies were done, two reexamining competitive and genetic aspects and one investigating the role of microsymbionts, specifically the nitrogen fixer *Rhizobium leguminosarum* biovar *trifolii.*

Biological nitrogen fixation is one of the most important factors determining productivity in pastures, and this will have a significant impact on growth of *T. repens* and consequently on the growth of neighboring grasses. Thus, in searching for the processes at work in clover–grass interactions, the third major partner of this association, the nodule bacterium *R. leguminosarum* biovar *trifolii*, must be considered. Productivity of *T. repens* depends on the precise association between plant cultivar and *Rhizobium* strain (Mytton, 1975; Turkington *et al.*, 1988). On average, a more productive symbiosis results when plants are nodulated with *Rhizobium* strains isolated from their own nodules than with *Rhizobium* strains isolated from different cultivars. In addition, the particular strain of *Rhizobium* used in hill-seeding trials had a significant impact on the growth of different *T. repens* cultivars—the outcome of competition between plant cultivars can be reversed by changing the strains of *Rhizobium* (Young and Mytton, 1983; Mytton and Hughes, 1984; Young *et al.*, 1986). The nodulating ability of *Rhizobium* strains may be altered by environmental conditions and soil type (Jones and Hardarson, 1979; Newbould *et al.*, 1982). The *T. repens–Rhizobium* symbiotic relationship is further complicated because Robinson (1969a,b) showed that individuals of *T. repens* tend to be nodulated by strains of *R. trifolii* that are more effective in fixing nitrogen. Masterson and Sherwood (1974) demonstrated that, when *T. subterraneum* and *T. repens* are presented with an array of *Rhizobium* strains, they tend to select the strain originally isolated from that species. The significance of root microorganisms in the ecosystem, their interactions with each other and with associated plants has been reviewed by Newman (1978), Gaskins *et al.* (1985), and Turkington *et al.* (1988).

As a first step in using these principles to help understand the competitive relations between neighbors in pastures and their community-structure consequences, J. D. Thompson, R. Turkington, and F. B. Holl (unpublished observations) conducted a series of studies to investigate various parts of the grass–clover–*Rhizobium* system. It is our thesis that grasses indirectly influence the growth of neighboring *T. repens* by their direct, or indirect, effect on soil microorganisms—this would fit into Connell's definition of apparent competition. Ramets of *T. repens* were collected in a 45-year-old pasture from patches dominated by either *D. glomerata, H. lanatus*, or *L. perenne*. At each collection site, tillers of the dominant grass were also collected (a matched pair) along with root nodules from the *T. repens*. A factorial experiment was done in sterilized soil in a greenhouse in which the three clover "types" (collection sites) were grown in all possible combinations with their three *Rhizobium*

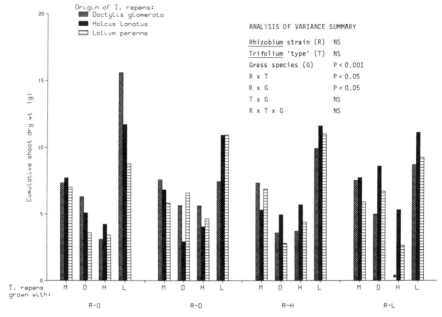

Figure 6 Cumulative shoot dry weight of individual *Trifolium repens* clones, sampled from patches dominated by three grass species, when grown in monoculture (M) and with three grasses, with three different *Rhizobium* strains, and an uninoculated control. R-O is uninoculated, and R-D, R-H, and R-L are *R. trifolii* strains isolated from *T. repens* growing in patches of *Dactylis glomerata*, *Holcus lanatus*, and *Lolium perenne*, respectively. One of the RL treatments (*) has a zero value. Data from J. D. Thompson, R. Turkington, and F. B. Holl (unpublished observations).

strains, in monoculture and in mixture with the three grasses. There was a significant difference in *T. repens* yield across treatments (Fig. 6), and the only significant main effect was the influence of the different grass species. There were also significant *T. repens* "type" × *Rhizobium* interaction effects, and a grass species × *Rhizobium* interaction; the grass × *Rhizobium* interaction is crucial to our arguments on individual-level patterns in the next section. These studies tend to add weight to the arguments concerning the divergence of *T. repens* populations in response to neighboring grasses. However, they also raise two other issues. First, the *T. repens* collected from the *H. lanatus* patch was higher yielding than *T. repens* from other patches (Fig. 6); this agrees with the first survey measurements by Evans (1986). Second, the growth of *T. repens*, regardless of its origin, is influenced by the strain of *Rhizobium* with which it has been inoculated; the relative abundance of various *Rhizobium* strains is influenced by the species of grass. Thus, because the *T. repens* in the Evans (1986) study was grown without competitors, there may have been no differential influence on the soil environment, so the microorganism populations in all plots may have been relatively similar. This would create conditions in which clovers, although from different origins, would be similar in their growth; hence, the disappearance of character differences between surveys 1 and 2 (Evans, 1986). The differences detected in the first survey presumably partially reflect a carryover effect from the original collection sites. However, the J. D. Thompson, R. Turkington, and F. B. Holl (unpublished observations) study showing that *T. repens* from *H. lanatus* patches tends to be the highest yielding regardless of neighboring grass or *Rhizobium* strain present lends some support for genetically based differences. There are still many unresolved issues concerning the genetic basis of the neighbor-specific patterns of differentiation observed in *T. repens*.

Another study involving British Columbian pastures and potential conditioning effects involved *T. repens* collected from pastures of 0, 3, 8, 27, and 46 years of age. Plants were grown under greenhouse conditions for 20 months to minimize the conditioning effects identified by Evans and Turkington (1988). During this time, genets were propagated in two different treatment regimes: without competitors and with the original cohabiting grass species. At the end of the propagation period, all genets from both treatments were transplanted into both pure stands of *D. glomerata* and pure stands of *L. perenne*. After 7 months the individual transplants were harvested for above-ground dry weight. In contrast to the Turkington and Harper (1979c) study, each clone did not grow best with the grass species with which it had originally cohabited (Fig. 7), indicating a lack of species-specific subpopulations in the British Columbian pastures.

B. Evolutionary Consequences

Due to apparent discrepancies between the studies from British Columbia and the original study by Turkington and Harper (1979c), the question of the evolution of competitors was reinvestigated. This was done by using Connell's (1980) proposed method in the old pasture in north Wales. This design can determine whether observed patterns of niche differentiation are due to competition or to some other cause. Ideally, one would prefer to observe the species both before and after contact, but in nature this is rarely possible. As an alternative, Connell (1980) proposed that we can observe the two species, the two presumed competitors, in areas where they cohabit (= sympatry) and in areas where they do not cohabit (= allopatry), assuming the former followed the latter and that present isolated conditions represent the precontact conditions. The terms "sympatry" and "allopatry" should only apply to populations that are within or not within, respectively, each other's breeding range. The

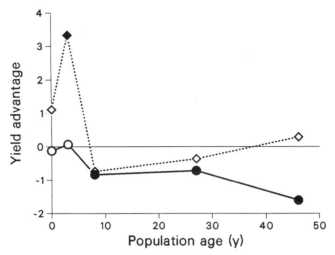

Figure 7 Comparison of proportional yield advantage of *Trifolium repens* from populations of different ages (0, 3, 8, 27, and 46 years), showing the advantage of native over alien
T. repens when grown with *L. perenne* (○) and *D. glomerata* (◇) under clipped, but ungrazed,
conditions at Vancouver, British Columbia. Filled symbols indicate a significant ($p < 0.05$,
Mann–Whitney U test) deviation from 0 (no advantage).

term "cohabit" (Harper *et al.*, 1961) is more appropriate than sympatric
for the situation described below, but to aid comparison with Connell's
(1980) paper, sympatric and allopatric will be used.

One pair of neighbors that was assumed to compete was *L. perenne* and
T. repens, and results from experiments involving these species are given
here. A site was established in which *L. perenne* and *T. repens* had 96 and
48% cover, respectively; this was the sympatric site. An allopatric site was
established in which *T. repens* was relatively abundant (40%), and the
presumed competitor, *L. perenne*, was at relatively low abundance (9%).
Various treatments using transplants between allopatric and sympatric
sites, replants, species removals, and controls, were carried out (Fig. 8).
Throughout, it is assumed that growth is related to niche width. As
corroboration with an earlier study (Turkington and Harper, 1979c),
differentiation was again demonstrated within the *T. repens* population
in response to the identity of its grass neighbor; each *T. repens* grew best
in its home site (i.e., 4 > 1 and 3 > 6 in Fig. 9 and Table 4). To determine
if competition is the most likely mechanism generating the observed
divergence, two conditions must be met. First, *T. repens* transplants from
the allopatric site to the sympatric site should grow better when the
presumed competitor, *L. perenne*, has been selectively removed from the
sympatric site compared to plots where *L. perenne* was not removed (i.e.,

Figs. 8 and 9, treatment 2 > 1). This tests for competition happening in the present. Second, in plots in the sympatric site where *L. perenne* has been selectively removed, *T. repens* transplants from the allopatric site should grow better than *T. repens* replants in the sympatric site (i.e., Figs. 8 and 9, treatment 2 > 5). The assumption underlying these predictions is that the two species living in sympatry will have diverged in their resource requirements and so will be relatively less competitive toward each other than their congeners in allopatric populations, i.e., sympatric *T. repens* should not be able to exploit resources made available by the removal of the *L. perenne* to the same extent as allopatric *T. repens* can.

It is not possible to establish that a characteristic is genetically controlled without at least a few generations of breeding experiments. However, appropriately controlled and replicated field experiments can provide strong circumstantial evidence. If we are to conclude that evolution has occurred rather than merely a plastic or conditioning response, two conditions must be met. First, with its natural competitor, *L. perenne*,

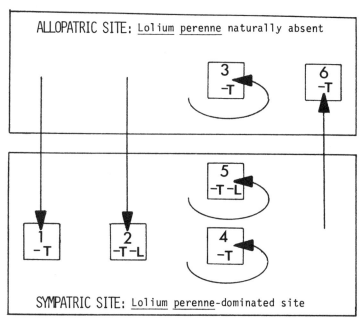

Figure 8 Outline of experimental design to test for the evolution of competitors. Treatment numbers are in the boxes. All treatments in which *Trifolium repens* or *Lolium perenne* was removed (−T, −L, respectively) had the indigenous populations of these species removed prior to the introduction of experimental *T. repens*. In all treatments, *T. repens* was collected from the base of the arrow and transplanted, or replanted, at the head of the arrow into the various treatments. Treatment numbers 1 through 6 are consistent with those used by Connell (1980).

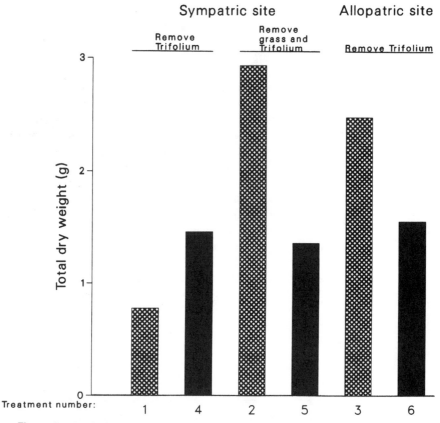

Figure 9 Total plot dry weight of *Trifolium repens* collected from an allopatric site ▨ and a sympatric site ■, and then transplanted or replanted into plots with different treatments (see Fig. 8 for details of treatments). Probability levels of all pairwise comparisons are given in Table 4a.

removed, the growth of the sympatric individuals of *T. repens* should not increase relative to its growth with *L. perenne* still present (i.e., Figs. 8 and 9, treatment 5 not different from 4). Second, sympatric individuals of *T. repens* should show no significant difference in growth when transplanted to the allopatric site compared to being replanted within the sympatric site (i.e., Figs. 8 and 9, treatment 6 should not be different from 4); but, if there is a difference, then growth in the allopatric site should be less than that of the natural allopatric population (i.e., Fig. 9, treatment 6 < 3). The assumptions underlying these predictions are that if a genetic change causing divergence had occurred in sympatry due to interspecific competition in the past, then, on removal of the competing species, the sympatric population should *not* quickly expand or shift its

Table 4 Pairwise *t*-Tests Compare Whether the Growth of
Trifolium repens is Different, or Equal, in Any
Two Treatments[a]

(a) All pairwise comparisons

1	1				
2	0.0043	2			
3	0.0436	0.5921	3		
4	0.0855	0.0943	0.3219	4	
5	0.0528	0.0332	0.2696	0.9345	5
6	0.0306	0.1121	0.2025	0.7243	0.5623

(b) Comparisons to show the evolution of competitors

Conditions to Demonstrate Competition		Conditions to Demonstrate a Probable Genetic Basis		
1 < 2	2 > 5	5 = 4	4 = 6, if not,	6 < 3
0.0043	0.0332	0.9345	0.7243	0.2025

[a] The treatments are numbered 1–6, and the values express the probability of no difference between means.

niche position back to what it presumably had been before evolution, or, if there is a shift, it should be as great as the allopatric population. The allopatric population is assumed to have the same niche position that the sympatric population used to have before evolution. If the sympatric population of *T. repens* did *not* change genetically after it had met and competed with the *L. perenne*, then, when *L. perenne* is experimentally removed, *T. repens* from the sympatric site would immediately bounce back to the niche position of *T. repens* from the allopatric site, the assumed former condition. So, the investigation is to compare the niche of *T. repens* from the sympatric site with and without *L. perenne*. If there has been a genetic change in *T. repens* from the sympatric site due to competition with *L. perenne*, then when *L. perenne* is removed the niche of the sympatric *T. repens* should either not change, or if it does, should remain narrower than the natural allopatric population of *T. repens*.

In this study, all conditions were met (Fig. 9 and Table 4), indicating that the population divergence observed in *T. repens* in association with *L. perenne* has been generated by competition in the past, is being maintained by present competition, and probably has an underlying genetic basis.

A simpler, but not so rigorous measure, can be made to address these questions. Using the aforementioned assumptions, we can compare the relative impact of a competitor, *L. perenne*, on both the allopatric *T. repens*

(2-1) and the sympatric *T. repens* (5-4). If the sympatric and allopatric *T. repens* show similar responses to the presumed competitor, then competition is not generating the observed pattern. If, however, the allopatric *T. repens* shows a much greater response to the absence of a competitor, (2-1) > (5-4), then competition is a major force generating the observed patterns, and present competition maintains them. In all cases, (2-1) was greater than (5-4), thus corroborating the more rigorous analyses. Two other concurrent experiments (Turkington, 1989) were done using *Agrostis capillaris* and *H. lanatus* as the presumed competitors. While the neighbor specificity patterns were again shown to have a "genetic" basis, competition in these two studies may not always have been the primary factor generating the divergence.

The north Wales pasture in this study is about 100 years old (Peters, 1980) and has large areas dominated by single grasses; the *L. perenne*-dominated site is about 150 m² (Turkington and Harper, 1979a; Thorhallsdottir, 1983). Here, during its lifetime, the individual *T. repens* will not extend its stolons through many different environments, but will sample its environment in a coarse-grained manner, and may spend many generations coexisting with one species of grass. In contrast, the old pasture in British Columbia is about 40 years old (Evans, 1986). This pasture has a similar botanical composition to the Welsh pasture but has very different patterns of species distribution. The pasture is a complex mosaic of grass patches where most patches are probably less than 1 m² (Evans, 1986). Here, an individual *T. repens* will extend its stolons through many different neighborhoods and sample the different biotic environments in a fine-grained way. Thus, one might expect to detect genetically based microevolutionary changes in the *T. repens* population in the Welsh pasture in response to different coarse-grained patches. In contrast, the British Columbia pasture population of *T. repens* should respond in a plastic manner to the fine-grained mosaic environment.

IV. Individual-Plant Patterns

To date, the individual-plant level has been largely ignored, but understanding of the events at this scale may be crucial to assessing which individual genotypes can continue to participate in the evolutionary game—it may or may not tell us much about larger patterns of community structure.

Turkington and Harper (1979c) demonstrated that individuals of *T. repens* can show great specificity to different species of grass neighbors. Aarssen and Turkington (1985b) looked at the same phenomenon, but using different genotypes of grass neighbors. They collected neighboring pairs of genotypes of *L. perenne* and *T. repens* from four different locations in a 40-year-old pasture. All 16 possible interspecific combinations of *L. perenne* and *T. repens* were planted together in pots, the mixture clipped regularly for 1 year, and the cumulative yield of each of the two species recorded. For each pair, the yield of *T. repens* was generally highest when grown in mixture with its natural *L. perenne* neighbor, but each grass genotype had quite low yield when grown with its natural clover neighbor (Fig. 10). If such local specialization is a genetic component of fitness, rather than an environmental conditioning, one would expect to find differences in local gene frequencies within a species. Gliddon and Trathan (1985) estimated the genetic variation within seven subpopulations of *T. repens* and *L. perenne* from the 100-year-old pasture in Wales. All 15 polymorphic loci studied in *T. repens* and 21 out of 23 in *L. perenne* showed significant heterogeneity of gene frequency, indicating that there was genetic differentiation between the samples. Their subsequent field experiments using transplants from four sites clearly showed that both *T. repens* (especially for number of survivors) and *L. perenne* (for dry weight) are highly locally specialized (Fig. 11). A final competition experiment using plant material from two of the sites showed a significant leading diagonal effect, and it is apparent that there is an interaction between *L. perenne* and *T. repens* genotypes.

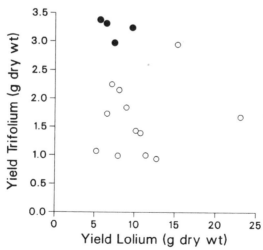

Figure 10 Yield of *Trifolium repens* and *Lolium perenne* when genotypes of the two species are grown in mixture. Genotypes of both species were collected as four pairs of physical neighbors and all four *T. repens* genotypes were grown in all combinations with the four *L. perenne* genotypes (16 mixtures). Natural pairs are designated by ● and mismatched pairs by ○. From data of Aarssen and Turkington (1985b).

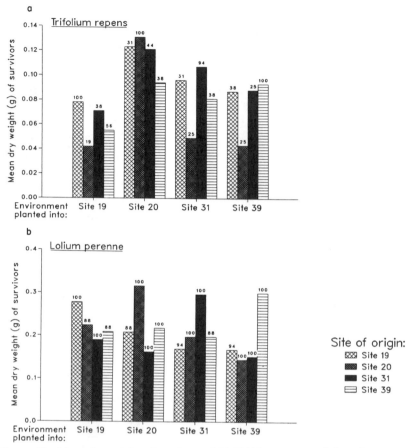

Figure 11   Mean dry weight of survivors of (a) *Trifolium repens* and (b) *Lolium perenne* collected from four sites and replanted into their site of origin and also transplanted into the other three sites. Numbers on top of bars are percentage survival values. Drawn from data of Gliddon and Trathan (1985).

Kelley and Clay (1987) used two grasses, *Danthonia spicata* and *Anthoxanthum odoratum,* to show that naturally co-occurring genotypes of these species differed in interspecific competitive ability. They collected eight different genotypes of each species from an area of species overlap in a 40-year-old field in North Carolina. The eight genotypes of *A. odoratum* were planted with each of the eight *D. spicata* genotypes, and each of the 64 combinations transplanted back into the same area of the field from which they were collected. The ramets were scored for survival, number of vegetative tillers, and number of reproductive tillers after 1 and 2 years. The competitive performance of a given genotype often de-

pended on the genotypic identity of the competing species, especially in
D. spicata, but there was not a clear principal diagonal effect. The results
from their study are compatible with a competitive explanation for the
maintenance of genetic variation and suggest that species interactions
have an important genetic component.

A. Ecological Significance

In an attempt to determine the role of microorganisms in the specializa-
tion phenomena described above, Chanway (1987) collected ramets of
T. repens from *L. perenne*-dominated areas in the old pasture used in the
Aarssen and Turkington (1985b) study. At each collection site, tillers of
L. perenne were also collected, along with root nodules from the *T. repens*.
A factorial experiment was done in sterilized soil in a greenhouse in
which the three factors were the origin of *T. repens* ramets, *L. perenne*
tillers, and *Rhizobium* strains—ramets, tillers, and strains from the same
collection site constituted a "matched" group. To corroborate the patterns observed by Aarssen and Turkington (1985b), we would have predicted a yield advantage to ramets of *T. repens* when growing with
matched *L. perenne* tillers; this effect was detected but it was not significant (Fig. 12). When *Rhizobium* from homologous sites is used to inocu-

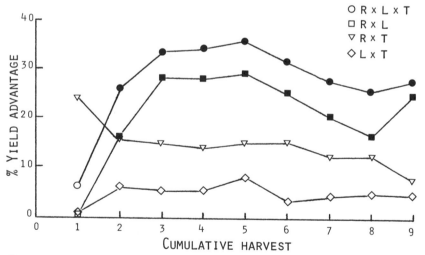

Figure 12 The influence of *Lolium perenne* genotype and *Rhizobium leguminosarum*
biovar *trifolii* strain on the dry weight yield of *Trifolium repens*. The yield advantage is a
comparative measure of the percentage increase in yield of *T. repens* in pots containing
matched combinations of *L. perenne* (L) and/or *Rhizobium* (R) compared with the mean yield
of all mismatched combinations. Solid symbols signify a significant yield advantage ($p <
0.05$). Data from C. P. Chanway, F. B. Holl, and R. Turkington (unpublished observations).

late the mixture (Fig. 12), the effect becomes very pronounced, with homologous groups outyielding nonhomologous groups by up to 35%. However, a *T. repens* yield advantage of up to 30% can also be achieved when only homologous *Rhizobium–L. perenne* combinations are used, regardless of the identity of the *T. repens*. This indicates that it is the specific associations of *Rhizobium–L. perenne* genotypes that have the greatest impact on *T. repens* yield, even though the *Rhizobium* is symbiotic with *T. repens*. It is noteworthy that a significant grass × *Rhizobium* interaction effect was also detected in a previously described study (J. D. Thompson, R. Turkington, and F. B. Holl, unpublished observations; Fig. 6) when different species of grass were used rather than different genotypes of the same grass. Both studies infer that some of the observed patterns in pastures are generated by grasses directly, or indirectly, influencing soil microorganism populations, which in turn influence the growth of their associated plants. Specifically, in these studies, *L. perenne* promotes the proliferation of those strains of *Rhizobium* that result in smaller *T. repens*, while *H. lanatus* indirectly promotes larger *T. repens*.

B. Evolutionary Consequences

Given Darwin's emphasis on the fate of individuals as the cornerstone of natural selection, it is surprising that we know so little about how the events occurring at this scale are relevant to influencing community structure. Numerous studies have monitored the fate of individuals in pastures, but mostly without regard for the population genetic consequences. For example, Charles (1961) followed the population size of *D. glomerata, L. perenne,* and *Phleum pratense* in a newly sown pasture and documented up to 90% mortality of individuals within the first year of sowing. In addition, McNeilly and Roose (1984) documented a decline from about 40 different genotypes of *L. perenne* per 0.25 m^2 in a 10-year-old pasture to 5 in 40-year-old pasture. The reason for the deaths of these individuals and the loss of variability may be related to many factors such as grazing, competition, pathogens, and variable abiotic factors, but it is not clear what role competition plays in this sorting process. It is also unclear how these events affect the evolution of community structure.

V. Conclusions

This chapter specifically focused on the community structure of pastures, a system that was originally chosen for study because of its apparent simplicity! It is evident that the structure of pastures is strongly influenced by inter- and intraspecific competition, grazing, abiotic factors, and by microorganisms, specifically *Rhizobium leguminosarum* biovar *trifolii*. Thus, we must assess which aspects of the community can, or cannot, be explained by competition, evaluate the relative impact of other factors on competition, and ultimately determine how this explains some aspects of community structure.

A. Competition

An attempt to provide a definitive statement on the role of competition in structuring pastures is difficult. Few competition studies have clearly and unequivocally concluded that competition is the major selective force in pasture communities, but it is clear that competition is of at least some importance and that it does contribute to community structure. As a first step then, it is useful to know that some community patterns are influenced and explained by competitive interactions. These patterns may or may not have a genetic basis and may or may not lead to evolutionary change.

Evolutionary theory (Lawlor and Maynard Smith, 1976; Pianka, 1983) predicts that inter- and intraspecific competition results in selection which decreases competition either by niche divergence or niche expansion, respectively. Alternative theories predict that interspecific competition leads to increased competitive ability through niche convergence or increased interference mechanisms. What few appropriate data are available from pastures tend to support the more traditional view of interspecific competition leading to niche differentiation. Pasture (and other) communities, however, have many complicating "third factors" (e.g., grazing, pathogens, mycorrhizae) which change or negate the structuring influence of competition. It is not easy to assess the relative impact of competition in pastures because of the difficulty in designing an experiment that provides unequivocal evidence, and because of the difficulty of studying competition in isolation. In addition, third factor effects are not easily observable, quantifiable, or separable.

B. Constraints on the Importance of Competition

Studying competition without consideration of these third factors is unrealistic, but nevertheless important, because they help us to understand and to assess the relative influence of these factors in generating departures from expectation. Competition for scarce resources is clearly important in pastures but it operates within the constraints set by environmental conditions, grazing, pathogens, mycorrhizae, and other microorganisms and it interacts simultaneously and complexly with them.

Competition experiments can be performed in greenhouses, but it is often not feasible, perhaps never feasible, to relate laboratory results to field situations because so many third factors have been controlled or ignored. It is quite daunting that the outcome of "Competition Experiments" can be altered, or even reversed, by adding third factors such as grazing (Berendse, 1985), mycorrhizae (Fitter, 1977), *Rhizobium* strains (Young and Mytton, 1983; Mytton and Hughes, 1984; Chanway, 1987), parasites (Dobson and Hudson, 1986), or viruses (Mackenzie, 1985). It is equally daunting that the results can be quite different depending on the plant genotypes used in the experiment (Aarssen and Turkington, 1985b; Kelley and Clay, 1987; Chanway, 1987).

These factors may nullify competition by preventing dominance, exclusion, or the establishment of equilibrium conditions. These factors may also impose additional constraints on the system or force component species to converge in similarity (Lawlor and Maynard Smith, 1976). For example, grazing apparently causes convergence in the morphologies of grasses, resulting in shorter and more prostrate plants in older pastures (Aarssen and Turkington, 1985c). Yet, within these constraints interspecific competition may still promote niche divergence.

Thus, it is necessary to integrate the relative importance (sensu Welden and Slauson, 1986) of intra- and interspecific competition with that of third factors in order to address community structure and such theoretical problems as species coexistence (Fowler, 1981) and community convergence (Antonovics, 1976; Bazzaz, this volume). Interspecific competition will not be, and should not be expected to be, the most important force in all ecosystems or indeed even in all pastures.

C. Integrating Levels of Pattern

In this chapter, we have focused attention on patterns at three different levels. It is difficult to assess how the events at one level will influence the patterns at another. Do patterns at the community level set the bounds within which smaller scale patterns are generated, or vice versa, or do patterns at different levels arise more or less independently? Evolution by natural selection is about the fate of individuals. The degree of competitiveness of individuals, along with their resistance or susceptibility to disease organisms, to grazers, and to other microorganisms, clearly determines their fate. As argued above, these third factors have an impact on winners and losers and ultimately influence which genotypes of which species continue to play the evolutionary game, and it is these individuals that ultimately become part of the patterns described at all scales. However, these events are probably only part of the evolutionary play that is taking place in a larger ecological theater (sensu Hutchinson, 1965) where abiotic conditions, along with grazing, define the ultimate limits of those that can enter the play in the first place. We would argue that abiotic factors screen all would-be actors, grazing eliminates many more, and competition in interaction with third factors directs the complex evolutionary play, in which there is an ever-changing cast in an endless drama with many tragedies.

VI. Summary

In this chapter, we focus attention on one type of grassland, the managed permanent pasture. Patterns in the pasture community are evident at different, but overlapping, levels: between species (community level), within species, and at the individual plant or genotype level. Each of these is considered in turn, beginning with a description of the observed patterns, followed by a consideration of the role of competition in influencing the structure of pasture communities and how the outcome of competition is influenced by various other factors such as grazing and microorganisms. We then assess the ecological and evolutionary consequences of these species interactions.

The chapter draws attention to (1) the dangers of inferring processes from patterns, (2) the problems of using common garden studies to demonstrate a genetic basis for morphological differences in plants, (3) inability to extrapolate from common garden to field situations, and (4) the need for rigorous methods to discriminate between past, present, and apparent competition. We conclude that competition does play a major role in structuring pasture communities, but only within the limits set by a variety of environmental/biotic constraints.

Acknowledgments

Most of the research reported here has been supported by the Natural Sciences and Engineering Research Council of Canada (R.T.) and a Killam predoctoral fellowship (L.A.M.). We are grateful to C. Chanway, R. Evans, B. Holl, R. Parish, and J. Thompson for giving us access to material not yet published; to Bill and Mary Chard for unrestricted access to their pastures; to N. Fowler, D. Goldberg, M. Hutchings, and two anonymous reviewers who gave many useful suggestions on an earlier draft of the text; and to Elena Klein, who prepared many of the figures and did much of the field work.

References

Aarssen, L. W. (1983). Ecological combining ability and competitive combining ability in plants: Toward a general evolutionary theory of coexistence in systems of competition. *Am. Nat.* **122,** 707–731.

Aarssen, L. W. (1988). "Pecking order" of four plant species from pastures of different ages. *Oikos* **51,** 3–12.

Aarssen, L. W., and Turkington, R. (1985a). Vegetation dynamics and neighbour associations in pasture-community evolution. *J. Ecol.* **73**, 585–603.

Aarssen, L. W., and Turkington, R. (1985b). Biotic specialization between neighbouring genotypes in *Lolium perenne* and *Trifolium repens* from a permanent pasture. *J. Ecol.* **73**, 605–614.

Aarssen, L. W., and Turkington, R. (1985c). Within-species diversity in natural populations of *Holcus lanatus, Lolium perenne,* and *Trifolium repens* from four different-aged pastures. *J. Ecol.* **73**, 869–886.

Aarssen, L. W., and Turkington, R. (1985d). Competitive relations among species from pastures of different ages. *Can. J. Bot.* **63**, 2319–2325.

Aarssen, L. W., Turkington, R., and Cavers, P. B. (1979). Neighbour relationships in grass–legume communities. II. Temporal stability and community evolution. *Can. J. Bot.* **57**, 2695–2703.

Abrams, P. (1983). The theory of limiting similarity. *Annu. Rev. Ecol. Syst.* **14**, 359–376.

Agren, G. I., and Fagerstrom, T. (1984). Limiting dissimilarity in plants: Randomness prevents exclusion of species with similar competitive abilities. *Oikos* **43**, 369–375.

Allen, E. B., and Foreman, R. T. T. (1976). Plant species removals and old-field community structure and stability. *Ecology* **57**, 1233–1243.

Antonovics, J. (1976). The input from population genetics: "The new ecological genetics." *Syst. Bot.* **1**, 233–245.

Arthur, W. (1982). The evolutionary consequences of interspecific competition. *Adv. Ecol. Res.* **12**, 127–187.

Bender, E. A., Case, T. J., and Gilpin, M. E. (1984). Perturbation experiments in community ecology: Theory and practice. *Ecology* **65**, 1–13.

Berendse, F. (1983). Interspecific competition and niche differentiation between *Plantago lanceolata* and *Anthoxanthum odoratum* in a natural hayfield. *J. Ecol.* **71**, 379–390.

Berendse, F. (1985). The effect of grazing on the outcome of competition between plant species with different nutrient requirements. *Oikos* **44**, 35–39.

Boyce, M. S. (1984). Restitution or r- and K-selection as a model of density-dependent natural selection. *Annu. Rev. Ecol. Syst.* **15**, 427–447.

Burdon, J. J. (1980a). Intra-specific diversity in a natural population of *Trifolium repens.* *J. Ecol.* **68**, 717–735.

Burdon, J. J. (1980b). Variation in disease-resistance within a population of *Trifolium repens.* *J. Ecol.* **68**, 737–744.

Burdon, J. J., and Harper, J. L. (1980). Relative growth rates of individual members of a plant population. *J. Ecol.* **68**, 953–957.

Cahn, M. G., and Harper, J. L. (1976). The biology of the leaf mark polymorphism in *Trifolium repens* L. 1. Distribution of phenotypes at a local scale. *Heredity* **37**, 309–325.

Chanway, C. P. (1987). "Plant/Bacteria Coadaptation in a Grass/Legume Pasture," Ph.D. thesis, 264 pp. Univ. of British Columbia, Vancouver, British Columbia, Canada.

Charles, A. H. (1961). Differential survival of cultivars of *Lolium, Dactylis,* and *Phleum. J. Br. Grassl. Soc.* **16**, 69–75.

Colwell, R. K., and Winkler, D. W. (1984). A null model for null models in biogeography. *In* "Ecological Communities: Conceptual Issues and Evidence" (D. R. Strong, D. Simberloff, L. G. Abele, and A. B. Thistle, eds.), pp. 344–359. Princeton Univ. Press, Princeton, New Jersey.

Connell, J. H. (1975). Some mechanisms producing structure in natural communities: A model and evidence from field experiments. *In* "Ecology and Evolution of Communities" (M. L. Cody and J. M. Diamond, eds.), pp. 460–490. Belknap, Cambridge, Massachusetts.

Connell, J. H. (1980). Diversity and the coevolution of competitors, or the ghost of competition past. *Oikos* **35**, 131–138.

Connell, J. H. (1983). On the prevalence and relative importance of interspecific competition: Evidence from field experiments. *Am. Nat.* **122,** 661–696.

den Boer, P. (1986). The present status of the competitive exclusion principle. *Trends Ecol. Evol.* **1,** 25–28.

de Wit, C. T., and van den Bergh, J. P. (1965). Competition between herbage plants. *Neth. J. Agric. Res.* **13,** 212–221.

Dobson, A. P., and Hudson, P. J. (1986). Parasites, disease and the structure of ecological communities. *Trends Ecol. Evol.* **1,** 11–15.

Donald, C. M. (1963). Competition among crop and pasture plants. *Adv. Agron.* **15,** 1–118.

Evans, D. R., Hill, J., Williams, T. A., and Rhodes, I. (1985). Effects of coexistence in the performances of white clover–perennial ryegrass mixtures. *Oecologia* **66,** 536–539.

Evans, R. C. (1986). "Morphological Variation in a Biotically Patchy Environment: Evidence from a Pasture Population of *Trifolium repens* L.," M.S. thesis, 80 pp. Univ. of British Columbia, Vancouver, British Columbia, Canada.

Evans, R. C., and Turkington, R. (1988). Maintenance of morphological variation in a biotically patchy environment. *New Phytol.,* in press.

Fitter, A. H. (1977). Influence of mycorrhizal infection on competition for phosphorus and potassium by two grasses. *New Phytol.* **79,** 119–125.

Fowler, N. (1981). Competition and coexistence in a North Carolina grassland. II. The effects of the experimental removal of species. *J. Ecol.* **69,** 843–854.

Fowler, N. L. (1984). Patchiness in patterns of growth and survival of two grasses. *Oecologia* **62,** 424–428.

Fowler, N. L. (1986). Density-dependent population regulation in a Texas grassland. *Ecology* **67,** 545–554.

Fowler, N. L., and Antonovics, J. (1981). Competition and coexistence in a North Carolina grassland. 1. Patterns in undisturbed grassland. *J. Ecol.* **69,** 825–861.

Frame, J. (1986). Agronomy of white clover. *Adv. Agron.* **40,** 1–88.

French, N. (ed.) (1979). "Perspectives in Grassland Ecology," 204 pp. Springer-Verlag, New York.

Gaskins, M. H., Albrecht, S. L., and Hubbell, D. H. (1985). Rhizosphere bacteria and their use to increase plant productivity: A review. *Agric. Ecosystems Environ.* **12,** 99–116.

Ghilarov, A. M. (1984). The paradox of the plankton reconsidered; or, why do species coexist. *Oikos* **43,** 46–52.

Gill, D. E. (1974). Intrinsic rate of increase, saturation density, and competitive ability. II. The evolution of competitive ability. *Am. Nat.* **108,** 103–116.

Gliddon, C., and Trathan, P. (1985). Interactions between white clover and perennial ryegrass in an old permanent pasture. *In* "Structure and Functioning of Plant Populations, 2" (J. Haeck and J. W. Woldendorp, eds.), pp. 161–169. North-Holland, Amsterdam.

Goldberg, D. E., and Werner, P. A. (1983). Equivalence of competitors in plant communities: A null hypothesis and a field experimental approach. *Am. J. Bot.* **70,** 1098–1104.

Grant, P. (1975). The classical case of character displacement. *Evol. Biol.* **8,** 237–337.

Grubb, P. J., Kelly, D., and Mitchley, J. (1982). The control of relative abundances in communities of herbaceous plants. *In* "The Plant Community as a Working Mechanism," (E. I. Newman, ed.), Br. Ecol. Soc. Spec. Publ. 1, pp. 79–97. Blackwell, Oxford, England.

Hairston, N. G., Nishikawa, K. C., and Stenhouse, S. L. (1987). The evolution of competing species of terrestrial salamanders: Niche partitioning or interference? *Evol. Ecol.* **1,** 247–262.

Hancock, J. F., and Wilson, R. E. (1976). Biotype selection in *Erigeron annuus* during old field succession. *Bull. Torrey Bot. Club* **103,** 122–125.

Harper, J. L., Clatworthy, J. N., McNaughton, I. H., and Sagar, G. R. (1961). The evolution and ecology of closely related species living in the same area. *Evolution* **15**, 209–227.

Haynes, R. J. (1980). Competitive aspects of the grass–legume association. *Adv. Agron.* **33**, 227–261.

Hils, M. H., and Vankat, J. L. (1982). Species removals from a first-year old-field plant community. *Ecology* **63**, 705–711.

Hutchinson, G. E. (1965). "The Ecological Theater and the Evolutionary Play." Yale Univ. Press, New Haven, Connecticut.

Jones, D. G., and Hardarson, G. (1979). Variation within and between white clover varieties in their preference for strains of *Rhizobium trifolii. Ann. Appl. Biol.* **92**, 221–228.

Joy, P., and Laitinen, A. (1980). "Breeding for Coadaptation between Red Clover and Timothy," Hankkija's Seed Publ. 13. Hankkija Plant Breeding Inst., Finland.

Kelley, S. E., and Clay, K. (1987). Interspecific competitive interactions and the maintenance of genotypic variation within two perennial grasses. *Evolution* **41**, 92–103.

Lawlor, L. R., and Maynard Smith, J. (1976). The coevolution and stability of competing species. *Am. Nat.* **110**, 79–99.

Lowe, J. F. (ed.) (1970). "White Clover Research," Occas. Symp. 6, 327 pp. Br. Grassl. Soc., Hurley, England.

MacArthur, R., and Levins, R. (1967). The limiting similarity, convergence and divergence of coexisting species. *Am. Nat.* **101**, 377–385.

Mackenzie, S. (1985). "Reciprocal Transplantation to Study Local Specialization and the Measure of Components of Fitness," Ph.D. thesis, 135 pp. Univ. College of North Wales, Bangor, Wales.

Martin, M. M., and Harding, J. (1981). Evidence for the evolution of competition between two species of annual plants. *Evolution* **35**, 975–987.

Masterson, C. L., and Sherwood, M. T. (1974). Selection of *Rhizobium trifolii* strains by white and subterranean clovers. *Ir. J. Agric. Res.* **13**, 91–99.

McNeilly, T., and Roose, M. L. (1984). The distribution of perennial ryegrass genotypes in swards. *New Phytol.* **98**, 503–513.

Moore, C. W. E. (1964). Distribution of grasslands. *In* "Grasses and Grasslands" (C. Barnard, ed.), pp. 182–205. Macmillan, New York.

Moore, P. D. M. (1982). Measuring competition in plant communities. *Nature (London)* **298**, 515.

Mytton, L. R. (1975). Plant genotype X *Rhizobium* strain interactions in white clover. *Ann. Appl. Biol.* **80**, 103–107.

Mytton, L. R., and Hughes, D. M. (1984). Inoculation of white clover with different strains of *Rhizobium trifolii* on a mineral hill soil. *J. Agric. Sci.* **102**, 455–459.

Newbould, P., Holding, A. J., Davies, G. J., Rangeley, A., Copeman, G. J. F., Davies, D. A., Frame, J., Haystead, A., Herriot, J. B. D., Holmes, J. C., Lowe, J. F., Parker, J. W. G., Waterson, H. A., Wildig, J., Wray, J. P., and Younie, D. (1982). The effect of *Rhizobium* inoculation on white clover in improved hill soils in the United Kingdom. *J. Agric. Sci.* **99**, 591–610.

Newman, E. I. (1978). Root microorganisms: Their significance in the ecosystem. *Biol. Rev.* **53**, 511–554.

Parish, R. (1987). "The Role of Disturbance in Permanent Pastures," Ph.D. thesis, 159 pp. Univ. of British Columbia, Vancouver, British Columbia, Canada.

Parrish, J. A. D., and Bazzaz, F. A. (1982). Competitive interactions in plant communities of different successional ages. *Ecology* **63**, 314–320.

Peters, B. (1980). "The Demography of Leaves in a Permanent Pasture," Ph.D. thesis, 102 pp. Univ. College of North Wales, Bangor, Wales.

Pianka, E. R. (1983). "Evolutionary Ecology," 3rd ed. Harper & Row, New York.

Pickett, S. T. A. (1980). Non-equilibrium coexistence of plants. *Bull. Torrey Bot. Club* **107**, 238–248.

Platt, W. J. (1975). The colonization and formation of equilibrium plant species associations on badger disturbances in a tall-grass prairie. *Ecol. Monogr.* **45**, 285–305.

Rabotnov, T. A. (1966). Peculiarities of the structure of polydominant meadow communities. *Vegetatio* **13**, 109–116.

Robinson, A. C. (1969a). Competition between effective and ineffective strains of *Rhizobium trifolii* in nodulation of *Trifolium subterraneum. Aust. J. Agric. Res.* **20**, 827–841.

Robinson, A. C. (1969b). Host selection for effective *Rhizobium trifolii* by red clover and subterranean clover in the field. *Aust. J. Agric. Res.* **20**, 1053–1060.

Roos, F. H., and Quinn, J. A. (1977). Phenology and reproductive allocation in *Andropogon scoparius* (*Gramineae*) populations in communities of different successional stages. *Am. J. Bot.* **64**, 535–540.

Sarukhan, J., and Harper, J. L. (1973). Studies on plant demography: *Ranunculus repens* L., *R. bulbosus* L. and *R. acris* L. I. Population flux and survivorship. *J. Ecol.* **61**, 675–716.

Schoener, T. W. (1983). Field experiments on interspecific competition. *Am. Nat.* **122**, 240–285.

Silvertown, J., and Law, R. (1987). Do plants need niches? Some recent developments in plant competition ecology. *Trends Ecol. Evol.* **2**, 24–26.

Slatkin, M. (1980). Ecological character displacement. *Ecology* **61**, 163–177.

Snaydon, R. W. (1978). Genetic changes in pasture populations. *In* "Plant Relations in Pastures" (J. R. Wilson, ed.), pp. 253–269. Commonw. Sci. Ind. Res. Org., Melbourne, Australia.

Snaydon, R. W. (1985). Aspects of the ecological genetics of pasture species. *In* "Structure and Functioning of Plant Populations, 2" (J. Haeck and J. W. Woldendorp, eds.), pp. 127–152. North-Holland, Amsterdam.

Thompson, J. D. (1980). Implications of different sorts of evidence for competition. *Am. Nat.* **116**, 719–726.

Thorhallsdottir, T. E. (1983). "The Dynamics of a Grassland Community with Special Reference to Five Grasses and White Clover," Ph.D. thesis, 207 pp. Univ. College of North Wales, Bangor, Wales.

Tilman, D. (1981). Tests of resource competition theory using four species of Lake Michigan algae. *Ecology* **62**, 802–815.

Turkington, R. (1989). The growth, distribution and neighbour relationships of *Trifolium repens* in a permanent pasture. V. The coevolution of competitors. *J. Ecol.* (in press).

Turkington, R., and Burdon, J. J. (1983). Biology of Canadian weeds. 57. *Trifolium repens* L. *Can. J. Plant Sci.* **63**, 243–266.

Turkington, R., and Harper, J. L. (1979a). The growth, distribution and neighbour relationships of *Trifolium repens* in a permanent pasture. 1. Ordination, pattern and contact. *J. Ecol.* **67**, 201–218.

Turkington, R., and Harper, J. L. (1979b). The growth distribution and neighbour relationships of *Trifolium repens* in a permanent pasture. II. Inter- and intra-specific contact. *J. Ecol.* **67**, 219–230.

Turkington, R., and Harper, J. L. (1979c). The growth, distribution and neighbour relationships of *Trifolium repens* in a permanent pasture. IV. Fine scale biotic differentiation. *J. Ecol.* **67**, 245–254.

Turkington, R., Holl, F. B., Chanway, C. P., and Thompson, J. D. (1988). The influence of microorganisms, particularly *Rhizobium*, on plant competition in grass–legume communities. *In* "Plant Population Ecology" (A. J. Davy, M. J. Hutchings, and A. R. Watkinson, eds.), pp. 343–366. Blackwell, Oxford, England.

Underwood, T. (1986). The analysis of competition by field experiments. *In* "Community

Ecology: Pattern and Process" (J. Kikkawa and D. J. Anderson, eds.), pp. 240–268. Blackwell, Melbourne, Australia.

Watkin, B. R., and Clements, R. J. (1978). The effects of grazing animals on pastures. *In* "Plant Relations in Pastures" (J. R. Wilson, ed.), pp. 273–289. Commonw. Sci. Ind. Res. Org., Melbourne, Australia.

Welden, C. W., and Slauson, W. L. (1986). The intensity of competition versus its importance: An overlooked distinction and some implications. *Q. Rev. Biol.* **61,** 23–44.

Wiens, J. A. (1977). On competition and variable environments. *Am. Sci.* **65,** 590–597.

Williams, O. B., and Roe, R. (1975). Management of arid grasslands for sheep: Plant demography of six grasses in relation to climate and grazing. *Proc. Ecol. Soc. Aust.* **9,** 142–156.

Wilson, J. R. (ed.) (1978). "Plant Relations in Pastures," 425 pp. Commonw. Sci. Ind. Res. Org., Melbourne, Australia.

Young, N. R., and Mytton, L. R. (1983). The response of white clover to different strains of *Rhizobium trifolii* in hill land reseeding. *Grass Forage Sci.* **38,** 13–39.

Young, N. R., Hughes, D. M., and Mytton, L. R. (1986). The response of white clover to different strains of *Rhizobium trifolii* in hill land reseeding: A second trial. *Plant Soil* **94,** 277–284.

16

The Role of Competition in Agriculture

S. R. Radosevich   **M. L. Roush**

I. Introduction

In this chapter, we consider agricultural communities to include all human-manipulated systems used to produce food, feed, and fiber, as well as systems of early forest regeneration and rangeland production. These production systems have similar economic goals and environmental and biological characteristics. Many characteristics of agricultural systems also are similar to those of natural ecosystems (Harper, 1977; Snaydon, 1980); however, agricultural communities are unique because they are manipulated and managed for single-species (sometimes multi-species) productivity.

Environmental and biotic factors are relatively homogeneous and synchronous in agricultural communities because of management and because crops and weeds often possess similar life forms, life histories, and genetic and phenotypic characteristics (Snaydon, 1980; Radosevich and Holt, 1984; Roush and Radosevich, 1985). Disturbance and resource supplementation reduce spatial and temporal variability of the agricultural environment. Because these factors are routinely manipulated in agricultural systems, gradients in environment or plant proximity may be examined easily. Thus, agricultural communities are special, simplified systems for investigating plant competition. Linkages between agricultural and plant ecology also help explain mechanisms of crop–weed competition, predict the consequences of weed infestations on cropland, and provide a biological basis for weed management.

Research conducted in agriculture often differs from that performed in natural systems in objectives, method of study, and, therefore, the interpretation of experimental results (Levins, 1973; Snaydon, 1980). Experiments in agriculture usually are conducted to determine how manipulations of biology or environment influence crop productivity. Furthermore, agricultural scientists are not necessarily concerned with understanding ecological processes, especially if the empirical results of their studies are reasonably definitive (Snaydon, 1980; Radosevich and Holt, 1984). This reliance on empirical, phenomenological research has allowed a narrow perspective of competition in agriculture.

In this chapter, we address this current perspective of competition in agriculture. One objective is to review past and present methods used to study competition in agriculture, and describe our present understanding of crop–weed associations on an ecological basis. A second objective is to examine how ecological studies and agricultural systems can be combined to investigate important processes in plant communities. In so doing, we present a general approach for investigation of the process and role of competition in agricultural systems that utilizes biomathematical models and an understanding of ecological processes.

II. Methods for Studying Plant Competition in Agriculture

Competition among plants and the dynamics of plant communities involve interactions of biological and environmental factors. Welden and

Slauson (1986) suggest that sufficient distinction has not been made between the intensity of competition and its importance in the dynamics of plant communities. Intensity integrates physiological and morphological responses of individual plants with the presence of neighbors. It describes the process of competition among individual plants in a population or community. Importance describes the role of competition in relation to other processes that also may influence the future productivity or species composition of a plant community. The distinction between the intensity of competition and its importance in influencing crop–weed associations is only now being recognized by segments of the agricultural research community. For example, most competition models in agriculture (Zimdahl, 1980; Cousens, 1985) only consider the degree of crop yield loss due to competition (intensity), without concern for its role on weed composition or future abundance (importance).

A. Intensity of Competition in Agriculture

Historically, competition studies performed in agriculture have documented levels of crop yield loss, rather than the causes or implications of the interactions among weed and crop plants. Empirical studies in agriculture have primarily been of two types: additive studies and substitutive (replacement series) experiments (Radosevich, 1987, 1988). Additive studies are considered useful because the crop loss from weed abundance under current cropping practices, and the value of specific weed control tactics, can be determined from them. Replacement series experiments maintain a constant total plant density (de Wit, 1960; Harper, 1977) and have primarily been useful for evaluating intercropping systems (Trenbath, 1976). However, both approaches place different emphasis on spatial (proximity) factors, which influence how such studies are interpreted (Harper, 1977; Connolly, 1986; Radosevich, 1987, 1988).

Zimdahl (1980) and Cousens (1985) have summarized numerous additive and substitutive experiments conducted in an array of cropping systems. Stewart *et al.* (1982) have provided a similar summary of experiments in young forest stands. Crop yield response to weed density or weed cover (Auld and Tisdell, 1988; Cousens, 1985; Radosevich, 1987, 1988) is generally described by a rectangular hyperbola or similar (e.g., exponential) function (Fig. 1). The results of such experiments, although predictable in a general sense, vary markedly among cropping systems, locations, and season of experimentation. Often, the intensity of competition can only be interpreted qualitatively. Thus, predictions of competition intensity in any agricultural system, and assessments of its importance on long-term crop–weed dynamics, are difficult using only empirical additive and substitutive approaches.

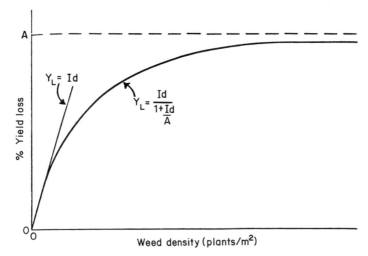

Figure 1   The rectangular hyperbolic model for relating yield loss to weed density. Y_L is percentage yield loss, A and I are the parameters that determine the shape of the curvilinear response of yield loss to weed density. From Cousens (1985).

Agricultural scientists often face a dilemma in accounting for the influence of spatial factors, especially total and relative plant density, on the outcome of their experiments. Crops usually are grown at a constant density, determined experimentally or intuitively to maximize economic yield, while weeds create conditions where both total and relative plant densities vary. Since both total plant density and species proportion influence the outcome and interpretation of competition experiments (de Wit, 1960; Harper, 1977; Spitters, 1983a,b; Joliffe *et al.*, 1984; Connolly, 1986; Radosevich, 1987, 1988; Roush *et al.*, 1988), it often is difficult to separate the effects of intra- and interspecific interactions in these experiments.

Carlson and Hill (1985) studied the influence of total and relative densities of wild oat (*Avena fatua*) and wheat on wheat yields (Fig. 2a), using several additive experiments conducted at different wheat densities. At any density of wheat, yields were always highest at low wild oat densities. However, the negative influence of wild oat diminished as the total density of the stand (crop + weed) increased. Because both total and relative densities of the species varied, it is difficult to differentiate between the effects of intra- and interspecific competition in this study. Carlson and Hill (1985) expressed the response of wheat as a function of the ratio of wild oat density to total plant density (Fig. 2b). This ratio diminished the impact of two simultaneously changing variables within the experiment by accounting for the influence of species proportion on

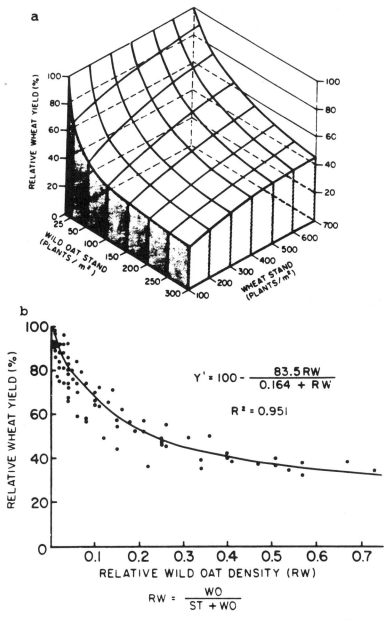

Figure 2 Data from Carlson and Hill (1985) showing (a) the expected relative wheat yields plotted as a function of wheat and wild oat plant densities, and (b) relative wheat yields as a function of the relative density of wild oat (RW). Yields are expressed as a percentage of yields attained in uninfested controls. WO, Wild oat density (plant m^{-2}); ST, wheat stand (plants m^{-2}).

wheat-yield response. The ratio also allowed predictions of wheat yield in relation to the abundance of both plant species.

Because of the joint influences of proximity factors in crop–weed competition, another experimental approach has been developed that systematically varies both total and relative plant densities (Watkinson, 1981; Spitters, 1983a,b; Firbank and Watkinson, 1985; Connolly, 1987; Radosevich, 1987, 1988). This approach provides a better basis for quantifying the intensity of competition than conventional additive or replacement experiments (Roush *et al.*, 1988). Analysis of competition using this approach is based on yield–density relationships (Fig. 3 and Table 1) (Shinozaki and Kira, 1956; Bleasdale and Nelder, 1960; Watkinson, 1980) for plants grown in monocultures. Watkinson (1981), Spitters (1983a,b), and Connolly (1987) proposed expansions of these yield–density relationships to include competition among multiple species.

Firbank and Watkinson (1985) have used this approach with an association of two plant species, fitting data to a general, nonlinear form of the yield–density equation (Watkinson, 1981). Studies in progress by Concannon and Radosevich (1987), Nichols *et al.*,[1] Shainsky and Radosevich (1987), Thill *et al.*,[1] and Westra *et al.*[1] are using this approach to examine the intensity of competition in two-species, crop–weed associations.

YIELD DENSITY RELATIONSHIPS

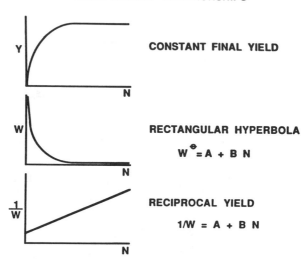

Figure 3 Diagrammatic representations of basic relationships between plant yield and density. Y is yield of a stand of plants (biomass area^{-1}), W is yield of an individual plant (biomass plant^{-1}), N is plant density, ϕ is an exponent that determines the curvilinear nature of individual plant responses, and A and B are the linear parameters of the models.

Table 1   Proposed Models that Expand Yield–Density Relationships from Monocultures to Mixtures[a]

Single Species (i)	Multiple Species (i and j)	Reference
$W_i = f(N_i)$	$W_i = f(N_i, N_j)$	Connolly (1987)
$1/W_i = B_{i0} + B_{ii}N$	$1/W_i = B_{i0} + B_{ii}N_i + B_{ij}N_j$	Spitters (1983a)
$W_i = W_{mi}(1 + a_iN_i)^{-b}$	$W_i = W_{mi}[1 + a(N_i + e_{ij}N_j)]^{-b}$	Watkinson (1981)

[a] W, Biomass per plant; W_m, maximum potential biomass per plant; N, density; B, regression coefficients, where $B_{i0} = 1/W_m$; B_{ii}, intraspecific competition; B_{ij}, interspecific competition; a, the area necessary to achieve W_m; b, efficiency of resource use; e, equivalency.

Miller and Werner (1987) and Roush and Radosevich (1987) have expanded the method for associations of four plant species.

The approach proposed by Watkinson (1981) and by Spitters (1983a,b) is especially useful for studies of crop and weed interactions because the effects of intra- and interspecific competition are separated by systematic variation of total plant density and species proportion. For example, Concannon and Radosevich (1987) systematically varied the densities of wheat and annual ryegrass (*Lolium multiflorum*) in monocultures and mixtures to create a matrix of total and relative densities that ranged from zero to 880 plants per m^2. Mean reciprocal yield of individuals and total stand yields were described using multiple linear regression models. The parameters from the reciprocal-yield models indicated that wheat responded more to variations in its own density (B_{ii}, intraspecific competition) than to the density of annual ryegrass (interspecific competition) (Table 2). In contrast, the influence of interspecific competition on ryegrass yields was more intense than the influence of intraspecific competition. The data suggest that the influence of approximately seven ryegrass neighbors was equivalent to a single wheat neighbor in determining wheat yield (Table 2). Thus, wheat productivity may be

Table 2   Multispecies Reciprocal-Yield Models for Interactions Between Spring Wheat and Italian Ryegrass (*Lolium multiflorum*)[a,b]

Species	$1/W = B_{i0} + B_{ii}N_i + B_{ij}N_j$	R^2	B_{ii}/B_{ij}
Wheat	$1/W = 10.72 + 1.18N_w + 0.17N_r$	0.90	6.70
Ryegrass	$1/W = 41.64 + 3.21N_r + 4.51N_w$	0.43	0.75

[a] From Concannon (1987).

[b] B_{i0} is the reciprocal of the theoretical maximum size of an individual, B_{ii} describes influences of intraspecific competition, B_{ij} describes influences of interspecific competition, and B_{ii}/B_{ij} predicts relative competitive ability of each species. $p < 0.01$ for B_{i0}, B_{ii}, and B_{ij} in each model.

optimized by adjusting densities of both wheat and ryegrass to minimize the joint influences of intraspecific and interspecific competition, rather than direct weed control tactics to reduce only interspecific effects from annual ryegrass. Since annual ryegrass competition is most detrimental to wheat yield at high weed and low crop densities, a threshold ratio of the crop and weed is suggested for biologically and economically optimum crop production.

In the past, most competition studies in agriculture have concentrated on total crop yields. However, reduced stand yields are the result of the plastic responses of individual plants to the presence of neighbors. When individual responses are of primary interest, a neighborhood approach to assess competition is appropriate (Harper, 1977; Goldberg and Werner, 1983; Radosevich, 1987, 1988). In neighborhood designs, "performance" of a target individual is determined as a function of the number, biomass, cover, aggregation, or distance of neighboring plants (Mack and Harper, 1977; Weiner, 1982; Goldberg and Werner, 1983; Watkinson *et al.*, 1983; Firbank and Watkinson, 1987). Gunsolus and Coble (1986) have used a neighborhood approach, which they call a "sphere of influence," to assess the competitive influence of individual weeds on crop productivity. Individual weed plants are grown with crop plants in additive-type experiments, and the influence of each weed individual on the yield of several crop plants sampled at various distances from the weed is determined.

The sphere of influence and neighborhood approaches may be more appropriate than conventional approaches to additive experiments, because individual responses to the presence of weeds can be determined without strict control of spatial factors. However, these approaches are not restricted to additive-type experiments. Both neighborhood and sphere of influence approaches assume that intraspecific crop effects and interspecific effects of crop plants on weed individuals of different species are constant. For sphere of influence studies, weeds are ideally restricted to low, widely spaced densities, because intraspecific weed interactions also could confound interpretations of experiments using this approach. In addition, difficulties in extrapolating yields of individual crop plants to stand yields may arise without better control of density and proportional factors in these experiments.

B. Importance of Competition in Agriculture

Welden and Slauson (1986) propose that the coefficient of determination (R^2) from a regression equation relating plant response to competition is a suitable measure of the importance of competition. They describe equations, derived from neighborhood experiments, in which the slope of the regression quantifies the intensity of competition on plant yield. The R^2 value for those equations suggests how important competition was, relative to all processes that influence plant yield (Welden and Slauson, 1986). However, interpretations assume that data have been fit to the most appropriate model. For example, if data are fit to a linear model when nonlinear techniques are most appropriate, R^2 values will be low, and underestimate the importance of competition. Values of R^2 may also overestimate the importance of competition when competition parameters used in regression models inadvertently include processes other than competition. For example, the influences of herbivory may interact with competition, so that competition parameters include the effects of both factors. Experimental design will also influence R^2 values.

A more definitive approach than estimates of coefficients of determination from regression equations is needed to separate the influence of competition from other processes in crop–weed associations. Such an approach requires a conceptual framework to define and organize the key processes and factors that influence the dynamics of crop–weed communities. Experiments then can be used to determine the influences of each process, while sensitivity analysis (Caswell and Werner, 1978) can quantitatively assess the importance of each process to the system. Maxwell *et al.* (1988) have successfully used this approach to describe the importance of intraspecific competition on leafy spurge (*Euphorbia esula*) abundance, and to predict the long-term outcome of several weed control strategies.

III. Process-Based Models for Competition in Agricultural Plant Communities

A recurring theme in this book has been the complexity and multiplicity of factors that mediate the influence of competition on plant communities. Figure 4 organizes the life histories of agricultural plant species into key ecological processes, e.g., germination, seed bank phenomena, growth, and reproduction. Each process in Fig. 4 is potentially important in determining weed densities, spatial and temporal distributions, the relative success of weed species, and the ultimate influence of weeds on crop productivity.

Although the relationships and interactions described in Fig. 4 appear to be reasonably straightforward, few studies have explored competition and community development of crop–weed associates by understanding variation in processes of germination, growth, and reproductive allocation. Furthermore, experiments have not been accomplished in agriculture to link management practices directly to either competition or plant community structure. Haas and Streibig (1982) provide a notable excep-

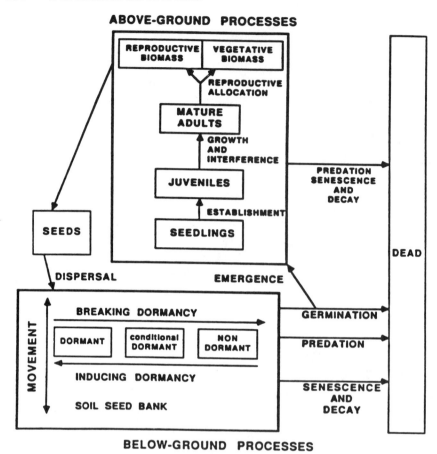

Figure 4   Conceptual model of key life history phases and processes of an annual weed community.

tion. Most factors and activities have been studied separately, however, for some crop and weed species.

A. Key Processes in Crop–Weed Competition and Community Dynamics

1. Emergence Time   The timing of plant emergence is an important factor in determining the outcome of competition in agriculture (Ross and Harper, 1972; Harper, 1977). Fischer and Miles (1973) developed a series of theoretical, stochastic models for competition between systematically arranged crop plants and randomly located weeds. Differential plant emergence times, rates of radial expansion (growth), and spatial arrangement all influenced competition among neighbors in these models. Depending on the density and spatial arrangement of the plants, Fischer and Miles (1973) predicted three- to eightfold increases in competitive advantage by combining early emergence and rapid growth in the weed species. Similarly, Firbank and Watkinson (1985) observed that emergence time and local crowding (proximity) accounted for up to 50% of the variation in performance of individual plants of *Agrostemma githago* L.

2. ***Growth Ability and Environment***   Plant growth rate has been proposed as a key plant trait for competitive success in disturbed, productive habitats typical of agriculture (Grime and Hunt, 1975; Grime, 1979). Roush and Radosevich (1985, 1987) described competitive interactions among weed species, and addressed plant–plant–environment interactions in weed communities using mathematical growth analysis techniques. In one set of experiments, equivalent relative growth rates (RGR) indicated potential competitive equivalency. However, a clear ranking in size, physiological, morphological, and allocation characteristics also suggested potential competitive hierarchies among the species (Fig. 5). In that study, equivalent and high RGRs were achieved among the species because physiological and morphological factors compensated in their influence on the ability of the plants to grow rapidly. Despite equivalent RGRs, the final sizes of the species varied markedly, due primarily to variation in the initial size of seedlings. The competitive hierarchy that was defined by the competition experiments (Fig. 6) was most closely related to plant size and the growth parameters of net assimilation rate (NAR) and leaf area ratio (LAR) (Table 3).

A second set of experiments was conducted for a similar ensemble of species in a contrasting environment (Roush and Radosevich, 1987, and in progress). These studies demonstrated consistent relationships between growth ability and competitiveness, as well as the importance of environment in determining how plant growth and competition are linked. The first experiment (Roush and Radosevich, 1985) was conducted in the hot, high-light intensity environment of the Central Valley of California; the second experiment (Roush and Radosevich, 1987) was conducted over 2 years in the cooler, lower-light intensity environment of the Willamette Valley in Oregon. In both environments, the competitive hierarchies among the species were consistent with hierarchies among similar growth parameters. However, the nature of the predictions shifted with changes in the environments (Table 4). In Davis, California, competitive ability was related to high net assimilation rates (physiological efficiency) and low leaf area ratios (morphological leafiness). In that environment, C_4 species out-competed C_3 species. In Corvallis, Oregon, this competitive advantage diminished, and competia

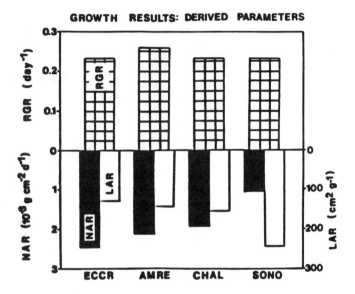

b

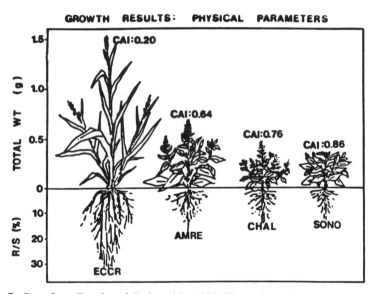

Figure 5 Data from Roush and Radosevich (1985) illustrating various measurements of the growth abilities of four annual weed species: ECCR, *Echinochloa crus-galli*; AMRE, *Amaranthus retroflexus*; CHAL, *Chenopodium album*; SONO, *Solanum nodiflorum*. The growth results include (a) directly measured, physical parameters (total WT = biomass in grams; root/shoot ratio, R/S = root biomass/shoot biomass × 100; canopy area index, CAI = leaf area/ground area), and (b) derived growth rates (relative growth rate, RGR = $d \ln \text{WT}/dT$; leaf area ratio, LAR = leaf area/WT; net assimilation rate, NAR = 1/leaf area × $d\text{WT}/dT$).

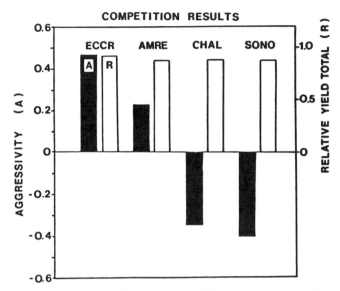

Figure 6 Data from Roush and Radosevich (1985) illustrating results of a replacement series experiment for measuring competition among four annual weed species: ECCR, *Echinochloa crus-galli*; AMRE, *Amaranthus retroflexus*; CHAL, *Chenopodium album*; SONO, *Solanum nodiflorum*. Aggressivity (a measure of relative competitive ability) and relative yield total were calculated as described by McGilchrist and Trenbath (1971).

tive ability was related more to leaf area than physiological efficiency (Table 4).

Pickett and Bazzaz (1978) demonstrated that environmental conditions influence competitive relationships and cause hierarchies among plant species grown in mixtures to be inconsistent. Although competitive

Table 3 Correlations among Competitive Ability (Aggressivity) and Growth Parameters Measured for Four Summer Annual Weeds[a,b]

	A	WT	NAR	LAR	HT	SEED WT	RGR
A	1.00						
WT	0.97	1.00					
NAR	0.81	0.87	1.00				
LAR	−0.73	−0.75	−0.95	1.00			
HT	0.69	0.77	0.70	−0.72	1.00		
SEED WT	0.39	0.47	0.46	−0.38	0.72	1.00	
RGR	0.33	0.28	0.38	−0.39	−0.26	−0.36	1.00

[a] From Roush and Radosevich (1985).

[b] The growth parameters dry weight (WT, seed WT), net assimilation rate (NAR), leaf area ratio (LAR), plant height (HT), and relative growth rate (RGR) were measured and derived as described by Hunt (1982). A, Aggressivity.

Table 4   Hierarchies of Growth Abilities and Competitive Ability among Four Summer Annual Weed Species in Two Environments[a,b]

	Davis, CA 1983	Corvallis, OR	
		1985	(1986)
Growth ability			
WT	E > A > C > S	C = A = E > L	C = E > A = L
NAR	E > A > C > S	C = A < E = L	C = A < E > L
LAR	E < A < C < S	C = A > E = L	C = A > E > L
RGR	E = A = C = S	C = A > E > L	A > C = E > L
Competitive ability	E > A > C > S	C = A > E = L	(C ≥ A > E ≥ L)

[a] From Roush and Radosevich (1987, 1988).

[b] The species were *Amaranthus retroflexus* (A), *Chenopodium album* (C), *Echinochloa crus-galli* (E), and *Lolium multiflorum* (L). The growth parameters were total dry weight (WT), net assimilation rate (NAR), leaf area ratio (LAR), and relative growth rate (RGR). Competitive ability was measured as aggressivity (McGilchrist and Trenbath, 1971) from a replacement series experiment at Davis, CA, and as relative competitive ability (Spitters, 1983a) from addition series experiments at Corvallis, OR.

relationships among weeds and crops suggest that site-specific or regional hierarchies of competitive ability among weed species and crops should exist (Obeid, 1965; Pickett and Bazzaz, 1978; Radosevich and Holt, 1984), consistent hierarchies of competition among plants have rarely been observed (Zimdahl, 1980; Roush and Radosevich, 1985, 1987). Climate, location, and management activities all contribute to competitive hierarchies among crop and weed species by varying resources, environmental conditions, and the degree of vegetative suppression (Aldrich, 1984; Grime, 1979; Radosevich and Holt, 1984).

Given the year-to-year variation in growing conditions, crop rotation patterns, and long-term changes in cultural practices, shifts in weed species composition have occurred, and should be expected to continue in agricultural systems (Salisbury, 1981; Haas and Striebig, 1982). Because these systems consist of species of generally similar life histories and competitive abilities (Baker, 1965; Grime, 1979; Roush and Radosevich, 1985, 1987), competitive advantage among the species is then determined by variation in proximity, germination, growth, and reproductive responses to environment and management, rather than by inherent differences in competitiveness. Predictions of competitive relationships and possible shifts in relative dominance among the species in a crop–weed community require an understanding of interactions among environment, proximity, and biological characteristics of the weed and crop.

3. Processes other than Competition   Besides the environmental and biological components of competition already discussed, other noncompetitive factors must be assessed to determine the relative importance of competition in agriculture. For example, numerous studies demonstrate that herbivory and competition interact to influence the structure and productivity of plant communities (e.g., Whittaker, 1978; Dirzo, 1984; Fowler and Rausher, 1985; Louda *et al.*, this volume). In addition, accurate assessment of the impacts of noncompetitive factors on crop–weed associations will allow a better understanding of how agricultural communities are organized. These processes include density-independent mortality, predation, senescence, seed bank dynamics, and noncompetitive forms of plant interference (e.g., allelopathy). Agricultural researchers often attempt to minimize the influence of noncompetitive factors on competition, or to assume a negligible role for them in the cropping system, e.g., the influence of herbivory may be lessened by controlling agricultural pests.

B. Models as Tools to Link Ecology with Crop–Weed Competition and Community Dynamics in Agriculture

Mathematical models can provide the functional framework to organize and identify processes of crop–weed communities and to focus experiments to better understand how weeds and crops respond to environment and management. For example, Shainsky and Radosevich (1987, 1988) have developed a model to describe the mechanisms of competition among Douglas fir and red alder in young forest stands. Douglas fir is of significant economic importance, while red alder has traditionally been considered as a weed tree in these plant communities. In past studies (Tarrant and Trappe, 1971; Binkley, 1983, 1984), both positive and negative interactions have resulted from associations of the two tree species. By using an addition series experiment (Radosevich, 1987, 1988), Shainsky and Radosevich (1987, 1988) demonstrated that responses for Douglas fir in relation to its own density ranged from negative to positive, depending on the density of red alder (Fig. 7).

The responses of tree size to competitive regime that were observed by Shainsky and Radosevich (1987, 1988) (Fig. 7) can be explained by examining the effects of species frequency on leaf area accumulation, and subsequent impacts on the light and water resources (Fig. 8). The density of each species had a positive influence on development of its own leaf area, but a negative influence on the leaf area of the other species. This response to density most significantly influenced the amount of light available to the understory species, Douglas fir. Both species also diminished soil moisture throughout the growing season, which resulted in less canopy development by red alder and relatively more light availability to the Douglas fir. Individuals of each species grew at different rates and attained different sizes depending on the availability of both resources as

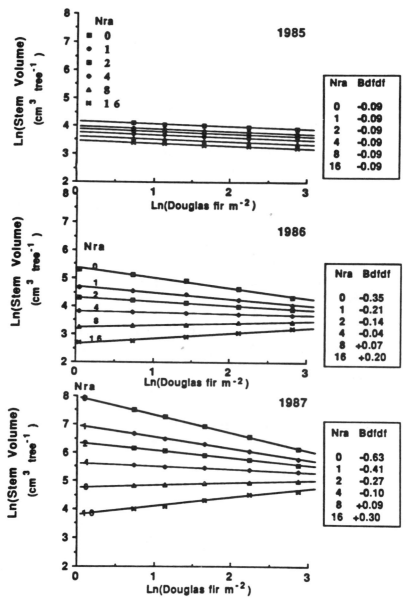

Figure 7 Response of Douglas fir stem volume to density of Douglas fir (Ndf) at different densities of red alder (Nra) in 3 years. Bdfrf is the slope of the isoline for each red alder density, which is the predicted influence of intraspecific competition among Douglas fir neighbors. From Shainsky and Radosevich (1987; Shainsky, 1988).

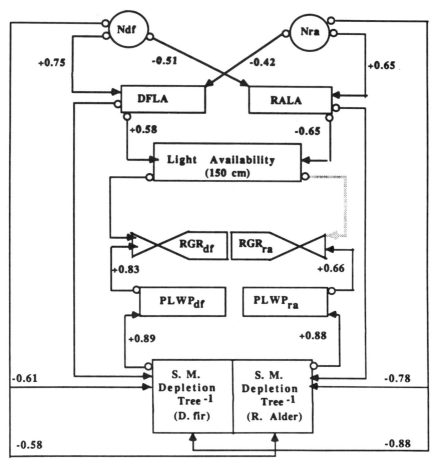

Figure 8 A conceptual model that illustrates interrelationships among factors of proximity, resources, plant growth, and moisture stress for Douglas fir and red alder mixtures. Ndf, Douglas fir density; Nra, red alder density, DFLA, Douglas fir leaf area; RALA, red alder leaf area; RGR_{df} and RGR_{ra}, relative growth rates of Douglas fir and red alder, respectively; $PLWP_{df}$ and $PLWP_{ra}$, predawn leaf water potential in August for Douglas fir and red alder, respectively. Simple correlations (r) between these parameters are indicated. Dotted lines represent potential linkages that were not directly evaluated. In Shainsky (1988).

mediated by plant density and frequency (Figs. 7 and 8). The model demonstrates that plant species may compete differently, and that availability of environmental resources depends on plant density, proportion, and size structure. The model also suggests that foresters may manage these species for optimal productivity by understanding the interrelationships between factors of tree proximity, biology, and environment.

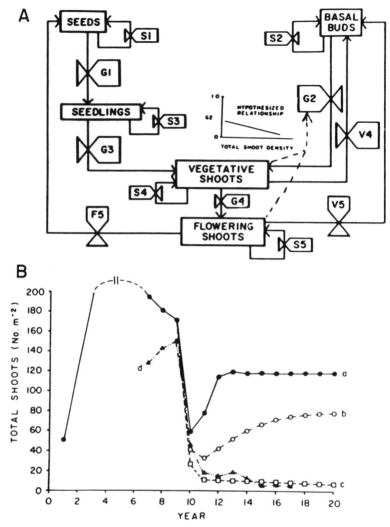

Figure 9 (A) Diagrammatic model of a leafy spurge population. Boxes represent stages in the life cycle, arrows indicate processes, and valve symbols represent rates for each process. Model transition parameters are S1, proportion of seeds that remain viable in the seed bank; S2, proportion of basal buds that remain viable; S3, proportion of seedlings that remain seedlings; S4, proportion of vegetative shoots (nonflowering mature) that remain vegetative; S5, proportion of flowering shoots that remain flowering; G1, proportion of seed that germinates to become seedlings; G2, proportion of basal buds that grow to vegetative shoots; G3, proportion of seedlings that become vegetative shoots; G4, proportion of vegetation shoots that become flowering shoots; F5, number of seeds produced per flowering shoot; V4, number of buds produced per vegetative shoot; V5, number of buds produced per flowering shoot. (B) Simulation of leafy spurge population, with density-dependent functions simulating the introduction of a foliage-feeding herbivore at year 10 that removes (a) 40, (b) 50, and (c) 60% of the stems. Also shown are (d) observed results of actual sheep feeding on leafy spurge (4). From Maxwell *et al.* (1988).

In the past, modeling efforts in agriculture have focused either on the intensity of competition alone (Zimdahl, 1980; Cousens, 1985) or on aspects of weed communities other than competition (Harper, 1977; Snaydon, 1980). Demographic models have been proposed for various plant species in agricultural systems (e.g., Sagar and Mortimer, 1976; Watson, 1985). Such models, although useful for describing the dynamics of weed populations to variation in management tactics, have virtually ignored competition as a process that may be important in determining weed species composition and abundance. When competition models are integrated into models that also address seed bank dynamics, plant growth, reproduction, and variation in environment and disturbance, the role of competition in the organization of agricultural plant communities can be elucidated (e.g., Maxwell *et al.*, 1988). These models also can provide farmers and other land managers with a tool to evaluate various management options.

Maxwell *et al.* (1988) expanded an existing demographic model that predicted the population dynamics of leafy spurge (*Euphorbia esula*) (Watson, 1985). The expansion included a mathematical relationship that introduced the effect of intraspecific competition among leafy spurge individuals into the model. Maxwell *et al.* (1987) demonstrated how the process of competition could be integrated into demographic models to predict accurately long-term stand dynamics of a weed (Fig. 9A), and provide assessments of the long-term effectiveness of several management tactics, e.g., grazing (Fig. 9B).

Ultimately, population process-based models that include competition must be perfected to predict how crop–weed communities will respond to variation caused by disturbance, biological manipulation, and environmental change. However, agriculturalists and ecologists must first understand the fundamental relationships that determine the intensity of competition in crop–weed associations and the importance of competition in organizing agricultural communities. Progress then can be made to implement those relationships into actual decision-making or economic models of specific crop–crop or crop–weed combinations.

IV. Summary

Agricultural systems are special, simplified plant communities that are characterized by human disturbance and high productivity. Because of their relative simplicity in comparison to natural systems, environmental and proximity factors in agricultural communities are manipulated easily. These traits should stimulate linkages between agriculture and plant ecology that will elucidate the mechanisms and implications of competition in plant communities, as well as provide an ecological basis for crop–weed management. Methods to study the intensity of competition in agriculture have been restricted primarily to additive or substitutive experiments. Recently, other experimental approaches have been developed that systematically vary total and relative plant densities, separate the influences of intra- and interspecific competition, and provide better quantification of competition in agricultural systems. These studies also contribute to the development of strategies for managing crop–weed communities. For example, economic weed thresholds can be developed from the outcome of such studies. The importance of competition for determining future species composition or system productivity also is not well understood by agricultural scientists. Realistic assessments of the importance of competition require a framework to define and organize ecological processes or factors that influence crop–weed dynamics. Process-based models provide a means to study competition, link the process of competition with other processes in crop–weed communities, and place it within a general context for understanding plant–plant–environment interactions. Several examples are provided to demonstrate how models can be used to link ecological principles with studies of crop–weed competition and plant community dynamics in agricultural systems.

Notes

1. Personal communication with Steven Simmons (University of Minnesota, St. Paul), Donald Thill (University of Idaho, Moscow), and Philip Westra (Colorado State University, Fort Collins), respectively.

References

Aldrich, R. J. (1984). "Weed–Crop Ecology: Principles in Weed Management," 465 pp. Breton, North Scituate, Massachusetts.

Auld, B. A., and Tisdell, C. A. (1988). Influence of spatial distribution of weeds on crop yield loss. *Plant Prot. Q.* **3,** 31.

Baker, H. G. (1965). Characteristics and modes of origin of weeds. *In* "The Genetics of Colonizing Species" (H. G. Baker and G. L. Stebbins, eds.), pp. 147–172. Academic Press, New York.

Binkley, D. (1983). Ecosystem production in Douglas-fir plantation: Interaction of red alder and site fertility. *For. Ecol. Manage.* **5,** 215–227.

Binkley, D. (1984). Douglas-fir stem growth per unit leaf area increased by interplanted sitka alder and red alder, *For. Sci.* **30,** 259–263.

Bleasdale, J. K. A., and Nelder, J. A. (1960). Plant population and crop yield. *Nature (London)* **188,** 342.

Carlson, H. L., and Hill, J. E. (1985). Wild oat (*Avena fatua*) competition with spring wheat: Plant density effects. *Weed Sci.* **33,** 176–181.

Caswell, H., and Werner, P. A. (1978). Transient behavior and life history analysis of teasel (*Dipsacus sylvestris* Huds.). *Ecology* **59,** 53–66.

Concannon, J. A. (1987). The effect of density and proportion on spring wheat and *Lolium multiflorum.* M.S. thesis, Oregon State University, Corvallis, Oregon.

Concannon, J. A., and Radosevich, S. R. (1987). Intra- and inter-specific effects of wheat and ryegrass. *Proc. West. Soc. Weed Sci.*

Connolly, J. (1986). On difficulties with replacement-series methodology. *J. Appl. Ecol.* **23,** 125–137.

Connolly, J. (1987). On the use of response models in mixture experiments. *Oecologia* **72,** 95–103.

Cousens, R. (1985). A simple model relating yield loss to weed density. *Ann. Appl. Biol.* **107,** 239–252.

de Wit, C. T. (1960). "On Competition." Inst. Biol. Scheik. Onderz. Landbouwg., Wageningen, The Netherlands.

Dirzo, R. (1984). Herbivory: A phytocentric overview. *In* "Perspectives in Plant Population Ecology" (R. Dirzo and J. Sarukhan, eds.), pp. 141–165. Sinauer, Sunderland, Massachusetts.

Firbank, L. G., and Watkinson, A. R. (1985). On the analysis of competition within two-species mixtures of plants. *J. Appl. Ecol.* **22,** 503–517.

Firbank, L. G., and Watkinson, A. R. (1987). On the analysis of competition at the level of the individual plant. *Oecologia* **71,** 308–317.

Fischer, R. A. and Miles, R. E. (1973). The role of spatial pattern in the competition between crop plants and weeds. A theoretical analysis. *Math. Biosci.* **18,** 335–350.

Fowler, N. L., and Rausher, M. D. (1985). Joint effects of competitors and herbivores on growth and reproduction in *Aristolochia reticulata. Ecology* **66,** 1580–1587.

Goldberg, D. E., and Werner, P. A. (1983). Equivalence of competitors in plant communities: A null hypothesis and a field experimental approach. *Am. J. Bot.* **70,** 1098–1104.

Grime, J. P. (1979). "Plant Strategies and Vegetation Processes," Chaps. 1 and 2. Wiley, New York.

Grime, J. P., and Hunt, R. (1975). Relative growth rate, its range and adaptive significance in a local flora. *J. Ecol.* **63,** 393–422.

Gunsolus, J. L., and Coble, H. D. (1986). The area of influence approach to measuring weed interference effects on soybean. *Abstr. Weed Sci. Soc. Am.* **26,** 25.

Haas, H., and Streibig, J. C. (1982). Changing patterns of weed distribution as a result of herbicide use and other agronomic factors. *In* "Herbicide Resistance in Plants" (H. M. LeBaron and J. Gressel, eds.). Wiley, New York.

Harper, J. L. (1977). "Population Biology of Plants." Academic Press, London.

Hunt, R. (1982). "Plant Growth Curves." Univ. Park Press, Baltimore, Maryland.

Joliffe, P. A., Minjas, A. N., and Runeckles, V. C. (1984). A reinterpretation of yield relationships in replacement series experiments. *J. Appl. Ecol.* **21,** 227–243.

Levins, R. (1973). Fundamental and applied research in agriculture. *Science* **181,** 523–524.

Mack, R., and Harper, J. L. (1977). Interference in dume annuals: Spatial pattern and neighborhood effects. *J. Ecol.* **65,** 345–363.

Maxwell, B. D., Wilson, M. V., and Radosevich, S. R. (1988). A population modeling approach for studying leafy spurge (*Euphorbia esula*). *Weed Technol.* **2,** 132–138.

McGilchrist, C. A., and Trenbath, B. R. (1971). A revised analysis of plant competition experiments. *Biometrics* **27,** 659–671.

Miller, T. E., and Werner, P. A. (1987). Competitive effects and responses in plants. *Ecology* **68,** 1201–1210.

Obeid, M. (1965). "Experimental Models in the Study of Interference in Plant Populations," Ph.D. thesis. University of Wales, Bangor, Wales.

Patterson, D. T. (1982). Effects of light and temperature on weed/crop growth and competition. *In* "Biometeorology in IPM" (J. L. Hatfield and I. J. Thomason, eds.). Academic Press, New York.

Pickett, S. T. A., and Bazzaz, F. A. (1978). Organization of an assemblage of early successional species on a soil moisture gradient. *Ecology* **59**, 1248–1255.

Radosevich, S. R. (1987). Methods to study interactions among crops and weeds. *Weed Technol.* **1**, 190–198.

Radosevich, S. R. (1988). Methods to study crop and weed interactions. *In* "Weed Management in Agroecosystem: Ecological Approaches" (M. A. Actieri and M. Liebman, eds.), pp. 121–144. CRC Press, Boca Raton, Florida.

Radosevich, S. R., and Holt, J. S. (1984). "Weed Ecology: Implications for Vegetation Management." Wiley, New York.

Ross, M. A., and Harper, J. L. (1972). Occupation of biological space during seedling establishment. *J. Ecol.* **68**, 919–927.

Roush, M. L., and Radosevich, S. R. (1985). Relationships between growth and competitiveness of four annual weeds. *J. Appl. Ecol.* **22**, 895–905.

Roush, M. L., and Radosevich, S. R. (1987). A weed community model of germination, growth and competition of annual weed species. *Abstr. Weed Sci. Soc. Am.* **27**, 147.

Roush, M. L., and Radosevich, S. R. (1988). Growth ability and competition in summerannual weed communities. *Abstr. Weed Sci. Soc. Am.* **28**, 255.

Roush, M. L., Radosevich, S. R., Wagner, R. G., Maxwell, R. G., and Petersen, T. D. (1988). A comparison of methods for measuring effects of density and proportion in plant competition experiments. *Weed Sci.* **37**, 268–275.

Sagar, G. R., and Mortimer, A. M. (1976). An approach to the study of the population dynamics of plants with special reference to weeds. *Ann. Appl. Biol.* **1**, 1–47.

Salisbury, E. J. (1961). "Weeds and Aliens." Collins, London.

Shainsky, L. J. (1988). Competitive interactions between Douglas-fir and red alder seedlings: growth analysis, resource use, and physiology. Ph.D. thesis. Oregon State University, Corvallis, Oregon.

Shainsky, L. J., and Radosevich, S. R. (1987). Competitive interactions between Douglas-fir (*Pseudotsuga menziesii* Mirb. Franco) and red alder (*Alnus rubra*) seedlings: Growth analysis, resource use, and physiology. *Abstr. Weed Sci. Soc. Am.* **27**, 139.

Shainsky, L. J., and Radosevich, S. R. (1988). Competition coefficients and their behavior through time. *Abstr. Ecol. Soc. Am.* **69**, 293.

Shinozaki, K., and Kira, T. (1956). Intraspecific competition among higher plants. VII. Logistic theory of the C–D effect. *J. Inst. Polytech., Osaka City Univ. Ser. D* **7**, 35–72.

Snaydon, R. W. (1980). Plant demography in agricultural systems. *In* "Demography and Evolution in Plant Populations" (O. T. Solbrig, ed.). Univ. of California Press, Berkeley, California.

Spitters, C. J. T. (1983a). An alternative approach to the analysis of mixed cropping experiments. 1. Estimation of competition effects. *Neth. J. Agric. Sci.* **31**, 1–11.

Spitters, C. J. T. (1983b). An alternative approach to the analysis of mixed cropping experiments. 2. Marketable yield. *Neth. J. Agric. Sci.* **31**, 143–155.

Stewart, R. E., Gross, L. L., and Honkala, B. H. (1982). Effects of competing vegetation on forest trees: A bibliography with abstracts. *U.S., For. Serv., Gen. Tech. Rep.* **WO-43.**

Tarrant, R. F., and Trappe, J. M. (1971). The role of *Alnus* in improving the forest environment. *Plant Soil, Spec. Vol.* pp. 335–348.

Trenbath, B. R. (1976). Plant interactions in mixed communities. *In* "Multiple Cropping"

(R. I. Papendick, P. A. Sanchez, and G. B. Triplett, eds.), Am. Soc. Agron. Spec. Publ. 27, pp. 129–169. Am. Soc. Agron, Madison, Wisconsin.

Watkinson, A. R. (1980). Density dependence in single-species populations of plants. *J. Theor. Biol.* **83,** 345–357.

Watkinson, A. R. (1981). Interference in pure and mixed populations of *Agrostemma githago. J. Appl. Ecol.* **18,** 967–976.

Watkinson, A. R., Lonsdale, W. M., and Firbank, L. G. (1983). A neighborhood approach to self-thinning. *Oecologia* **56,** 381–384.

Watson, A. K. (1985). Integrated management of leafy spurge. *In* "Leafy Spurge" (A. K. Watson, ed.), pp. 93–104. Weed Sci. Soc. Am. Champaign, Illinois.

Weiner, J. (1982). A neighborhood model of annual–plant interference. *Ecology* **63,** 1237–1241.

Welden, C. W., and Slauson, W. L. (1986). The intensity of competition versus its importance: An overlooked distinction and some implications. *Q. Rev. Biol.* **61,** 23–44.

Whittaker, J. B. (1978). Invertebrate grazing, competition and plant dynamics. *In* "Population Dynamics Symposium" (R. M. Anderson, B. D. Turner, and L. R. Taylor, eds.), Br. Ecol. Soc. 20. Blackwell, Oxford, England.

Zimdahl, R. L. (1980). "Weed Crop Competition: A Review." Int. Plant Prot. Cent., Corvallis, Oregon.

III

The Impact of Herbivores, Parasites, and Symbionts on Competition

17

The Mediation of Competition by Mycorrhizae in Successional and Patchy Environments

Edith B. Allen   **Michael F. Allen**

I. Introduction

The growing evidence that mycorrhizal fungi have numerous physiological effects on individual plants has prompted a great deal of speculation, and some recent research, on their importance in community processes such as competition. This chapter examines the ways in which mycorrhizae may influence competition, and the conditions under which mycor-rhizae may be important to competition.

Mycorrhizal fungi form associations with the roots of approximately 90% of the terrestrial plant species that have been examined (e.g., Harley and Smith, 1983), so their inclusion in experiments may explain patterns and mechanisms of competition that were formerly unknown. The association is mutualistic, whereby the fungus obtains carbon from the plant, and the plant in return receives nutrients that are transported via the hyphal network.

The three major groups of mycorrhizae are discussed, including vesic-ular–arbuscular (VA) mycorrhizae, ectomycorrhizae, and ericoid mycor-rhizae. VA mycorrhizae are the most abundant of the endomycorrhizae, and are the most abundant worldwide of any of the groups. They are associations of certain species of zygomycetous fungi with both woody and herbaceous plants, and form internal structures for carbon storage (vesicles) and nutrient and carbohydrate exchange between fungus and plant (arbuscules).

Ectomycorrhizae, as their name implies, form hyphal structures that are external to the plant cells, and are associations of higher fungi (basid-iomycetes, ascomycetes) with woody plants. Ericoid mycorrhizae are as-sociations of higher fungi that form both internal and external hyphal structures in the roots of members of the family Ericaceae. We refer the reader to Harley and Smith (1983) or Read (1983) for a more detailed review of the biology of mycorrhizae.

While some of the mechanisms differ among the three groups (Read, 1983), they have each been demonstrated to cause physiological changes in many plant species. The kinds of physiological changes include in-creased rates of growth and seed production, increased water and nutri-ent uptake, increased drought stress tolerance, changes in hormonal balance, and a number of morphological and anatomical changes.

For mycorrhizae to change the competitive balance between neighbor-ing plants they must have different physiological effects on those plants. The recognition that plants exhibit different degrees of mycotrophy (= fungal feeding) came early with the work of Stahl (1900), who divided plants into nonmycotrophic, facultatively mycotrophic, and obligately mycotrophic categories. While the plants vary in their dependence on the fungi, the fungi are generally obligate mutualists (Lewis, 1973). In addition, the fungi exhibit little specificity in their plant associations, except for some species of ectomycorrhizal fungi that only infect certain species of plants (e.g., Molina and Trappe, 1982). The significance of this fact to competition is that neighboring plants may often be infected with the same fungal species or group of species (Read *et al.*, 1985), and different plant species are known to have different physiological responses to the same fungal species (e.g., Graw *et al.*, 1979; Allen *et al.*, 1984).

Among the three groups of fungi, plants that form associations with ectomycorrhizal and ericaceous species have large growth responses to infection fall on the obligately mycorrhizal end of the continuum of mycotrophy (Read *et al.*, 1985; Janos, 1987). These are largely gymnosperms and some woody angiosperms in temperate regions, and some woody angiosperms in the tropics and subtropics (Harley and Harley, 1987; Newman and Reddell, 1987). Species with ericoid mycorrhizae are largely confined to the order Ericales, which become dominant in temperate heathlands (Read, 1983). Early transplant experiments of ectomycorrhizal conifers from a variety of habitats demonstrated that without mycorrhizae the plants simply died, or they grew at such slow rates that they would be unlikely to survive with competition in the field (e.g., Kessell, 1927; Anonymous, 1931; Hatch, 1936; Briscoe, 1959). The VA mycorrhizal plants, on the other hand, are either facultatively or obligately mycotrophic (Janos, 1980a,b). Nonmycotrophic species, those which have never been observed to form mycorrhizae, are most often systematically related to the VA mycorrhizal species (Pendelton and Smith, 1983; Harley and Harley, 1987; Newman and Reddell, 1987). In some cases, the nonmycotrophic species are conspecific with mycorrhizal ones, as for the nonmycotrophic *Atriplex rosea* (Allen, 1984) and the VA mycorrhizal *Atriplex gardneri* (Allen, 1983).

Plants with different degrees of mycotrophy form a continuum from the least to the most responsive to mycorrhizal fungi. This continuum has been noted along successional sequences, where species in the early seral stages were mostly nonmycorrhizal, followed by facultative and obligate species (Janos, 1980a). Lack of mycotrophy is perhaps to be expected in early succession, as disruption of the soil, either natural or man-made, reduces or eliminates mycorrhizal inoculum (e.g., Powell, 1980; Allen and Allen, 1980; Allen *et al.*, 1987). The greatest potential for competitive interactions with mycorrhizae may be in seral communities where species with different degrees of mycotrophy exist as neighbors. The next section describes the degrees of mycotrophy that may be found in different seral biomes, and how mycorrhiza-mediated competition may be more important in some than in other biomes.

II. Mycorrhizae in Successional Biomes

Terrestrial biomes are often classified by measurements of temperature and precipitation (e.g., MacMahon, 1981), but a more important consideration for mycorrhizal response than temperature is soil nutrients. We

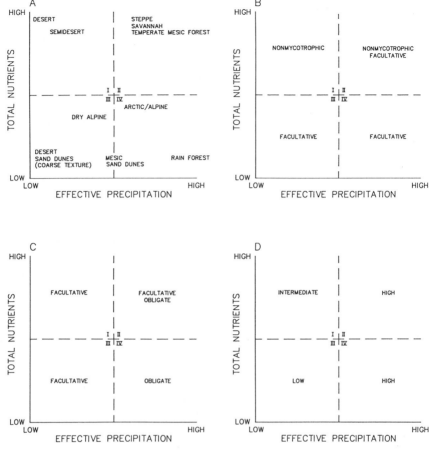

Figure 1   The relationship between effective precipitation and total nutrients for different biomes (A), degree of mycotrophy in early succession (B), degree of mycotrophy in late succession (C), and importance of micorrhizae in regulating competition in seral biomes (D).

have classified biomes from which we and others have data on mycorrhizae according to nutrient and soil moisture gradients (Fig. 1A). Effective rather than total precipitation is used for the *x* axis so that only water available for plant transpiration is considered. For instance, sandy soils even in mesic environments may have little available moisture because of deep infiltration.

Total rather than available nutrients are used because mycorrhizal fungi have enzymes that solubilize nutrients from the pool of total nutrients that are otherwise not available to plants. Enzymes that increase N uptake have been known for some time from ectomycorrhizal fungi

(Melin, 1953). There is more recent evidence that ericoid mycorrhizae also release enzymes for increased N uptake, which is critical to plant growth in the low-available-nutrient peat soils in which ericaceous plants are often found (Bajwa and Read, 1986). VA mycorrhizae are known to produce alkaline phosphatases for P uptake (Allen *et al.*, 1981; Gianinazzi-Pearson and Gianinazzi, 1983), and may increase the uptake of N as well (Ames *et al.*, 1984). The biomes cannot be arranged precisely along the nutrient axis because, of course, there is a great deal of variation in the soil nutrients of any one biome. In addition, the nutrient axis is represented in a synthetic form which combines the growth-limiting elements, especially N and P. While a separate axis for each element might lead to some differences in the precise placement of each biome, data on total nutrients of all biomes are not available in any case. Therefore, the graph is divided into four quadrants to show the approximate location of each biome.

The degree of mycotrophy of plants in each biome changes with seral stage, with plants that are less dependent on mycorrhizal fungi being most abundant during early succession. Nonmycotrophic species are most important in nutrient-rich biomes (Fig. 1B quadrants I and II), including colonizing annuals in such families as the Amaranthaceae, Chenopodiaceae, Brassicaceae, and Zygophyllaceae (Pendleton and Smith, 1983; Allen and Knight, 1984; Harley and Harley, 1987). Lack of mycotropy is to be expected most frequently in nutrient-rich biomes with labile, inorganic forms of nutrients because, without mycorrhize, these plants probably have a reduced capacity for nutrient uptake, especially organically bound nutrients. Disruption of the soil increases the proportion of inorganic nutrients (e.g., Vitousek and Reiners, 1975), and hence the ability of nonmycotrophic species to colonize. Nonmycotrophic species do exist in nutrient-poor biomes, but they are normally not the most important colonizing species and they may be restricted to patches of high-nutrient status. For instance, the nonmycotrophic annual *Salsola kali* is a common pioneer species in semideserts and grasslands (quadrants I and II), but it also grows in both inland (Utah) and coastal (Massachusetts) sand dunes representing the dry, nutrient-poor quadrant III of Fig. 1A. However, in the inland dunes it was found only in stabilized foredunes where *Psoralea lanceolata*, a colonizing legume, was abundant. In the coastal dunes it grew along with other nonmycotrophic annuals (*Atriplex arenaria*, *Cakile edentulata*), not in the nutrient-poor, shifting dunes but rather at the high tide line where oceanic deposition of debris may have contributed to the higher soil P (Allen and Allen, unpublished observations). Nonmycotrophic species are also likely to be less important in quadrant IV. For instance, early succession in the tropical rain forest is often dominated by woody facultative species (Richards, 1952;

Bazzaz and Picket, 1980), and nonmycotrophic species may be restricted to relatively nutrient-rich agricultural soils in the tropics (Janos, 1987).

A point of controversy exists in designating species as nonmycotrophic. Some species that were once thought to be nonmycotrophic (Gerdemann, 1968) have since been found with mycorrhizae after taking seasonality (Allen, 1983) and root depth (Virginia *et al.*, 1986) into account. For instance, sedges in a flooded field formed the early seral stage of Janos' (1980a) study on Barro Colorado Island. These were nonmycorrhizal upon examination, but sedges in nonflooded environments ranging from the alpine (Read and Haselwandter, 1981; Allen *et al.*, 1987; Lesica and Antibus, 1986) to semiarid and mesic grassland (Davidson and Christensen, 1977; Harley and Harley, 1987) do form mycorrhizae. Mycorrhizal fungi do not withstand flooding, probably because of oxygen depletion (Allen and St. John, 1982). Thus sedges should be considered facultatively mycorrhizal, not nonmycotrophic. Since many published studies on mycorrhizal occurrence are one-time observations, further observation is needed before positively designating a species as nonmycotrophic.

Observations of colonizing species in other biomes of the nutrient-poor quadrants III and IV indicate that facultatively mycorrhizal species are most important during early succession. Each of the species found in disturbed alpine communities in Montana was considered to be facultatively mycorrhizal because each occurred without mycorrhizae at least once during the multisite, 2-year survey (Allen *et al.*, 1987). All of the colonizing species from the Mount Saint Helens volcanic ash (quadrant IV) were facultatively mycorrhizal (Allen, 1987).

During late succession, the nonmycotrophic colonizing species diminish in importance and are replaced by facultative and obligate species (Fig. 1C). The mycorrhizal species from biomes with low precipitation (quadrants I and III) are considered to be facultative, as many have been grown with and without infection in the field and greenhouse (e.g., Reeves *et al.*, 1979; Lindsey, 1984). However, the usually obligately mycorrhizal genus *Pinus* is known from mesic sandy soils with relatively low effective precipitation (e.g., Buchholz and Motto, 1981). The obligate nature of some late seral mycorrhizal plants from the tropics (quadrant IV) has been verified by Janos (1980b), and obligate ericoid mycorrhizae are found in arctic and alpine tundra (Read, 1983). Contrary to the pattern suggested by Fig. 1C, some late seral trees in the tropics have also been reported without mycorrhizae (reviewed by Janos, 1987; St. John, 1980), but other workers found all observed tropical species with infection (e.g., Högberg, 1982). Given the importance of the mycelial network in nutrient uptake in poor tropical soils (Stark and Jordan, 1978; Janos, 1987), and the transient nature of mycorrhizal infection (Allen *et al.*,

1987), it seems prudent to repeat any one-time observations before concluding that the plants actually do not form mycorrhizae (Newman and Reddell, 1987).

Late seral species in quadrant II (high precipitation, high nutrients) may be facultative or obligate, depending on the biome. Although it appears to be contradictory for a plant to require mycorrhizae in an environment high in nutrients and water, these resources result in high plant productivity and hence competitive interactions. Resources are limiting in a competitive environment. Trees of temperate zones, whether coniferous or deciduous, are generally considered to be obligate (Ruehle and Marx, 1979). By contrast, the late seral grasses in mesic steppe are facultatively mycorrhizal, existing with and without mycorrhizal infection, although some authors consider them obligate because of their large growth response to mycorrhizae (Daniels Hetrick *et al.*, 1987).

A growth form classification of degree of mycotrophy becomes apparent from the preceding discussion, since certain plant growth forms are associated with particular biomes and mycorrhizal groups. In general, herbaceous species tend to be nonmycotrophic or facultative, while woody species are facultative or obligate (Table 1).

The biome approach shows that few biomes have dominant representatives of all three forms of mycotrophy during succession. Some adventive nonmycotrophic species may exist in nutrient-rich patches in sand dunes or tropical forest, as discussed above, but the only vegetation types that have dominant representatives of all three forms are probably semiarid or subhumid dwarf tree types. An example of this would be savannah with obligately mycorrhizal *Acacia* (Hoffman and Mitchell, 1986) or *Pinus–Juniperus* woodland in the western United States.

The importance of mycorrhizae in competitive relationships in each biome may depend on the co-occurrence of species that exhibit different degrees of mycotrophy (Fig. 1D). In biomes where facultatively mycorrhizal species are dominant during all stages of succession (quadrant III),

Table 1   The Relationship between Growth Form, Degree of Mycotrophy, and Mycorrhizal Group

Growth Forms	Degree of Mycotrophy	Mycorrhizal Group
Forbs		
Ruderals	Nonmycotrophic, facultative	VA mycorrhizal
Later seral	Facultative	VA mycorrhizal
Grasses	Facultative	VA mycorrhizal
Shrubs	Facultative, obligate	VA, ericoid, and ectomycorrhizal
Trees	Facultative, obligate	VA, ericoid, and ectomycorrhizal

competition should be less important than where two or all three forms of mycotrophy exist (quadrants I, II, and IV). Competition is given as intermediate in the high-nutrient, low-precipitation quadrant (I) because a dry environment may in itself cause competition to be less important (Fowler, 1986). Thus, the importance of mycorrhizae in mediating competition in any biome is dependent on the variability of response to infection by plants, coupled with changing levels of mycorrhizal inoculum (as during succession). Specific examples of competition with mycorrhizae are given below.

III. Competition and Mycorrhizae in Field and Greenhouse Experiments

To date, there are a number of published experiments on the effects of mycorrhizal fungi on competition, and most of these have been done in the greenhouse. This section is a review and discussion of these experiments, all done using VA mycorrhizal plants.

Some of the competition experiments focused on competition between species of different seral stages in Wyoming sagebrush steppe (Allen and Allen, 1984; Benjamin and Allen, 1987). When these species are arranged according to a successional cronosequence, their physiological responses in monoculture are related to this sequence, with greater response for late than early seral species (Table 2). Interestingly, the non-mycotrophic species (*Salsola kali* and *Atriplex rosea*) had decreased dry mass and stomatal conductance with inoculation, even though no mycorrhizal association was formed. The possible mechanisms for this negative effect of VA mycorrhizal fungi on nonmycotrophic species is reviewed in Allen and Allen (1988).

Although competition experiments have not been done for the late seral shrub species of Table 2, the earlier seral species were tested in the greenhouse using a de Wit replacement design (Table 3). The crowding coefficients (de Wit, 1960) that were calculated for aboveground biomass indicate that the addition of mycorrhizal inoculum to a mixture of mycotrophic perennial *Agropyron* spp. with nonmycotrophic annuals did not generally reverse the competitive outcome, but did confer a greater advantage in terms of increased biomass to the mycorrhizal neighbor. This increased biomass may prove beneficial in the long term in the field. The mycotrophic annual grass *Bromus tectorum* gained a slight benefit in mixtures with *Agropyron dasystachyum* or *Agropyron spicatum* with inoculation, even though it had reduced growth (Schwab and Loomis, 1987) or little response in monoculture (Table 3). Conversely, *Hordeum jubatum* had a growth response to infection in monoculture, but was at a slight disad-

Table 2 Relative Responses of Plant Species from Different Seral Stages in Wyoming Sagebrush Grassland to VA Mycorrhizal Infection[a]

Species	Dry Mass	CO_2 Exchange	H_2O Conductance	P Concentration	Source
Salsola kali	−, 0		−, 0	0	Allen and Allen (1984, 1988)
Atriplex rosea	−, 0		−		Allen (1984)
Bromus tectorum	0		0	0	Allen (1984)
Hordeum jubatum	+		+	0	Benjamin and Allen (1987)
Agropyron dasystachyum	+, −, 0		+, 0	0	Allen and Allen (1986), Benjamin and Allen (1987)
Agropyron smithii	+, 0	+	+	+, 0	Allen et al. (1984), Allen and Allen (1984)
Bouteloua gracilis	+, 0	++	++	+	Allen et al. (1981), Allen et al. (1984)
Artemisia tridentata	+++			+	Call and McKell (1985), Lindsey (1984)
Atriplex canescens	+++			+	Call and McKell (1985), Lindsey (1984), Aldon (1975)

[a] +, Increase; −, decrease; 0, no change.

Table 3   Crowding Coefficients (k_{12}) Calculated for Above-Ground Biomass in Mycorrhizal and Nonmycorrhizal Treatments[a]

		k_{12}		
Species 1	Species 2	NM	M	Source
Agropyron smithii	*Salsola kali*	0.68	0.90	Allen and Allen (1984)
Bouteloua gracilis	*S. kali*	0.30	0.38	
Agropyron dasystachyum	*S. kali*	0.30	1.10	Benjamin and Allen (1987)
A. dasystachyum	*Atriplex rosea*	1.80	2.57	
A. dasystachyum	*Bromus tectorum*	0.68	0.59	
A. dasystachyum	*Hordeum jubatum*	2.07	3.02	
Agropyron spicatum	*B. tectorum*	0.27	0.16	Schwab and Loomis (1987)
Lolium perenne	*Trifolium repens*	14.65	1.47	Hall (1978)
L. perenne	*T. repens*	2.33[b]	2.01[b]	

[a] If $k_{12} > 1$, species 1 has the biomass advantage, if $k_{12} < 1$, species 2 has the advantage. *Salsola kali* and *Atriplex rosea* are nonmycotrophic. M, Mycorrhizal; NM, nonmycorrhizal.
[b] Fertilized with 108 kg P/ha.

vantage in mixture. These results indicate that response to mycorrhizae in monoculture and mixture are not necessarily correlated. Some indirect mechanisms, rather than direct competitive interactions, may be involved, such as changes in species composition of fungi or percent root infection in mixtures compared to monocultures. Further research is needed to sort out the mechanisms involved.

A replacement experiment was also performed to test competition between *Lolium perenne* and *Trifolium repens* (Hall, 1978; Table 3). The addition of phosphorus had an effect on the crowding coefficient that was similar to inoculation, an expected result since mycorrhizae are known to increase the ability of plants to take up limited soil phosphorus (e.g., Allen *et al.*, 1981). The huge imbalance that favored the grass in the nonmycorrhizal mixture was corrected by inoculation. This suggests the hypothesis that in some situations mycorrhizae may improve diversity by increasing evenness. This hypothesis is further discussed in the studies on multispecies competition below.

Two additional competition experiments on *Lolium* and *Trifolium* that were done using mixtures only showed conflicting results (Table 4). Unfertilized mixtures of *Trifolium* had higher M/NM (mycorrhizal/nonmycorrhizal) ratios than *Lolium* in the experiments by Hall (1978) and by Crush (1974), but *Trifolium* M/MN ratios of unfertilized mixtures were lower in Buwalda's (1980) experiment. An interaction with nitrogen nutrition may explain the difference, as Crush's *Trifolium* was inoculated with *Rhizobium,* Hall's plants were growing in the field and were presumably naturally inoculated, but Buwalda's *Trifolium* plants were growing in sterile soil without the benefit of *Rhizobium.* Even with high soil N levels,

Table 4   Biomass Ratio of Mycorrhizal to Nonmycorrhizal (M/NM) Treatments[a]

| | M/NM | | | |
	Lolium perenne	*Trifolium repens*	*Holcus lanatus*	Source
Monocultures	0.88		1.11	Fitter (1977)
Mixtures	0.54		1.39	Fitter (1977)
Monocultures	1.06	3.30		Hall (1978)
Mixtures	0.77	23.80		Hall (1978)
Monocultures[b]	1.10	0.88		Hall (1978)
Mixtures[b]	1.02	0.95		Hall (1978)
Mixtures	0.94	2.90		Crush (1974)
Mixtures	2.40	0.70		Buwalda (1980)
Mixtures[c]	1.30	2.80		Buwalda (1980)

[a] Insufficient data were given to calculate crowding coefficients as in Table 3. Data from Hall (1978) are recalculated in this form for comparison.
[b] Fertilized with 108 kg P/ha.
[c] Fertilized with 280 mg P per 600 ml soil.

mycorrhizae did not enable Buwalda's *Trifolium* to overcome competition from *Lolium*.

An additional experiment on grass–grass competition was performed by Fitter (1977; Table 4), who showed that the competitive imbalance between *Lolium perenne* and *Holcus lanatus* in the potting mixture he used was due in part to reduced growth of mycorrhizal *Lolium* in monoculture. As shown in the studies on nonmycotrophic species (Table 3), mycorrhizal fungi may have detrimental effects on the biomass of some species. In such a case, the shift in balance in a competition experiment may not be due to competition at all, but rather to the direct effects of mycorrhizae on plants in the mixture.

Several multispecies competition studies have been carried out to determine how mycorrhizal fungi may structure plant communities. In an artificial community of mycotrophic and nonmycotrophic garden flowers, mycorrhizae shifted the balance toward the mycotrophic plants (Yocum, 1983; Table 5). However, with the addition of P fertilizer, the absence of mycorrhizal inoculum did not reduce the proportion of mycorrhizal plants. In another experiment on pasture plants, a mixture of forbs and grasses shifted toward a greater percent composition of forbs, although diversity and evenness were not changed (Grime *et al.*, 1987; Table 6). This suggests the hypothesis that mycorrhizae are instrumental in maintaining forbs in grass-dominated communities, as suggested by Hall's (1978) *Trifolium* field study. Finally, in a field experiment, a mixture of tropical woody species had greater evenness when mycorrhizal (Janos, 1981). It is apparent from these data that mycorrhizae cause shifts in plant communities, shifts that may occur under natural field

Table 5 Shift in Frequency of Nonmycotrophic and Mycotrophic Garden Flower Species with and without Mycorrhizal Inoculum, and with and without Added Phosphorus[a,b]

	Without P		With P	
	NM	M	NM	M
Nonmycotrophic	73.4 (3)	55.4 (3)	59.6 (4)	55.3 (4)
Mycotrophic	26.4 (4)	44.7 (4)	41 (5)	45.3 (4)

[a] Data from Yocum (1983).

[b] Values in parentheses are number of species per pot. M and NM, with and without mycorrhizal inoculum, respectively.

conditions where mycorrhizal inoculum has been reduced by disturbance. It is, however, premature to hypothesize that mycorrhizae increase the diversity or evenness of plant communities. This would be the case in a community where a high proportion of species have a large response to mycorrhizae, while a small proportion have little or no response and are successful without inoculum. A task for mycorrhizal researchers is to identify the physiological responses of more species within communities to mycorrhizal infection, so that we can begin to make predictions about the relationships of mycorrhizae and plant species diversity.

Each of these experiments was done in a situation where competition was an important mechanism in structuring species mixtures. However, this is not always the case. In a series of five field weed removal ×

Table 6 Shift in Grass and Forb Dry Mass and Composition with Mycorrhizal and Nonmycorrhizal Treatments in a Multispecies Experiment[a,b]

	Dry Mass (g) (No. of Species)		Composition (%)	
	NM	M	NM	M
Grasses	1168 (7)	1023 (7)	96.4	86.6
Forbs	44 (12)	158 (12)	3.6	13.4
Total	1212 (19)	1181 (19)		

[a] Data from Grime *et al.* (1987).

[b] All forb species had increased dry mass with infection except for the two nonmycotropic species, which had slightly decreased mass. Grass species had increased, decreased, or no change in mass with infection. M and NM, with and without mycorrhizal inoculum, respectively.

Table 7   Proposed Relationships among Mycorrhizae and
Competition during Succession

Importance of Competition in Succession	Importance of Mycorrhizae in Succession
Strong (competition)	+
Weak (tolerance)	0
Very weak (facilitation)	−

inoculation experiments in a semiarid, nutrient-rich environment (see
Fig. 1A), competition was important only in one of the field sites (Allen
and Allen, 1986, 1988, and unpublished observations). The removed
plants were primarily nonmycotrophic chenopods and brassicas and the
mycorrhizal *Bromus tectorum,* but the response of the remaining perennial grasses varied from increases (competition) to no change (tolerance)
and decreases (facilitation) (Table 7). In a species removal experiment an
increase in the remaining plants implies competition, no change in the
remaining plants implies no interaction or tolerance, and a decrease
implies facilitation of the remaining plants by the removed plants. The
importance of mycorrhizae in regulating the rate of succession may be
related to the importance of competition versus facilitation between species of different seral stages. Facilitation occurred on the harshest, windi-
est sites where a reduction of the weeds, either by removal or by inoculation, also meant loss of a litter cover that increased snow capture (Allen
and Allen, 1988). Inoculation caused reduced percent cover and density
of the nonmycotrophic weeds that facilitated grass establishment.

Although the initial effect of inoculation was to decrease the establishment of later seral grasses on this harsh site, mycorrhizae may still have
long-term benefits for this and the other sites. Physiological measurements showed that some of the grass species had decreased stomatal
resistance during drought, delayed phenology, and decreased leaf mortality with infection (Allen and Allen, 1986, and unpublished observations). Longer term observations of succession would be needed to sort
out these conflicting effects.

IV. Hyphal Connections in Patchy Environments

We have emphasized primarily mycorrhizal relationships in seral environments where inoculum density and plant species are changing. In
addition, smaller scale disturbances are important in any environment
(Pickett and White, 1985), so soil disturbance, inoculum loss, colonization by species with low dependence on mycorrhizae, and the potential

for competition which is mediated by mycorrhizae are continual processes. However, mycorrhizae may have a role even where disturbance is unimportant. Soil nutrients are naturally heterogeneous in their distribution (e.g., Allen and MacMahon, 1985). This section describes how mycorrhizae may transport nutrients from patches of high soil nutrients between plants interconnected by hyphae, and how this may affect competition between plants that grow in these patches.

The transport of nutrients via mycorrhizal hyphae was first demonstrated by injecting a pulse of labeled P onto hyphal tips of ecto- (Kramer and Wilber, 1949) and VA mycorrhizal (Hattingh *et al.*, 1973) plants. Since that time, a number of researchers have also injected ^{32}P into plant shoots and roots and detected the label in nearby plants of the same and other species (Woods and Brock, 1964; Chiarello *et al.*, 1982; Francis *et al.*, 1986; Newman and Ritz, 1986). Whether the ^{32}P was transported via hyphae or leaked into the soil and was picked up by roots of an adjacent plant is still under debate (e.g., Newman and Ritz, 1986). By contrast, Finlay and Read (1986) did not detect labeled P in the acceptor plant at all. There has been a great deal of speculation about how the potential sharing of nutrients among plants might structure plant communities, including reduced competition and greater similarity of plant niches.

The potential importance of hyphal connections between plants is related to the quantity of nutrients that may be transported through the hyphae, and the quantity of nutrients is related to gradients between the nutrient sources and sinks (Finlay and Read, 1986). Hyphal connections have been microscopically observed by a number of researchers, but the total number relative to the number of hyphae that enter the soil instead is unknown. In each of the experiments the amount of labeled P that reached an acceptor plant was several orders of magnitude less than the amount injected into the donor plant, so the number of connections are probably relatively low. Longer term growth experiments showed that this low level transport actually increased the P content of an acceptor plant (Francis *et al.*, 1986). Whether the P moved only via hyphae, or into the soil and was then picked up by a neighboring plant, has not yet been conclusively demonstrated. What is apparent from some of these experiments is that a plant which receives a nutrient pulse or is growing with a higher nutrient concentration than a neighbor may give some of these nutrients up to a neighboring plant.

Hyphal transport, or nutrient movement via root leakage, may be operative in natural communities with nutrient patches. An example of such a nutrient patch is from shrub "islands of fertility" in semiarid shrub steppe (Fig. 2; Allen and MacMahon, 1985). Soil P concentrations are higher under shrubs, but shrub roots reach into the nutrient-poor

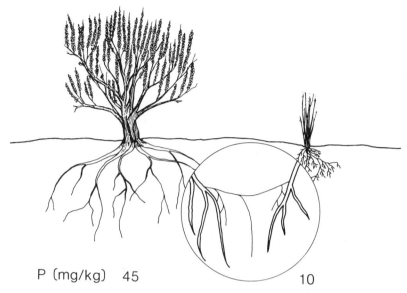

P (mg/kg)   45   10

Figure 2 Mycorrhizal hyphal connections between a shrub (*Artemisia tridentata*) and a grass. The soil under the shrub has higher bicarbonate-extractable P than the grass, possibly setting up a nutrient gradient. P data from Allen and MacMahon (1985).

interspaces dominated by grasses. Other species, such as the dominant grasses, occur in these interspaces where they may interact with the shrub roots. Two gradients of P movement can be constructed from this scenario, depending on the concentrations of soil P relative to the needs of the plant. A hypha which extends into the soil may absorb a phosphate ion and transport it to the hyphal junction between the two plants. The plant which receives the phosphate is that which has a greater demand for it, i.e., that plant which can set up a greater depletion gradient for phosphate to move in its direction. The shrub is likely to be this recipient more frequently, as it has already shown an ability to concentrate P in its rhizosphere. Under such conditions, it would not be energetically advantageous for the grass to maintain a hyphal connection with the shrub.

The second gradient of P movement would be from the shrub to the grass, perhaps during time periods of low shrub growth and high grass growth. This could occur only if the shrub experiences luxury uptake of excess P that can be transported via hyphae or leaked back out of the roots for grass uptake. The extent to which this occurs in natural plant communities is not known, and researchers differ in their opinions on whether hyphal connections will supply enough nutrients to affect the nutrient balance of neighboring plants. Competition experiments need to be done to show whether the nutrient donation of a plant with luxury consumption will overcome any competitive effects on the acceptor plant.

V. Other Considerations

A. Proteoid Roots

The Proteaceae family is an interesting exception to mycorrhizal formation that does not fit into our classifications of mycotrophy (Fig. 1). No mycorrhizal infection has been reported in any members of this family, which is limited to South Africa and Australia (Lamont, 1981). Even so, they grow in nutrient-poor Mediterranean-type climates, and may be early or late seral species of fire-adapted vegetation. They may be able to survive in low-nutrient soils by forming "proteoid" roots, which have increased root branching in response to nutrient patches in poor soils. Up to a 16-fold increase in root length was reported in experimentally created nutrient patches (Lamont *et al.*, 1984). Mycorrhizae were present in the earliest land plants (e.g., Pirozynski and Malloch, 1975; Stubblefield *et al.*, 1987) but this family has evolved a completely different mechanism to cope with limiting soil resources. Communities with Proteaceae should provide unique opportunities for research on the importance of mycorrhizae in competition, as they are typically interspersed with other species that do form mycorrhizae (Lamont, 1981).

B. Shifts in Fungal Species

The interest in mycorrhizal fungi has led to recent research on the structure and function of the fungal community itself. Identifications of mycorrhizal spores across gradients and seral chronosequences have shown shifts in spore species that are sometimes as great as the shifts in plant species (e.g., Anderson *et al.*, 1984; Allen *et al.*, 1987). In addition, species composition and abundance of mycorrhizal fungi change between seasons and years (Allen *et al.*, 1987; M. F. Allen and C. F. Friese, unpublished observations). These changes are important to community level processes because different species of fungi are known to confer different physiological responses to the same plant species, including different rates of growth, nutrient uptake, and water uptake (e.g., Allen and Boosalis, 1983; Stahl and Smith, 1984; Trappe, 1977; Sinclair and Marx, 1982). Further work is needed to determine how mycorrhizal fungal composition may affect plant composition.

VI. Conclusions

The available evidence indicates that mycorrhizae may be involved in regulation of competition wherever species with different responses to mycorrhizae are neighbors. There have been suggestions that they increase diversity and evenness by allowing otherwise less-competitive species to coexist. They also may influence succession where competition between species of different seral stages is important, as different seral stages exhibit different degrees of mycotrophy in many biomes.

Most published competition experiments did not consider mycorrhizae, and were often run in sterilized greenhouse potting soils. While the results of these experiments still have theoretical value, they probably seldom explain patterns of competition in the field. Mycorrhizal inoculation procedures need to involve little more than using nonsterile field soil, and can be easily incorporated into future experiments if realism is a goal.

We do not mean to infer that mycorrhizal fungi are necessarily the most important factor influencing the outcome of competition between plant species. In a series of experiments between the nonmycotrophic *Salsola kali* and the facultatively mycorrhizal *Agropyron smithii*, mycorrhizae had no greater influence on crowding coefficients than temperature or watering regime (Table 8). Other biotic factors are also important, as Grime *et al.* (1987) showed, mycorrhizae were as important as herbivory (clipping) in determining plant composition. The importance of mycorrhizae to competition will vary between plant species and environments.

Table 8 Crowding Coefficients Calculated for Above-Ground Dry Mass in Competition Experiments between *Agropyron smithii* (Species 1) and *Salsola kali* (Species 2)[a]

	k_{12}	Source
NM	0.68	Allen and Allen (1984)
M	0.90	
Dry	0.51	Allen (1982a)
Wet	1.14	
Warm	1.18	Allen (1982b)
Cool	1.45	

[a] $k_{12} > 1$ indicates that species 1 has the biomass advantage (see Fig. 2). In the moisture and temperature experiments, *A. smithii* was mycorrhizal. The mycorrhizal experiment was run at a warm temperature and intermediate moisture regime.

Their incorporation will help determine mechanisms of competition, for example, increased arbuscular development in a more competitive plant species (Caldwell *et al.,* 1985), and will explain one more portion of the unexplained variance in competition experiments.

VII. Summary

The increasing evidence that mycorrhizae improve nutrient and water uptake by plants, coupled with experimental studies on the effects of mycorrhizae on plant competition, have given rise to a great deal of speculation on their possible role in species composition in plant communities. The arguments in support of a role for mycorrhizae in competition are based on one logical premise: for mycorrhizal fungi to change the competitive balance between neighboring species, they must supply nutrients and water to their host plants at different relative rates. If two neighboring plants have resource acquisition rates with mycorrhizae that are simply higher and proportionally equal, the expected outcome would be greater growth but no change in the competitive ability of each plant. There is sufficient evidence on physiological effects of mycorrhizae on plant growth to state that mycorrhizae do have different effects on different species.

Three general groups of plants can be distinguished based on their different physiological responses to mycorrhizae: they may be nonmycotrophic, or facultatively or obligately mycotrophic. These groups generally are thought to exist in different seral stages, where nonmycotrophic plants are colonizing species followed by facultative and obligate species. Thus, the greatest possibility for the mediation of competition between plants may exist in seral communities, where groups of plants with these different physiological responses may exist as neighbors.

Not all environments have representatives of all three groups during succession, and in fact, some successions may begin and end with facultative species. We present pictorial models to show how combinations of nutrients and water in different environments may determine which groups are present during early and late succession. The importance of mycorrhizae in mediating competition may depend on the presence of more than one of the three groups.

Mycorrhizae may influence competition between two species in the same seral stage. Again, it is likely that there are subtle differences in the physiological responses that plants have to mycorrhizae for competition to occur. Nutrient "sharing" has been demonstrated for plants of the same and different species when they are linked by hyphae, and has caused some researchers to conclude that we must rethink our notions of plant interactions in communities. Hyphal nutrient transport may be important when nutrient gradients exist between plants (e.g., in patchy soils). Whether competition will eventually occur between plants connected by hyphae is dependent on the rate of depletion of soil nutrients and the nutrient demand of the plants.

Differences in the fungal community composition may also affect competition. Mycorrhizal fungal inoculum composition, density, and root infection vary with season, climatic fluctuation, and small-scale disturbance, so their contributions to plant interactions will also vary. The effects of mycorrhizae on competition are no greater than other variables, e.g., moisture or temperature. However, their inclusion is recommended where realism is a goal, and they may explain previously unknown mechanisms of competition.

Acknowledgments

We thank David Read, Keith Clay, and Mike Austin for reviewing the manuscript. This synthesis was supported by USDA grants 83-CRCR-1-1229 and 85-CRSR-2-2719 and NSF grant BSR 83-17358.

References

Aldon, E. F. (1975). Endomycorrhizae enhance survival and growth of four-wing saltbush on coal mine spoils. *USDA For. Res. Note* **RM-294,** 1–4.

Allen, E. B. (1982a). Germination and competition of *Salsola kali* with native C_3 and C_4 species under three temperature regimes. *Bull. Torrey Bot. Club* **109,** 39–46.

Allen, E. B. (1982b). Water and nutrient competition between *Salsola kali* and two native grass species (*Agropyron smithii* and *Bouteloua gracilis*). *Ecology* **63,** 732–741.

Allen, E. B. (1984). VA mycorrhizae and colonizing annuals: Implications for growth, competition and succession. *In* "VA Mycorrhizae and Reclamation of Arid and Semiarid Lands" (S. E. Williams and M. F. Allen, eds.), Univ. Wyoming Agric. Exp. Stn. Sci. Rep. SA1261, pp. 41–51. Univ. of Wyoming, Laramie, Wyoming.

Allen, E. B., and Allen, M. F. (1980). Natural reestablishment of vesicular–arbuscular mycorrhizae following stripmine reclamation in Wyoming. *J. Appl. Ecol.* **17,** 139–148.

Allen, E. B., and Allen, M. F. (1984). Competition between plants of different successional stages: Mycorrhizae as regulators. *Can. J. Bot.* **62,** 2625–2629.

Allen, E. B., and Allen, M. F. (1986). Water relations of xeric grasses in the field: Interactions of mycorrhizas and competition. *New Phytol.* **104,** 559–571.

Allen, E. B., and Allen, M. F. (1988). Facilitation of succession by the nonmycotrophic colonizer *Salsola kali* (Chenopodiaceae) on a harsh site: Effects of mycorrhizal fungi. *Am. J. Bot.* **75,** 257–266.

Allen, E. B., and Knight, D. H. (1984). The effects of introduced annuals on secondary succession in sagebrush–grassland, Wyoming. *Southwest. Nat.* **29,** 407–421.

Allen, E. B., Chambers, J. C., Connor, K. F., Allen, M. F., and Brown, R. W. (1987). Natural reestablishment of mycorrhizae in disturbed alpine ecosystems. *Arct. Alp. Res.* **19,** 11–20.

Allen, M. F. (1983). Formation of vesicular–arbuscular mycorrhizae in *Atriplex gardneri* (Chenopodiaceae): Seasonal response in a cold desert. *Mycologia* **75**, 773–776.

Allen, M. F. (1987). Reestablishment of mycorrhizas on Mount St. Helens: Migration vectors. *Trans. Br. Mycol. Soc.* **88**, 413–417.

Allen, M. F., and Boosalis, M. G. (1983). Effects of two species of VA mycorrhizal fungi on drought tolerance of winter wheat. *New Phytol.* **93**, 67–76.

Allen, M. F., and MacMahon, J. A. (1985). Impact of disturbance on cold desert fungi: Comparative microscale dispersion patterns. *Pedobiologia* **28**, 215–224.

Allen, M. F., and St. John, T. V. (1982). Dual culture of endomycorrhizae. *In* "Methods and Principles of Mycorrhizal Research" (N. C. Schenck, ed.), pp. 85–90. Am. Phytopathol. Soc., St. Paul, Minnesota.

Allen, M. F., Sexton, J. C., Moore, T. S., and Christensen, C. (1981). Influence of phosphate source on vesicular–arbuscular mycorrhizae of *Bouteloua gracilis*. *New Phytol.* **87**, 687–694.

Allen, M. F., Allen, E. B., and Stahl, P. D. (1984). Differential niche response of *Bouteloua gracilis* and *Pascopyrum smithii* to VA mycorrhizae. *Bull. Torrey Bot. Club* **111**, 361–365.

Ames, R. N., Porter, L. K., St. John, T. V., and Reid, C. P. P. (1984). Nitrogen sources and "A" values for vesicular–arbuscular and non-mycorrhizal sorghum grown at three rates of ^{15}N-ammonium sulphate. *New Phytol.* **97**, 269–276.

Anderson, R. C., Liberta, A. E., and Dickman, L. A. (1984). Interaction of vascular plants and vesicular–arbuscular mycorrhizal fungi across a soil moisture–nutrient gradient. *Oecologia* **674**, 111–117.

Anonymous (1931). Establishing pines. Preliminary observations on the effects of soil inoculation. *Rhod. Agric. J.* **28**, 185–187.

Bajwa, R., and Read, D. J. (1986). Utilization of mineral and amino N sources by the ericoid mycorrhizal endophyte *Hymenoscyphus ericae* and by mycorrhizal and non-mycorrhizal seedlings of *Vaccinium*. *Trans. Br. Mycol. Soc.* **87**, 269–277.

Bazzaz, F. A., and Pickett, S. T. A. (1980). Physiological ecology of tropical succession: A comparative review. *Annu. Rev. Ecol. Syst.* **11**, 287–310.

Benjamin, P. K., and Allen, E. B. (1987). The influence of VA mycorrhizal fungi on competition between plants of different successional stages in sagebrush–grassland. *Proc. North Am. Conf. Mycorrhizae, 7th* p. 144.

Briscoe, C. B. (1959). Early results of mycorrhizal inoculation of pine in Puerto Rico. *Caribb. For.* **July–December**, 73–77.

Buchholz, K., and Motto, H. (1981). Abundance and vertical distributions of mycorrhizae in plains and barrens forest soils from the New Jersey Pine Barreno. *Bull. Torrey Bot. Club* **108**, 268–271.

Buwalda, J. G. (1980). Growth of a clover–ryegrass association with vesicular arbuscular mycorrhizas. *N.Z. J. Agric. Res.* **23**, 379–383.

Caldwell, M. M., Eissenstat, D. M., Richards, J. H., and Allen, M. F. (1985). Competition for phosphorus: Differential uptake from dual-isotope-labeled soil interspaces between shrub and grass. *Science* **229**, 384–386.

Call, C. A., and McKell, C. M. (1985). Endomycorrhizae enhance growth of shrub species in processed oil shale and disturbed native soil. *J. Range Manage.* **38**, 258–261.

Chiarello, N., Hickman, J. C., and Mooney, H. A. (1982). Endomycorrhizal role for interspecific transfer of phosphorus in a community of annual plants. *Science* **217**, 941–943.

Crush, J. R. (1974). Plant growth responses to vesicular–arbuscular mycorrhizas. VII. Growth and nodulation in some herbage legumes. *New Phytol.* **73**, 743–749.

Daniels Hetrick, B. A., Gerschefske Kitt, D., and Thompson Wilson, G. (1987). Effects of drought stress on growth response in corn, sudan grass, and big bluestem to *Glomus etunicatum*. *New Phytol.* **105**, 403–410.

Davidson, D. E., and Christensen, M. (1977). Root-microfungal and mycorrhizal associations in a shortgrass prairie. *In* "The Belowground Ecosystem: A Synthesis of Plant-Associated Processes" (J. K. Marshall, ed.), Range Sci. Dep. Sci. Ser. 26, pp. 379–389. Colorado State University, Fort Collins, Colorado.

de Wit, C. T. (1960). On competition. *Versl. Landbouwkd. Onderz.* **66,** 1–82.

Finlay, R. D., and Read, D. J. (1986). The structure and function of the vegetative mycelium of ectomycorrhizal plants. II. The uptake and distribution of phosphorus by mycelial strands interconnecting host plants. *New Phytol.* **103,** 157–165.

Fitter, A. H. (1977). Influence of mycorrhizal infection on competition for phosphorus and potassium by two grasses. *New Phytol.* **79,** 119–125.

Fowler, N. (1986). The role of competition in plant communities in arid and semiarid regions. *Annu. Rev. Ecol. Syst.* **17,** 89–110.

Francis, R., Finlay, R. D., and Read, D. J. (1986). Vesicular–arbuscular mycorrhiza in natural vegetation systems. IV. Transfer of nutrients in inter- and intra-specific combinations of host plants. *New Phytol.* **102,** 103–111.

Gerdemann, J. W. (1968). Vesicular–arbuscular mycorrhiza and plant growth. *Annu. Rev. Phytopathol.* **6,** 397–418.

Gianinazzi-Pearson, V., and Gianinazzi, S. (1983). The physiology of vesicular–arbuscular mycorrhizal roots. *Plant Soil* **71,** 197–209.

Graw, D., Moawad, M., and Rehm, S. (1979). Untersuchungen zur Wirts- und Wirkungsspezifität der VA-Mykorriza. *Z. Acker- Pflanzenbau* **148,** 85–98.

Grime, J. P., Mackey, J. M. L., Hillier, S. H., and Read, D. J. (1987). Mechanisms of floristic diversity: Evidence from microcosms. *Nature (London)* **328,** 420–422.

Hall, I. R. (1978). Effects of endomycorrhizas on the competitive ability of white clover. *N.Z. J. Agric. Res.* **21,** 509–515.

Harley, J. L., and Harley, E. L. (1987). A checklist of mycorrhiza in the British flora. *New Phytol.* **105** (Suppl.), 1–102.

Harley, J. L., and Smith, S. E. (1983). "Mycorrhizal Symbiosis." Academic Press, New York.

Hatch, A. B. (1936). The role of mycorrhizae in afforestation. *J. For.* **34,** 22–29.

Hattingh, M. J., Gray, L. E., and Gerdemann, J. W. (1973). Uptake and translocation of ^{32}P-labelled phosphate to onion roots by endomycorrhizal fungi. *Soil Sci.* **116,** 383–387.

Hoffman, M. T., and Mitchell, D. T. (1986). The root morphology of some legume species in the south-western Cape and the relationship of vesicular–arbuscular mycorrhizas with dry mass and phosphorus content of *Acacia saligna* seedlings. *S. Afr. J. Bot.* **52,** 316–320.

Högberg, P. (1982). Mycorrhizal associations in some woodland and forest trees and shrubs in Tanzania. *New Phytol.* **92,** 407–415.

Janos, D. P. (1980a). Mycorrhizae influence tropical succession. *Biotropica* **12,** 56–64.

Janos, D. P. (1980b). Vesicular–arbuscular mycorrhizae affect lowland tropical rainforest plant growth. *Ecology* **61,** 151–162.

Janos, D. P. (1981). VA mycorrhizae increase productivity and diversity of tropical tree communities. *Proc. North Am. Conf. Mycorrhizae, 5th* p. 18.

Janos, D. P. (1987). VA mycorrhizas in humid tropical ecosystems. *In* "Ecophysiology of VA Mycorrhizal Plants" (G. R. Safir, ed.), pp. 107–134. CRC Press, Boca Raton, Florida.

Kessell, S. L. (1927). Soil organisms. The dependence of certain pine species on a biological soil factor. *Emp. For. J.* **6,** 70–74.

Kramer, P. J., and Wilbur, K. M. (1949). Absorption of radioactive phosphorus by mycorrhizal roots of pine. *Science* **110,** 8–9.

Lamont, B. B. (1981). Specialized roots of non-symbiotic origin in heathlands. *In* "Heathlands and Related Shrublands. Ecosystems of the World: 9" (R. L. Specht, ed.), pp. 183–195. Elsevier, Amsterdam.

Lamont, B. B., Brown, G., and Mitchell, D. T. (1984). Structure, environmental effects on their formation, and function of proteoid roots in *Leucadendron laureolum* (Proteaceae). *New Phytol.* **97,** 381–390.

Lesica, P., and Antibus R. K. (1986). Mycorrhizae of alpine fell-field communities on soils derived from crystalline and calcareous parent materials. *Can. J. Bot.* **64,** 1691–1697.

Lewis, D. H. (1973). Concepts in fungal nutrition and the origin of biotrophy. *Biol. Rev.* **48,** 261–278.

Lindsey, D. L. (1984). The role of vesicular–arbuscular mycorrhizae in shrub establishment. *In* "VA Mycorrhizae and Reclamation of Arid and Semiarid Lands" (S. E. Williams and M. F. Allen, eds.), Univ. Wyoming Agric. Exp. Stn. Sci. Rep. SA1261, pp. 52–69. Univ. of Wyoming, Laramie, Wyoming.

MacMahon, J. A. (1981). Successional processes: Comparisons among biomes with special reference to probable roles of and influences on animals. *In* "Forest Succession: Concept and Application" (D. West, H. Shugart, and D. Botkin, eds.), pp. 277–304. Springer-Verlag, New York.

Melin, E. (1953). Physiology of mycorrhizal relations in plants. *Annu. Rev. Plant Physiol.* **4,** 325–346.

Molina, R. J., and Trappe, J. M. (1982). Patterns of ectomycorrhizal host specificity and potential amongst Pacific Northwest conifers and fungi. *For. Sci.* **28,** 423–457.

Newman, E. I., and Reddell, P. (1987). The distribution of mycorrhizas among families of vascular plants. *New Phytol.* **106,** 745–751.

Newman, E. I., and Ritz, K. (1986). Evidence on the pathways of phosphorus transfer between vesicular–arbuscular mycorrhizal plants. *New Phytol.* **104,** 77–87.

Pendleton, R. L., and Smith, B. N. (1983). Vesicular–arbuscular mycorrhizae of weedy and colonizer plant species at disturbed sites in Utah. *Oecologia* **59,** 296–301.

Pickett, S. T. A., and White, P. S. (eds.) (1985). "The Ecology of Natural Disturbance and Patch Dynamics." Academic Press, Orlando, Florida.

Pirozynski, K. A., and Malloch, D. W. (1975). The origins of land plants: A matter of mycotrophism. *Biosystems* **6,** 153–164.

Powell, C. L. (1980). Mycorrhizal infectivity of eroded soils. *Soil Biol. Biochem.* **12,** 247–250.

Read, D. J. (1983). The biology of mycorrhiza in the Ericales. *Can. J. Bot.* **61,** 985–1004.

Read, D. J., and Haselwandter, K. (1981). Observations on the mycorrhizal status of some Alpine plant communities. *New Phytol.* **88,** 341–352.

Read, D. J., Francis, R., and Finlay, R. D. (1985). Mycorrhizal mycelia and nutrient cycling in plant communities. *In* "Ecological Interactions in Soil" (A. H. Fitter, ed.), pp. 193–217. Blackwell, Oxford, England.

Reeves, F. B., Wagner, D., Moorman, T., and Kiel, J. (1979). The role of endomycorrhizae in revegetation practices in semi-arid West. I. A comparison of incidence of mycorrhizae in severely disturbed versus natural environments. *Am. J. Bot.* **66,** 6–13.

Richards, P. W. (1952). "The Tropical Rain Forest: An Ecological Study." Cambridge Univ. Press, London.

Ruehle, J. L., and Marx, D. H. (1979). Fiber, food, fuel, and fungal symbionts. *Science* **206,** 419–422.

Schwab, S. M., and Loomis, P. A. (1987). VAM effects on growth of grasses in monocultures and mixtures. *Proc. North Am. Conf. Mycorrhizae, 7th* p. 264.

Sinclair, W. A., and Marx, D. H. (1982). Evaluation of plant response to inoculation. A. Host variables. *In* "Methods and Principles of Mycorrhizal Research" (N. C. Schenck, ed.), pp. 165–174. Am. Phytopathol. Soc. Press, St. Paul, Minnesota.

Stahl, E. (1900). Der Sinn der Mycorrhizen-bildung. *Jahrb. Wiss. Bot.* **34,** 539–667.

Stahl, P. D., and Smith, W. K. (1984). Effects of different geographic isolates of *Glomus* on the water relations of *Agropyron smithii*. *Mycologia* **76,** 261–267.

Stark, N. M., and Jordan, C. F. (1978). Nutrient retention by the root mat of an Amazonian rain forest. *Ecology* **59**, 434–437.

St. John, T. V. (1980). A survey of micorrhizal infection in an Amazonian rain forest. *Acta Amazonica* **10**, 527–533.

Stubblefield, S. P., Taylor, T. N., and Trappe, J. M. (1987). Fossil mycorrhizae: A case for symbiosis. *Science* **237**, 59–60.

Trappe, J. M. (1977). Selection of fungi for ectomycorrhizal inoculation in nurseries. *Annu. Rev. Phytopathol.* **15**, 203–222.

Virginia, R. A., Jenkins, M. B., and Jarrell, W. M. (1986). Depth of root symbiont occurrence in soil. *Biol. Fertil. Soils* **2**, 127–130.

Vitousek, P. M., and Reiners, W. A. (1975). Ecosystem succession and nutrient retention: A hypothesis. *BioScience* **25**, 376–381.

Woods, F. W., and Brock, K. (1964). Interspecific transfer of ^{45}Ca and ^{32}P by root systems. *Ecology* **45**, 886–889.

Yocum, D. H. (1983). The costs and benefits to plants forming mycorrhizal associations. PhD Dissertation. State University of New York, Stony Brook, New York.

18

The Impact of Parasitic and Mutualistic Fungi on Competitive Interactions among Plants

Keith Clay

I. Introduction

Ecologists have long been concerned with the biological and physical forces influencing the structure and dynamics of communities. This interest has been expressed in the debate over the importance of interspecific competition versus other biological interactions or abiotic factors

(Hairston *et al.*, 1960; Connell, 1983; Schoener, 1983; Hairston, 1986). However, questions on the importance of single factors in ecology obfuscate the complex interactions that occur among species. Communities and ecosystems consist of species at different trophic levels that interact simultaneously as competitors, predators, pathogens, and mutualists. Competitive interactions among species may be changed by the actions of predators (Connell, 1961; Morin, 1983), herbivores (Tansley and Adamson, 1925; Lubchenco, 1978; Bentley and Whittaker, 1979; Windle and Franz, 1979; Dirzo, 1984), and microorganisms (Burdon and Chilvers, 1977; Fitter, 1977; Hall, 1978; Burdon *et al.*, 1984; Clay, 1984).

Much of the previous empirical and theoretical work on competition has focused on the degree of niche overlap and the limiting similarity of coexisting species (May, 1981; Pianka, 1981). The morphological similarity of plants in many communities and their requirements for the same limited resources ensure some niche overlap (Fowler, 1981; Goldberg and Werner, 1983), yet many species appear to coexist indefinitely. Aarssen (1983) suggested that, even with substantial niche overlap, species can coexist given equivalent competitive abilities. Turkington and Aarssen (1984) provide evidence that selection for balanced competitive abilities occurs in pasture communities. Alternatively, the influence of other trophic levels may prevent or retard the development of competitive hierarchies and the exclusion of competitive subordinates from the community. No plant communities exist independently of other trophic levels. Competitive interactions among plant species therefore could be influenced by organisms from these different trophic levels, yet there are relatively few studies that explicitly examine the effects of viruses, bacteria, fungi, and animals on competitive interactions among plants. Beneficial or detrimental interactions with other organisms may profoundly affect the competitive dynamics of plant communities.

The purpose of this chapter is to review the role that fungi, both parasitic and mutualistic, play in mediating competitive interactions among plants and to consider how plant communities may be structured by fungi. Fungal infection can affect the realized niche of host plants and the host's competitive ability in those habitats. Such changes can result from direct effects on plant physiological and biochemical processes and from indirect effects on the relationship of host plants with biotic (e.g., herbivores and pathogens) and abiotic (e.g., mineral resources) features of their environment. The influence of mycorrhizal fungi on competition is considered elsewhere in this volume (see Allen and Allen). Unlike mycorrhizal fungi, which tend to infect all plants in a population (often all species in a community), the parasitic and nonmycorrhizal mutualistic fungi considered here tend to infect plants on an individual basis, which can have quite different consequences for plant populations and communities.

II. Mutualism and Parasitism

Fungal associates of plants can be divided arbitrarily into parasites and mutualists based on their effect on the fitness of host plants relative to uninfected plants. However, these dichotomous terms encompass a continuum of interactions that may not be fixed; a parasite under some circumstances can be a mutualist under others (Harley, 1968). Fungal parasites of plants encompass a broad array of fungi with a range of effects on their hosts (Burdon, 1987). According to Lewis (1988), there are three types of plant–fungal mutualisms (lichens, mycorrhizae, and endophytic, leaf-inhabiting fungi) that include a more narrow range of fungal species.

Considering plant competitive relations, in the simplest case, infection by a parasitic fungus reduces competitive ability while infection by a mutualistic fungus enhances it. Consider two species A and B that are equivalent competitors in the absence of fungal infection (Fig. 1). If B is infected by a parasitic fungus its yield in pure stands and its relative yield in interspecific competition (de Wit, 1960) is reduced (Fig. 1). The relative yield of A, the competing species, is increased as a result. The opposite situation exists when B is infected by a mutualistic fungus; its yield in pure stand and relative yield in mixture is increased compared to uninfected plants (Fig. 1). Simultaneously, the relative yield of A in mixture is reduced.

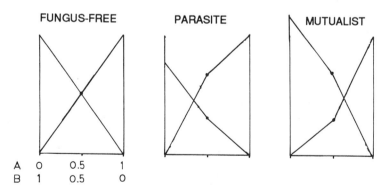

Figure 1   de Wit replacement series diagram of competition between species A and B. In the absence of fungal infection the two species are equivalent competitors, but the relative competitive abilities are altered when B is infected by a parasitic or mutualistic fungus.

III. Effects of Fungi on Plant Competition

A. Direct Physiological Changes in Hosts

Fungi are heterotrophic microorganisms that obtain their nutritional requirements directly from their host, often damaging host tissues in the process. Plants infected by parasitic fungi are weakened compared to uninfected plants of the same species, resulting in reduced growth and competitive abilities. Experimental studies of the effect of fungal parasites on plant competitive ability are rather few but illustrative. Grove and Williams (1975) found that the competitive inferiority of the Australian weed *Chondrilla juncea* in mixtures with *Trifolium subterraneum* was greatly exacerbated in the presence of the rust *Puccinia chondrillina*, which is specific to *C. juncea*. In competition experiments carried out in the greenhouse, barley and wheat were grown together in a replacement series in the presence or absence of powdery mildew (Burdon and Chilvers, 1977). Without disease, barley tended to exclude wheat, but in the presence of powdery mildew, the competitive advantage of barley was reduced and became roughly equivalent to wheat. In another study, Burdon *et al.* (1984) compared the intraspecific competitive abilities of two genotypes of *C. juncea*, one susceptible to *P. chondrillina* and the other resistant. In the absence of the rust, the susceptible line had a slight competitive advantage, but when rust was present, the resistant line was the significantly better competitor (Fig. 2). Thus, the presence of rust reversed the competitive outcome between the two lines. In experiments with *Senecio vulgaris* and the rust *Puccinia lagenophorae*, Paul and Ayres

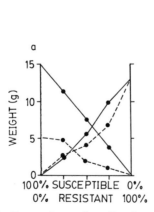
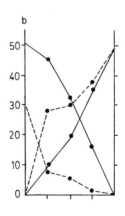

Figure 2 Data redrawn from Burdon *et al.* (1984) showing competitive interactions between two genotypes of *Chondrilla juncea* that are either susceptible or resistant to the rust *Puccinia chondrillina*. The solid lines represent the outcome of competition when the rust was absent, and dotted lines the outcome when the rust was present. There were two harvests, one of rosettes (a) and one of flowering plants (b).

(1986) found that infected plants exhibited significantly reduced growth in monocultures but that this disadvantage was increased further in mixtures with uninfected plants.

Reductions in seed or biomass production in agricultural and natural communities with increasing disease incidence can result from direct damage to the host and/or reduced competitive ability of hosts with neighboring conspecifics and other plants. Damage to individual crop plants may have little effect on total yield per unit area if surrounding uninfected plants compensate with increased growth. In contrast, natural plant communities are characterized by higher species diversity and greater genotypic diversity within populations. Plants surrounding infected individuals are likely to be of a different species or genotype. Compensatory growth of neighbors is therefore likely to lead to proportional increase of other species or different genotypes of the same species.

Mutualistic fungi can also influence plant competitive interactions. Mycorrhizal fungi typically enhance the ability of their hosts to obtain limiting resources from the soil. Fitter (1977) and Hall (1978) examined the effect of mycorrhizal fungi on plant competition. In the absence of VA mycorrhizae the two grasses *Lolium perenne* and *Holcus lanatus* were equivalent competitors, but when both plants were infected by mycorrhizal fungi, *L. perenne* was severely repressed by *H. lanatus* compared to its yield in mixture without mycorrhizae (Fitter, 1977). Hall (1978) found that infection with VA mycorrhizae had a much greater stimulatory effect on growth of *Trifolium repens* under conditions of interspecific competition than in monocultures. The role of mycorrhizae in plant competition is considered in greater depth elsewhere in this volume (see Allen and Allen).

The two other types of fungal mutualists with plants are lichens and leaf-inhabiting endophytic fungi (Lewis, 1988). Because lichens are an obligate symbiosis it is inappropriate to consider the effect of fungi on lichen competition. On the other hand, endophytic fungi, particularly clavicipitaceous endophytes infecting grasses, can have significant effects on the competitive abilities of their hosts. These fungi (tribe Balansiae, Clavicipitaceae, Ascomycetes) are systemic in their host plants (Diehl, 1950). There are five genera (*Atkinsonella, Balansia, Balansiopsis, Epichloe,* and *Myriogenospora*) encompassing about 20 species (Diehl, 1950; Luttrell and Bacon, 1977), plus asexual, or conidial, forms (anamorphs) that are given separate names from the sexual forms (telomorphs). The fungi infect at least 80 host genera and hundreds of species, including many important food, forage, turf, and weed grasses (Clay, 1986a). Some endophytes are completely internal and asexual and are transmitted by vegetative growth of hyphae into the developing ovules and seeds of the host while others produce sexual fruiting bodies on inflorescences or leaves of infected plants, which are often rendered sterile (Sampson, 1933; Neill, 1941; Clay, 1986a).

Hosts typically exist in mixed populations where they are often larger than uninfected plants (Bradshaw, 1959; Harberd, 1961; Clay, 1984, 1986b). In the greenhouse, infected perennial ryegrass (*Lolium perenne*) and tall fescue (*Festuca arundinacea*) produced significantly more tillers and biomass after 6, 10, and 14 weeks of growth (Clay, 1987a; see also Latch *et al.*, 1985). Forage production of endophyte-infected tall fescue and perennial ryegrass pastures was greater than in uninfected pastures (Mortimer and di Menna, 1983; Read and Camp, 1986). Endophyte-infected grasses are more common in older populations (Diehl, 1950; Large, 1952; Clay, 1986a; Lewis and Clements, 1986; Latch *et al.*, 1987), suggesting that infection builds up by contagious spread to new plants or by competitive displacement of uninfected plants. However, contagious spread is unusual (and impossible in some species) while increased vigor of infected plants is common (Clay, 1986a). Therefore, the more likely hypothesis is that infection provides a competitive advantage to host plants.

Several lines of evidence indicate that endophyte-infected grasses are better competitors than uninfected grasses. Following anecdotal observations, Bradshaw (1959) suggested that plants of *Agrostis tenuis* infected by *Epichloe typhina* were at a competitive advantage over uninfected plants in grazed pastures because of adaptive morphological changes in host plants, including suppression of flowering, denser tillering, and a more prostrate, creeping growth form. In a study of the grass *Danthonia spicata* infected by *Atkinsonella hypoxylon*, Clay (1984) and Kelley and Clay (1987) compared directly the competitive interactions of infected and uninfected ramets with the commonly co-occurring (and uninfected) grass *Anthoxanthum odoratum*. When two ramets of the same *Danthonia* genotype were grown together, there was little difference between infected and uninfected genotypes but, in competition with *Anthoxanthum*, the infected *Danthonia* genotypes consistently outperformed uninfected genotypes (Table 1). In another study, infected clones of *Danthonia* had significantly higher survival and growth than uninfected clones when planted into a dense sward of natural vegetation, further supporting the idea that infected plants have greater competitive ability (Clay, 1984; Antonovics *et al.*, 1987).

In a more recent experiment performed in a growth chamber, endophyte-infected and uninfected seedlings of tall fescue were planted in pure stands and mixtures in small pots where one central target individual was surrounded by five competing plants. In addition, target seedlings of tall fescue were surrounded with seedlings of uninfected perennial ryegrass. The plants were harvested after biomass per pot reached

Table 1 Comparison of Infected and Uninfected *Danthonia spicata* Genotypes[a,b]

	Intragenotypic		Interspecific	
	Infected (28)	Uninfected (48)	Infected (109)	Uninfected (197)
Growth				
Year 1	5.41 + 0.68	4.92 + 0.57	3.82 + 0.28*	2.90 + 0.16
Year 2	6.04 + 0.88	6.13 + 1.01	4.51 + 0.63*	2.44 + 0.17
Total	11.45 + 1.39	11.06 + 1.40	8.33 + 0.85*	5.35 + 0.29
Reproductive output				
Year 1	1.78 + 0.28*	0.82 + 0.17	1.27 + 0.12*	0.48 + 0.04
Year 2	2.90 + 0.52	2.68 + 0.60	2.72 + 0.36*	1.44 + 0.11
Total	4.68 + 0.72	3.50 + 0.68	3.98 + 0.44*	1.91 + 0.14

[a] From Kelley and Clay (1987).
[b] Intragenotypic treatment consisted of two ramets of the same genotype in one pot, while interspecific treatment consisted of one ramet of *D. spicata* and one ramet of *Anthoxanthum odoratum* in the same pot. Growth and reproductive output represent the numbers of vegetative tillers and reproductive tillers, respectively, at the sample date divided by the number of vegetative tillers at the start of the experiment. Asterisks denote significant differences between infected and uninfected genotypes ($p <$ 0.05). Number of pots in each treatment given in parentheses.

an asymptote. The performance of infected tall fescue was greater than that of uninfected plants in most comparisons, including competition with ryegrass (Table 2). Under conditions of intraspecific competition, target plants of uninfected tall fescue produced significantly less biomass when surrounded by infected competitors compared to uninfected competitors. However, when target plants of tall fescue were endophyte-

Table 2 Mean Weights (+ One Standard Error) of One Target and Five Surrounding, Competing Plants of *Festuca arundinacea* and *Lolium perenne* Grown in Competition[a]

Target	Competitor	Mean Target Weight (mg)	Mean Competitor Weight (mg)
Intraspecific competition			
NI Fescue	NI Fescue	63.5 + 3.3[a]	271.7 + 8.3[ab]
NI Fescue	I Fescue	50.7 + 3.2[b]	276.5 + 7.7[a]
I Fescue	I Fescue	59.4 + 3.0[ab]	290.0 + 5.1[a]
I Fescue	NI Fescue	52.8 + 3.2[b]	252.9 + 8.0[b]
Interspecific competition			
NI Fescue	Ryegrass	51.3 + 2.4[b]	238.1 + 4.9[a]
I Fescue	Ryegrass	57.4 + 2.8[a]	211.5 + 5.5[b]

[a] Experiment described in text. I and NI refer to infected and uninfected, respectively. $N = 47$ pots for each treatment. Within a column (intra- and interspecific competition considered separately), different letters indicate the means are significantly different as determined by one-way ANOVA.

infected, they produced significantly less biomass when surrounded by uninfected plants compared to when they were surrounded by infected plants, although infected target plants still produce more biomass on average than surrounding uninfected competitors (Table 2). Under conditions of intraspecific competition with perennial ryegrass, target plants of infected fescue produced significantly more biomass than uninfected target plants. Moreover, the surrounding ryegrass plants produced significantly less biomass when the target fescue plant was infected than when it was uninfected (Table 2). The results of this experiment suggest that infected plants are at a competitive advantage both in inter- and intraspecific mixtures.

B. Effects on Other Trophic Levels

In addition to changes of host physiologies that result directly in altered competitive ability of infected plants, fungi can induce or select for other changes in host plants that may indirectly affect competitive ability. Fungal infection can have important consequences for the interaction of the host plant with organisms at different trophic levels (Clay, 1987b), which themselves can alter plant competitive relationships. For example, prior infection by smut fungi increases the susceptibility of various grains to infection by rusts (Fischer and Holton, 1957). Conversely, infection by mycorrhizal fungi can reduce the probability of attack of several root pathogenic fungi (Dehne, 1982). Perennial ryegrass infected by the fungal endophyte *Acremonium lolii* was more resistant to infection by the rust *Puccinia coronata* than uninfected plants in growth chamber inoculation trials (unpublished observations). Simultaneous infections by several pathogenic fungi could severely weaken the host plant and reduce its ability to persist in a competitive community; alternatively, induced resistance to pathogen infection caused by a mutualistic fungus could provide further advantage to an infected plant.

Fungal infection can alter the likelihood of herbivory in many plants (Clay, 1987b), and herbivory can affect competitive interactions (Louda *et al.*, this volume). For example, Lewis (1984) found that rust-infected leaves of sunflower were significantly more attractive to grasshoppers than uninfected leaves. Alternatively, cotton infected by verticillium wilt resulted in poor growth of spider mites compared to uninfected plants (Karban *et al.*, 1987). Similarly, mycorrhizal soybeans resulted in reduced growth of two lepidopteran larvae compared to nonmycorrhizal soybeans (Rabin and Pacovsky, 1985). A particularly well-documented example of fungi deterring herbivory is clavicipitaceous fungal endophytes of grasses. Infected forage grasses can result in poisonings of cattle and other domestic animals (Bailey, 1903; Nobindro, 1934; Bacon *et al.*, 1977; Siegel *et al.*, 1987; White, 1987), and probably affect nondomestic mammalian herbivores. Infected plants also exhibit increased resistance to insect herbivory (Clay *et al.*, 1985; Hardy *et al.*, 1986; Cheplick and Clay, 1988). Prestidge *et al.* (1982) found that endophyte-infected perennial ryegrass plots sustained less damage and supported fewer numbers of Argentine stem weevils compared to uninfected plots. In laboratory studies fall armyworm larvae had reduced survival, lower weight gains, and increased developmental time when fed leaves from grasses infected by several clavicipitaceous endophytes compared to uninfected grasses (Clay *et al.*, 1985; Cheplick and Clay, 1988). The greater resistance of infected plants to herbivory appears to result from alkaloids produced by the fungi and present in host tissues (Bacon *et al.*,1986; Lyons *et al.*, 1986). Significant differences in herbivory among individuals within a population or among species resulting from differential endophyte-infection could alter competitive relationships.

The relationship between herbivory and competitive ability is controversial. While some authors have suggested that plants benefit by being eaten, most data indicate that plant fitness and competitive ability are reduced by herbivory (Bentley and Whittaker, 1979; Windle and Franz, 1979; McNaughton, 1983; Belsky, 1986; Westoby, 1986). The competitive consequences of fungal infection-induced changes in herbivory depend on the balance between the advantages or disadvantages of infection and herbivory. This balance is likely to be dynamic, depending on many factors that vary in time and space. If the competitive dominant is preferentially fed upon, then less competitive species can increase, and species diversity is promoted, but if the less competitive species is preferred then it is at a double disadvantage (Crawley, 1983). Interactions among competitive ability, fungal infection, and herbivory (or secondary pathogen infection) almost certainly will be missing in controlled environments unless these factors are built in to the experiments. In field situations these interactions can occur unbeknownst to the researcher.

C. Cost of Resistance

Infection by fungal parasites can result in selection for resistance loci that have pleiotropic effects on plant fitness or that are linked with loci that affect fitness and competitive ability (Harlan, 1976). The cost of resistance can be defined as the reduction in fitness of a resistant plant, compared to a susceptible plant, in the absence of the parasite (Fig. 3a). However, in the presence of the parasite, the resistant plant has relatively higher fitness and competitive ability compared to the susceptible plant (Fig. 3b). This is, of course, an oversimplification that could be considerably more complex depending on the nature of the resistance and its relationship with other plant characteristics. For example, constitutive resistance represents a constant cost to the plant, while facultative,

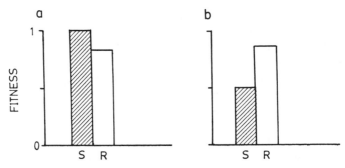

Figure 3   Mean fitness of susceptible (S) and resistant (R) genotypes when a pathogen is absent (a) or present (b). The reduction in fitness of the resistant genotype, compared to the susceptible genotype, in the absence of a pathogen is the cost of resistance.

or induced, resistance in response to a specific attack entails costs primarily during the period when defenses are mobilized, with little effect on plant competitive ability at other times. Variation in disease incidence can drive oscillations in plant populations in the proportions of resistant genotypes and, hence, the proportions of more versus less competitive genotypes. In mutualistic associations, resistance is a detrimental trait reducing plant fitness and so should not be expected to develop.

What evidence do we have for a competitive cost to pathogen resistance? There are few studies of natural plant populations. In the legume *Amphicarpea bracteata* there are genotypes resistant to various strains of the host-specific pathogen *Synchytrium decipiens*, but in greenhouse growth trials there was no evidence for reduced growth of resistant genotypes compared to susceptible forms (M. Parker, personal communication). In contrast, Burdon *et al.* (1984) observed lower pure stand yields and relative yields of rust-resistant *Chondrilla juncea* compared to susceptible plants in the absence of the rust (see also Fig. 2). There are several examples from the agricultural literature demonstrating a cost to pathogen resistance expressed as reduced growth and yield in competitive communities (Harlan, 1976). However, in some crops resistance alleles are known to actually increase yields, while in others the effect on yield may depend on the genetic background of the host plant (Frey and Browning, 1971; Brinkman and Frey, 1977). Additional research, especially with natural plant populations, is needed to determine whether a cost to pathogen resistance is a common phenomenon. The relative lack of empirical evidence primarily reflects the paucity of studies, not that there is no cost of resistance.

Plants resistant to herbicides and heavy metal-contaminated soils provide a useful analogy with pathogen-resistant plants. Several studies have documented a competitive cost of resistance in plants occurring on heavy metal-contaminated soils and in weeds subject to herbicide treatments (Cook *et al.*, 1972; Hickey and McNeilly, 1975; Warwick and Black, 1981). Experiments conducted on normal soils or in the absence of herbicides indicated that resistant plants were significantly less competitive than susceptible plants (see also Snaydon, 1971; Conrad and Radosevich, 1979; Weaver and Warwick, 1982), although the situation with heavy metal-tolerant plants has recently been questioned (MacNair, 1987). Plants resistant to insect attack also can exhibit reduced competitive ability or growth rate due to the energetic expenditures for chemical or physical defenses (Berenbaum *et al.*, 1986).

D. Apparent Competition

A shared parasite can cause "apparent competition" between hosts when no direct competition is taking place. Competition historically has referred to either direct interference or exploitation of shared resources. The concept of apparent competition was first introduced by Holt (1977) in the context of predator–prey relationships. A predator can produce indirect negative interactions that mimic competition between two prey species. An increase in the predator population that results from consuming prey species A results in increased predation of prey species B, and vice versa. The topic of apparent competition is considered in greater detail by Connell (this volume).

Apparent competition, as used in this chapter, is an indirect negative interaction between plants mediated by fungi. As such, it bears some similarity to exploitation competition, where plants interact indirectly through depletion of resources or through secretion of allelopathic compounds. However, unlike these traditionally accepted forms of "real" competition, apparent competition is dependent on the presence of additional species other than the competing species in question. While some researchers may feel that this is a semantic point, not a biological one, it nevertheless provides an operational definition for experimentally approaching apparent versus "real" competition. If future research shows that apparent competition is a widespread phenomenon, we will need to be more precise in our terminology and experimental design.

Consider a plant parasite that infects several co-occurring species. If two species are equally susceptible to infection by the same fungus and one species has a higher frequency of infection, individuals of the other species will then be more likely to become infected. This could lead to a decrease in their numbers due to death or reduced seed production. Thus, the abundance of the two species could be inversely related even though there may be no direct interaction or utilization of shared resources. With respect to mutualism, consider two species that both form mutualistic associations with the same fungus. If the growth of one species is limited by lack of fungal inoculum, this may reduce the probability of infection of the second species, in turn decreasing its potential growth. Alternatively, if inoculum is not limiting, an increasing frequency of infected plants of both species will tend to bring them into more direct competitive interaction by increasing their growth and requirements for limiting resources.

Apparent competition mediated by fungi will be limited to species or populations susceptible to infection by the same fungus. Some fungi require two different hosts to complete their life cycles, e.g., heteroecious rusts. Increases in one host result in increased inoculum and greater infection of the other host, decreasing its vigor and yield. Similarly, congeners or members of the same family are often susceptible to the same fungus and they typically have similar resource requirements. Both "real" and apparent competition should be expected to be more prevalent in related than in unrelated species.

Colonization events often could be influenced by apparent competition. An invader infected by a fungal parasite with low virulence can transmit the fungus to the resident species, where the fungus has a higher virulence. The resident species may be driven to local extinction, apparently by competition with the invader, but actually by the fungal parasite. The devastation of endemic American peoples with the introduction of novel diseases by Europeans and the extinction of Hawaiian avifauna with the introduction of blood parasites from introduced birds provide dramatic examples of this phenomenon (Anderson and May, 1986). Similar devastations of plant species are known (e.g., American chestnut, elm), but the original host of the pathogen is not filling the vacant niche left by the deaths of resident trees. In the case of chestnut, its extinction resulted in a major reorganization of the eastern North American forest community. Alternatively, resident species harboring a fungal parasite of low virulence can transmit it to invading species where it has greater virulence, reducing the probability of successful introductions, apparently by competitive exclusion. In his review, Harlan (1976) cites several examples of failed introductions of crop varieties into new areas because of resident pathogens that had less detrimental effects on the locally adapted crop varieties; Anderson (1972) and Freeland (1983) provide similar examples for animal systems. Coexistence of related species may depend on a certain degree of differentiation such that shared fungal parasites are few. Where parasites have significant negative effects on all hosts they can obscure competitive interactions and promote coexistence. The stable coexistence of codominant eucalypts in south-eastern Australia may be due to the presence of host-specific fungi and insects that prevent competitive exclusion (Burdon and Chilvers, 1974).

IV. Factors Affecting Infection Frequency

The significance of fungi on competitive relations among plants will be greater where a large proportion of plants in the population are infected and where infection has substantial effects on plant fitness. For fungi that systemically infect their hosts, the frequency of infection can be accurately estimated by the proportion of infected plants. The situation with fungi causing localized infections is more complicated. The proportion of infected plants, the proportion of infected leaves (or flowers, fruits, etc.) on an infected plant, and the number of lesions per leaf are all important parameters for estimating the incidence and severity of fungal infection (Campbell, 1986). Often, both host and fungal populations are dynamic so that the frequency of infection can vary dramatically over time.

The density and distribution of host plants influences the probability that infection occurs and spreads to uninfected plants. The frequency of infection is often positively correlated with the density of host plants (Burdon and Chilvers, 1982; Augspurger and Kelly, 1984), as is the intensity of competition (Clay and Shaw, 1981). Increasing host density results in a greater amount of biomass or leaf surface per unit area capable of intercepting inoculum and a reduction in the distance between plants that inoculum must travel (Burdon and Chilvers, 1982). High density may increase humidity and decrease light and wind speed in the immediate vicinity of the plants (Antonovics and Levin, 1980), which can favor infection by many pathogens (Weber, 1973). There may be a density threshold below which pathogens cannot persist (Alexander, 1984) and above which disease does not increase monotonically (Burdon and Chilvers, 1976).

The spatial distribution of plants and the frequency of the species within a community can influence the spread and frequency of fungal pathogens. Epidemics are most prevalent in agronomic situations where dense, evenly spaced, single-species communities occur, as opposed to natural, multispecies plant communities (Harper, 1977). Southern fusiform rust of pines, a rare occurrence in undisturbed forest communities, reaches epidemic proportions in southern pine plantations (Squillace, 1976). Clumped distributions result in lower levels of spread than even distributions, as shown experimentally by Burdon and Chilvers (1976).

The size of clumps also is important. Jennersten *et al.* (1983) found that the infection rate of *Viscaria vulgaris* by *Ustilago violacea* was lower in smaller, more isolated patches and higher in larger, less isolated patches. Janzen (1970) suggested that the wide dispersal and low density of conspecific tropical forest trees may reduce pest pressure. The results of Augspurger and Kelly (1984) were consistent with this idea. Accordingly, infection levels in a given host species may be lowest where many species coexist and highest when one species dominates a community.

Other factors inherent in the host plants themselves will influence the frequency of fungal infection. The genetic makeup of the host population can constrain the frequency and intensity of damage by pathogenic fungi. The genetic variability found in virtually all plant (and pathogen) populations will often result in degrees of resistance and/or damage among genotypes (Burdon *et al.*, 1983; Moseman *et al.*, 1983; Parker, 1985), and this genotypic mixture can change rapidly over time (Murphy *et al.*, 1982; Burdon *et al.*, 1981). The physiological status of plants, including prior exposure to the same fungus, will often affect the susceptibility of infection or the severity of the disease response. For example, drought stress can increase susceptibility to pathogens (Francke-Grossman, 1967). Competition itself can be stressful to plants, making them more susceptible to infection, which in turn can further reduce their competitive ability. Conversely, susceptibility to pathogen attack can be reduced by prior exposure to the same pathogen, unrelated pathogens, or insects (McIntyre and Miller, 1978; Dehne, 1982; Karban *et al.*, 1987).

Properties of fungal populations will also influence infection frequency. Most important are the density and distribution of inoculum sources, and the dispersal of inoculum and its efficiency in infecting new plants (Van der Plank, 1975; Campbell, 1986). Moisture and temperature conditions especially can influence the latter (Jones, 1986), and these can vary greatly over one growing season and between habitats. As with hosts, there is an important genetic component such that given fungal strains may be capable of infecting only a subset of species or genotypes within the community.

Similar factors probably operate in mutualistic associations between plants and fungi (see Allen and Allen, this volume). However, obligately seed-borne endophytes with no contagious spread are a special case. Here, only the relative number of seeds produced by infected versus uninfected plants ultimately determines the frequency of infection in populations (Clay, 1987a), i.e., the frequency of infection is determined totally by the fitness and competitive ability of infected plants. In general, fungi with limited powers of contagious transmission should not be detrimental to their hosts, otherwise they would go extinct.

V. Community Consequences

The significance of fungal infection on plant–plant interactions in communities ultimately arises from altered host physiology, with consequences for intra- and interspecific interactions. Plants infected by fungal pathogens tend to be weakened in competition with uninfected plants. Over time such plants will gradually be replaced by uninfected ones in competitive communities. The uninfected plants might be resistant genotypes of the same species or individuals of other species not afflicted by the same pathogen.

Several researchers have pointed out that plant disease can favor diversity in communities because host-specific pathogens become more significant as a community becomes dominated by a single species (Chilvers and Brittain, 1972; Harper, 1977). Burdon and Chilvers (1977) suggest three requirements must be met to prevent competitive exclusion by the dominant species: at least some of the pathogens are host-specific, pathogen numbers increase with host numbers, and high densities of pathogens cause damage sufficient to reduce host density. All of these requirements are frequently met in natural systems. In mixed communities, the most common species often will tend to be the best competitor and occur at the highest frequency and density, where it is most vulnerable to pathogens. Infection of the competitive dominant allows inferior competitors to persist and increase in the community. Given a diverse community and a multiplicity of pathogens capable of attacking any one or all plants species, frequency-dependent selection will prevent continuous dominance by one species and enhance species diversity and the persistence of rare species.

Intraspecific diversity may also be favored by fungal pathogens by maintaining polymorphism for disease resistance (Burdon, 1982; Dinoor and Eshed, 1984; Parker, 1985). Resistance may not become permanently fixed in populations if there is a substantial cost of resistance. If there are populations where all plants are resistant, the pathogen cannot persist, favoring susceptible plants with greater competitive ability (Harlan, 1976). The large amounts of genetic polymorphism present in populations may reflect selection by parasites and pathogens favoring rare genotypes (Haldane, 1949; Clarke, 1976). Thus, pathogenic fungi will tend to maintain species diversity within communities and genetic diversity within populations by the same processes.

Infection of plants by mutualistic fungi appears to be a prerequisite for survival and species persistence in many communities. Mycorrhizae are important in successional communities (see Allen and Allen, this volume). While early successional species are often nonmycorrhizal and midsuccessional species may be facultatively mycorrhizal, later successional, climax species tend to be obligately mycorrhizal in many biomes. Long-term persistence depends on forming mutualistic associations with fungi; nonmycorrhizal plants may be competitively excluded from successional communities. Similar patterns are present in endophyte-infected grasses (Large, 1952; Clay, 1986a; Watkin and Clements, 1978). With increasing age of the community, a larger proportion of species and individuals within species are endophyte-infected, reflecting their competitive superiority. Thus, mutualistic associations with fungi are often necessary for the persistence of plant species but parasitic fungi prevent the dominance of communities by one or several species.

VI. Conclusions and Future Research Directions

Both mutualistic and pathogenic fungi can play an important role in plant communities by affecting the competitive relationships among species and individuals. However, many fungi are not obvious, being microscopic and often internal, and their effects on host plants are often difficult to measure. As emphasized by Burdon (1987) in his recent book, the biological significance of pathogens in natural plant populations and communities is poorly understood, in contrast to the role of disease in agricultural communities. Given the meager data base now available, there are several areas of research that need to be pursued before many of the ideas presented here can be evaluated as being either conjecture or general principles. In particular, the question of whether fungi can mediate competition among plants will require additional field and greenhouse research with a greater diversity of plants and fungi than cited here. Second, the question of whether fungi do mediate competitive interactions calls for careful manipulations of plant and fungal populations in the field. We still know little about the numbers and frequency of pathogens (and mutualists) specific to different plants in natural communities. Are dominant species more frequently infected by pathogens or mutualists than less common species? Subtle and indirect effects of fungi on competition among plants (e.g., multitrophic level interactions, apparent competition) may be hard to evaluate but nevertheless may have a significant impact on plant communities. How frequently does infection with one fungus affect the probability of herbivory and infection by other fungi? The genetic correlates of susceptibility or resistance to pathogens need to be critically evaluated in light of their possible effect on competitive ability. Can plants be resistant to only a few pathogens if each entails a separate cost of resistance? In short, the same research effort that has generated the wealth of information on plant–insect interactions needs to be directed toward plant–fungal interactions.

Questions of whether competition, predation, mutualism, or parasitism are more important in natural plant communities ignore the essential interactive nature of these processes. The impact of fungi on the competitive interactions between plants is complex and includes both direct effects and a number of important indirect effects. Fungal infection often occurs in a frequency or density-dependent manner, with a large stochastic component. Because often only one or a few species are affected by a given fungus, the presence of fungi may hinder the development of competitive hierarchies and equilibria, and lead to increased species diversity within and among plant communities. To understand competitive interactions among plants at a single trophic level more fully, it is necessary to consider the role of other trophic levels.

References

Aarssen, L. (1983). Ecological combining ability and competitive combining ability in plants: Towards a general evolutionary theory of coexistence in systems of competition. *Am. Nat.* **122,** 707–731.

Alexander, H. M. (1984). Spatial patterns of disease induced by *Fusarium moniliforme* var. *subglutinans* in a population of *Plantago lanceolata*. *Oecologia* **62,** 141–143.

Anderson, R. M. (1972). The ecological relationship of meningeal worm and native cervids in North America. *J. Wildl. Dis.* **8,** 304–310.

Anderson, R. M., and May, R. M. (1986). The invasion, persistence, and spread of infectious diseases within animal and plant communities. *Philos. Trans. R. Soc. London, Ser. B* **314,** 533–570.

Antonovics. J., and Levin, D. A. (1980). The ecological and genetic consequences of density-dependent regulation in plants. *Annu. Rev. Ecol. Syst.* **11,** 411–452.

Antonovics, J., Clay, K., and Schmitt, J. (1987). The measurement of small-scale environmental heterogeneity using clonal transplants of *Anthoxanthum odoratum* and *Danthonia spicata*. *Oecologia* **71,** 601–607.

Augspurger, C. K., and Kelly, C. K. (1984). Pathogen mortality of tropical tree seedlings: Experimental studies of the effects of dispersal distance, seedling density, and light conditions. *Oecologia* **61,** 211–217.

Bacon, C. W., Porter, J. K., Robbins, J. D., and Luttrell, E. S. (1977). *Epichloe typhina* from toxic tall fescue grasses. *Appl. Environ. Microbiol.* **34,** 576–581.

Bacon, C. W., Lyons, P. C., Porter, J. K., and Robbins, J. D. (1986). Ergot toxicity from endophyte-infected grasses: A review. *Agron. J.* **78,** 106–116.

Bailey, V. (1903). Sleepy grass and its effects on horses. *Science* **17,** 392–393.

Belsky, A. J. (1986). Does herbivory benefit plants: A review of the evidence. *Am. Nat.* **127,** 870–892.

Bentley, S., and Whittaker, J. B. (1979). Effects of grazing by a chrysomelid beetle, *Gastrophysa viridual*, on competition between *Rumex obtusifolius* and *Rumex crispus*. *J. Ecol.* **67,** 79–90.

Berenbaum, M. R., Zangerl, A. R., and Nitao, J. K. (1986). Constraints on chemical coevolution: Wild parsnips and the parsnip webworm. *Evolution* **40,** 1215–1228.

Bradshaw, A. D. (1959). Population differentiation in *Agrostis tenuis* Sibth. II. The incidence and significance of infection by *Epichloe typhina*. *New Phytol.* **58**, 310–315.

Brinkman, M. A., and Frey, K. J. (1977). Growth analysis of isoline-recurrent parent grain yield differences in oats. *Crop Sci.* **17**, 426–430.

Burdon, J. J. (1982). The effect of fungal pathogens on plant communities. *In* "The Plant Community as a Working Mechanism" (E. I. Newman, ed.), pp. 99–112. Blackwell, Oxford.

Burdon, J. J. (1987). "Diseases and Plant Population Biology." Cambridge Univ. Press, Cambridge, England.

Burdon, J. J., and Chilvers, G. A. (1974). Fungal and insect parasites contributing to niche differentiation in mixed species stands of eucalypt saplings. *Aust. J. Bot.* **22**, 103–114.

Burdon, J. J., and Chilvers, G. A. (1976). The effect of clumped planting patterns on epidemics damping-off disease in cress seedlings. *Oecologia* **23**, 17–29.

Burdon, J. J., and Chilvers, G. A. (1977). The effect of barley mildew on barley and wheat competition in mixtures. *Aust. J. Bot.* **25**, 59–65.

Burdon, J. J., and Chilvers, G. A. (1982). Host density as a factor in plant disease ecology. *Annu. Rev. Phytopathol.* **20**, 143–166.

Burdon, J. J., Groves, R. H., and Cullen, J. M. (1981). The impact of biological control on the distribution and abundance of *Chondrilla juncea* in south-eastern Australia. *J. Appl. Ecol.* **18**, 957–966.

Burdon, J. J., Oates, J. D., and Marshall, D. R. (1983). Interactions between *Avena* and *Puccinia* species. I. The wild hosts: *Avena barbata* Pott. ex Link, *A. fatua* Durieu. *J. Appl. Ecol.* **20**, 571–584.

Burdon, J. J., Groves, R. H., Kaye, P. E., and Speer, S. S. (1984). Competition in mixtures of susceptible and resistant genotypes of *Chondrilla juncea* differentially infected with rust. *Oecologia* **64**, 199–203.

Campbell, C. L. (1986). Interpretation and uses of disease progress curves for root diseases. *In* "Plant Disease Epidemiology: Population Dynamics and Management" (K. J. Leonard and W. E. Fry, eds.), pp. 38–54. Macmillan, New York.

Cheplick, G. P., and Clay, K. (1988). Acquired chemical defences in grasses: The role of fungal endophytes. *Oikos* **52**, 309–318.

Chilvers, G. A., and Brittain, E. G. (1972). Plant competition mediated by host-specific parasites—A simple model. *Aust. J. Biol. Sci.* **25**, 749–756.

Clarke, B. (1976). The evolution of genetic diversity. *Proc. R. Soc. London, Ser. B* **205**, 453–474.

Clay, K. (1984). The effect of the fungus *Atkinsonella hypoxylon* (Clavicipitaceae) on the reproductive system and demography of the grass *Danthonia spicata*. *New Phytol.* **98**, 165–175.

Clay, K. (1986a). Grass endophytes. *In* "Microbiology of the Phyllosphere" (N. J. Fokkema and J. van den Heuvel, eds.), pp. 188–204. Cambridge Univ. Press, Cambridge.

Clay, K. (1986b). Induced vivipary in *Cyperus virens* and the transmission of the fungus *Balansia cyperi*. *Can. J. Bot.* **64**, 2984–2988.

Clay, K. (1987a). Effects of fungal endophytes on the seed and seedling biology of *Lolium perenne* and *Festuca arundinacea*. *Oecologia* **73**, 358–362.

Clay, K. (1987b). The effect of fungi on the interaction between host plants and their herbivores. *Can. J. Plant Pathol.* **9**, 380–388.

Clay, K., and Shaw, R. (1981). An experimental demonstration of density dependent reproduction in a natural population of *Diamorpha smallii*, a rare annual. *Oecologia* **51**, 1–6.

Clay, K., Hardy, T. N., and Hammond, A. M., Jr. (1985). Fungal endophytes of grasses and their effects on an insect herbivore. *Oecologia* **66**, 1–6.

Connell, J. H. (1961). Effects of competition, predation by *Thais lapillus*, and other factors on natural populations of the barnacle *Balanus balanoides*. *Ecol. Monogr.* **31**, 61–104.

Connell, J. H. (1983). On the prevalence and relative importance of interspecific competition: Evidence from field experiments. *Am. Nat.* **122**, 661–696.

Conrad, S. G., and Radosevich, S. R. (1979). Ecological fitness of *Senecio vulgaris* and *Amaranthus retroflexus* biotypes susceptible or resistant to atrazine. *J. Appl. Ecol.* **16**, 171–177.

Cook, S. C. A., Lefebvre, C., and McNeilly, T. (1972). Competition between metal tolerant and normal plant populations on normal soil. *Evolution* **26**, 366–372.

Crawley, M. J. (1983). "Herbivory: The Dynamics of Animal–Plant Interactions." Univ. of California Press, Berkeley, California.

Dehne, H. W. (1982). Interaction between vesicular–arbuscular mycorrhizal fungi and plant pathogens. *Phytopathology* **72**, 1115–1119.

de Wit, C. T. (1960). On competition. *Versl. Landbouwkd. Onderz.* **66**, 1–82.

Diehl, W. W. (1950). *Balansia* and the Balansiae in America. *U.S. Dep. Agric., Agric. Monogr.* **4**, 1–82.

Dinoor, A., and Eshed, N. (1984). The role and importance of pathogens in natural plant communities. *Annu. Rev. Phytopathol.* **22**, 443–466.

Dirzo, R. (1984). Herbivory: A phytocentric overview. *In* "Perspectives on Plant Population Ecology" (R. Dirzo and J. Sarukhan, eds.), pp. 141–165. Sinauer, Sunderland, Massachusetts.

Fischer, G. W., and Holton, C. S. (1957). "Biology and Control of the Smut Fungi." Ronald Press, New York.

Fitter, A. H. (1977). Influence of mycorrhizal infection on competition for phosphorus and potassium by two grasses. *New Phytol.* **79**, 119–125.

Fowler, N. (1981). Competition and coexistence in a North Carolina grassland: II. The effects of the experimental removal of species. *J. Ecol.* **69**, 843–854.

Francke-Grossman, H. (1967). Ectosymbiosis in wood-inhabiting insects. *In* "Symbiosis" (S. M. Henry, ed.), pp. 142–206. Academic Press, New York.

Freeland, W. J. (1983). Parasites and the coexistence of animal host species. *Am. Nat.* **121**, 223–236.

Frey, K. J., and Browning, J. A. (1971). Association between genetic factors for crown rust resistance and yield in oats. *Crop Sci.* **11**, 757–760.

Goldberg, D. E., and Werner, P. A. (1983). Equivalence of competitors in plant communities: A null hypothesis and a field experimental approach. *Am. J. Bot.* **70**, 1098–1104.

Groves, R. H., and Williams, J. D. (1975). Growth of skeleton weed (*Chondrilla juncea* L.) as affected by growth of subterranean clover (*Trifolium subterraneum*) and infection by *Puccinia chondrillina* Bubak & Syd. *Aust. J. Agric. Res.* **26**, 975–983.

Hairston, N. G. (1986). Species packing in *Desmognathus* salamanders: Experimental demonstration of predation and competition. *Am. Nat.* **127**, 266–291.

Hairston, N. G., Smith, F. E., and Slobodkin, L. B. (1960). Community structure, population control and competition. *Am. Nat.* **94**, 421–425.

Haldane, J. B. S. (1949). Disease and evolution. *Ric. Sci. Suppl.* **19**, 68–76.

Hall, I. R. (1978). Effects of endomycorrhizas on the competitive ability of white clover. *N.Z. J. Agric. Res.* **21**, 509–515.

Harberd, D. J. (1961). Note on choke disease of *Festuca rubra*. *Scott. Plant Br. Stn. Rep.* **1961**, 47–51.

Hardy, T. N., Clay, K., and Hammond, A. M., Jr. (1986). The effect of leaf age and related factors on endophyte-mediated resistance to fall armyworm (Lepidoptera : Noctuidae) in tall fescue. *J. Environ. Entomol.* **15**, 1083–1089.

Harlan, J. R. (1976). Diseases as a factor in plant evolution. *Annu. Rev. Phytopathol.* **14,** 31–51.

Harley, J. L. (1968). Fungal symbiosis. *Trans. Br. Mycol. Soc.* **51,** 1–11.

Harper, J. L. (1977). "Population Biology of Plants." Academic Press, New York.

Hickey, D. A., and McNeilly, T. (1975). Competition between metal tolerant and normal plant populations: A field experiment on normal soil. *Evolution* **29,** 458–464.

Holt, R. D. (1977). Predation, apparent competition, and the structure of prey communities. *Theor. Pop. Biol.* **12,** 197–229.

Janzen, D. H. (1970). Herbivores and the number of tree species in tropical forests. *Am. Nat.* **104,** 501–528.

Jennersten, O., Nilsson, S. G., and Wastljung, U. (1983). Local plant populations as ecological islands: The infection of *Viscaria vulgaris* by the fungus *Ustilago violacea*. *Oikos* **41,** 391–395.

Jones, A. L. (1986). Role of wet periods in predicting foliar diseases. *In* "Plant Disease Epidemiology: Population Dynamics and Management" (K. J. Leonard and W. E. Fry, eds.), pp. 87–100. Macmillan, New York.

Karban, R. Adamchek, R., and Schnathorst, W. C. (1987). Induced resistance and interspecific competition between spider mites and a vascular wilt fungus. *Science* **235,** 678–680.

Kelley, S. E., and Clay, K. (1987). Interspecific competitive interactions and the maintenance of genotypic variation within the populations of two perennial grasses. *Evolution* **41,** 92–103.

Large, E. C. (1952). Surveys for choke (*Epichloe typhina*) in cocksfoot seed crops, 1951. *Plant Pathol.* **1,** 23–28.

Latch, G. C. M., Hunt, W. F., and Musgrave, D. R. (1985). Endophytic fungi affect growth of perennial ryegrass. *N.Z. J. Agric. Res.* **28,** 165–168.

Latch, G. C. M., Potter, L. R., and Tyler, B. F. (1987). Incidence of endophytes in seeds from collections of *Lolium* and *Festuca* species. *Ann. Appl. Biol.* **111,** 59–64.

Lewis, A. C. (1984). Plant quality and grasshopper feeding: Effects of sunflower condition on preference and performance in *Melanoplus differentialis*. *Ecology* **65,** 836–843.

Lewis, D. H. (1988). Evolutionary aspects of mutualistic associations between fungi and photosynthetic organisms. *In* "Evolutionary Biology of the Fungi" (A. D. M. Rayner, C. M. Brasier, and D. Moore, eds.), pp. 161–178. Cambridge Univ. Press, Cambridge, England.

Lewis, G. C., and Clements, R. O. (1986). A survey of ryegrass endophyte (*Acremonium loliae*) in the U.K. and its apparent ineffectuality on a seedling pest. *J. Agric. Sci.* **107,** 633–638.

Lubchenco, J. (1978). Plant species diversity in a marine intertidal community: Importance of herbivore food preference and algal competitive abilities. *Am. Nat.* **112,** 23–39.

Luttrell, E. S., and Bacon, C. W. (1977). Classification of *Myriogenospora* in the Clavicipitaceae. *Can. J. Bot.* **55,** 2090–2097.

Lyons, P. C., Plattner, R. D., and Bacon, C. W. (1986). Occurrence of peptide and clavine ergot alkaloids in tall fescue grass. *Science* **232,** 487–489.

MacNair, M. R. (1987). Heavy metal tolerance in plants: A model evolutionary system. *Trends Ecol. Evol.* **2,** 354–359.

May, R. M. (1981). Models for two interacting populations. *In* "Theoretical Ecology: Principles and Practices" (R. M. May, ed.), pp. 78–104. Blackwell, Oxford, England.

McIntyre, J. L., and Miller, P. M. (1978). Protection of tobacco against *Phytopthora parasitica* var. nicotianae by cultivar-nonpathogenic races, cell-free sonicates, and *Pratylenchus penetrans*. *Phytopathology* **68,** 235–239.

McNaughton, S. J. (1983). Compensatory plant growth as a response to herbivory. *Oikos* **40,** 329–336.

Morin, P. J. (1983). Predation, competition, and the composition of larval anuran communities. *Ecol. Monogr.* **53,** 119–138.

Mortimer, P. H., and di Menna, M. E. (1983). Ryegrass staggers: Further substantiation of a *Lolium* endophyte aetiology and the discovery of weevil resistance of ryegrass pastures infected with *Lolium* endophyte. *Proc. N.Z. Grassl. Assoc.* **44,** 240–243.

Moseman, J. G., Nevo, E., and Zohary, D. (1983). Resistance of *Hordeum spontaneum* collected in Israel to infection with *Erysiphe graminis hordei. Crop Sci.* **23,** 1115–1119.

Murphy, J. P., Helsel, D. B., Elliot, A., Thro, A. M., and Frey, F. J. (1982). Compositional stability of an oat multiline. *Euphytica* **31,** 33–40.

Neill, J. C. (1941). The endophytes of *Lolium* and *Festuca. N.Z. J. Sci. Technol.* **23,** 185–193.

Nobindro, U. (1934). Grass poisoning among cattle and goats in Assam. *Indian Vet. J.* **10,** 235–236.

Parker, M. A. (1985). Local population differentiation for compatibility in an annual legume and its host-specific fungal pathogen. *Evolution* **39,** 713–723.

Paul, N. D., and Ayres, P. G. (1986). Interference between healthy and rusted groundsel (*Senecio vulgaris* L.) within mixed populations of different densities and proportions. *New Phytol.* **104,** 257–269.

Pianka, E. R. (1981). Competition and niche theory. *In* "Theoretical Ecology: Principles and Applications" (R. M. May, ed.), pp. 167–196. Blackwell, Oxford, England.

Prestidge, R. A., Pottinger, R. P., and Barker, G. M. (1982). An association of *Lolium* endophyte with ryegrass resistance to Argentine stem weevil. *Proc. N.Z. Weed Pest Control Conf., 35th* pp. 199–222.

Rabin, L. B., and Pacovsky, R. S. (1985). Reduced larva growth of two Lepidoptera (Noctuidae) on excised leaves of soybean infected with a mycorrhizal fungus. *J. Econ. Entomol.* **78,** 1358–1363.

Read, J. C., and Camp, B. J. (1986). The effect of fungal endophyte *Acremonium coenophialum* in tall fescue on animal performance, toxicity, and stand maintenance. *Agron. J.* **78,** 848–850.

Sampson, K. (1933). The systematic infection of grasses by *Epichloe typhina* (Pers.) Tul. *Trans. Br. Mycol. Soc.* **18,** 30–47.

Schoener, T. W. (1983). Field experiments on interspecific competition. *Am. Nat.* **122,** 240–285.

Siegel, M. C., Latch, G. C. M., and Johnson, M. C. (1987). Fungal endophytes of grasses. *Annu. Rev. Phytopathol.* **25,** 293–315.

Snaydon, R. W. (1971). An analysis of competition between plants of *Trifolium repens* L. populations collected from contrasting soils. *J. Appl. Ecol.* **8,** 687–697.

Squillace, A. E. (1976). Geographic patterns of fusiform rust infection in loblolly and slash pine plantations. *USDA For. Serv. Res. Note* **SE-232,** 1–4.

Tansley, A. G., and Adamson, R. S. (1925). Studies on the vegetation of the English chalk. III. The chalk grasslands of the Hampshire–Sussex border. *J. Ecol.* **13,** 177–223.

Turkington, R., and Aarssen, L. W. (1984). Local-scale differentiation as a result of competitive interactions. *In* "Perspectives on Plant Population Ecology" (R. Dirzo and J. Sarukhan, eds.), pp. 107–127. Sinauer, Sunderland, Massachusetts.

Van der Plank, J. E. (1975). "Principles of Plant Infection." Academic Press, New York.

Warwick, S. I., and Black, L. (1981). The relative competiveness of atrazine resistant and susceptible populations of *Chenopodium album* and *C. strictum. Can. J. Bot.* **59,** 689–693.

Watkin, B. R., and Clements, R. J. (1978). The effects of grazing animals on pastures. *In* "Plant Relations in Pastures" (J. R. Wilson, ed.), pp. 273–289. Commonw. Sci. Ind. Res. Org., Melbourne, Australia.

Weaver, S. E., and Warwick, S. I. (1982). Competitive relationships between atrazine resistant and susceptible populations of *Amaranthus retroflexus* and *A. powellii* from southern Ontario. *New Phytol.* **92,** 131–139.

Weber, G. F. (1973). "Bacterial and Fungal Diseases of Plants in the Tropics." Univ. of Florida Press, Gainesville, Florida.

Westoby, M. (1986). Mechanisms influencing grazing success for livestock and wild herbivores. *Am. Nat.* **128,** 940–941.

White, J. F. (1987). The widespread distribution of endophytes in the Poaceae. *Plant Dis.* **71,** 340–342.

Windle, P. N., and Franz, E. H. (1979). Plant population structure and aphid parasitism: Changes in barley monocultures and mixtures. *J. Appl. Ecol.* **16,** 259–268.

19

Herbivore Influences on Plant Performance and Competitive Interactions

Svaťa M. Louda   Kathleen H. Keeler

Robert D. Holt

I. Introduction

Both competition and herbivory can affect plant abundance and distribution (e.g., Harper, 1977; Whittaker, 1979; Crawley, 1983; Sih *et al.*, 1985; Fowler, 1986). However, it is still not clear when, how, or how often these two processes interact to determine plant community structure or dynamics. Theory predicts that consumers, including herbivores, will have major effects on their resource populations; the character of these effects depends on many factors, including feeding preferences, refuges, differential growth rates, differential recruitment rates, and competition among co-occurring plants (e.g., Crawley, 1983; Jeffries and Lawton, 1985; Holt, 1985; Maschinski and Whitham, 1989).

Leaf, root, and seed damage by herbivores is common and well documented (e.g., Harper, 1977; Edwards and Wratten, 1980; Hodgkinson and Hughes, 1982; Crawley, 1983, 1988a; Hendrix, 1988), as is selective consumption among co-occurring plants (e.g., Janzen, 1971; Morrow, 1977; Coley, 1983; Denno and McClure, 1983; Dirzo, 1985; Brown, 1985; Louda *et al.*, 1987a; Joern, 1989). But the significance of these observations remain controversial (e.g., Belsky, 1986; McNaughton, 1986). Fox and Morrow (1986), however, effectively argue that the effect of herbivory should be related to its differential impact on competing species and may be independent of the absolute amount of damage inflicted.

In this chapter, we consider three questions: (1) To what extent does herbivory affect plant growth and resource acquisition? (2) Will herbivory modify the intensity or alter the outcome of resource-mediated competitive interactions? and (3) When will such effects be most marked? By herbivory we mean consumption of living plant tissues, including grazing, browsing, defloration, seed predation, parasitism, and disease. Our examples are generally drawn, however, from the interactions with which we are most familiar: insects feeding on foliage and seeds.

Herbivory can decrease growth and fecundity, stimulate compensatory regrowth, or cause mortality (Harper, 1977; Crawley, 1983). So, herbivory might influence competitive interactions: (1) by changing a plant's relative ability to acquire limited resources, or (2) by eliminating the plant as a competitor. We develop theory and review evidence bearing on both effects. Crawley (1983, p. 8) has suggested, "The principal effect of herbivores on plant species richness acts not through the animals eating plants to extinction (although this can happen), but through their feeding modifying the competitive abilities of *one* plant species with another." We suggest that the impact of herbivory on competitive interactions among plants is and should be greatest in general when environmental constraints limit plant response and compensatory regrowth after selective consumption on one competitor. However, the available data are as yet insufficient for a definitive evaluation of this hypothesis. We conclude by suggesting methods for such studies.

II. Herbivory in Models of Competition

"Competition" generally means reciprocal negative interactions among individuals or populations. In ecological discussions, the term competition usually refers to negative interactions that arise from direct interference or, more indirectly, from the preemptive exploitation of limiting resources (Fig. 1). In multispecies communities, however, alternative indirect pathways leading to reciprocal negative interactions are possible, including shared natural enemies (Holt, 1977, 1984) and mixed mutual-

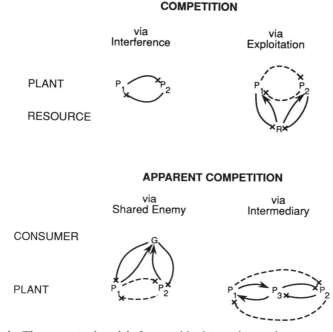

Figure 1 The conceptual model of competitive interactions and apparent competition. Adapted from Holt (1977, 1984) and Connell (this volume). The × indicates negative effect, whereas the arrow indicates the direction of a positive effect. Solid lines represent *direct effects,* and dashed lines represent *indirect effects.* Two directly competing plant species, P_1 and P_2 are represented. P_3 is added to represent a plant for which facilitation occurs, such as by a mutualistic association. R represents a limiting resource, and G represents a generalist herbivore that is a shared natural enemy of P_1 and P_2.

ism–competition systems (Connell, this volume); these alternatives are called "apparent competition" (Fig. 1).

In this chapter we schematically depict how herbivory can be included in the conceptual models corresponding to the disparate varieties of competition (Fig. 2). Furthermore, we examine how the impact of selective herbivory might influence competitive interactions. Specifically, we use a graphical model to examine two factors that will govern the impact of asymmetrical, selective herbivory on competing plants. The first factor is the benefit gained by the nonconsumed species. This benefit will vary inversely with the capacity of the consumed species to compensate for its losses. The second factor is the relationship of herbivore dynamics to plant dynamics; the effect on coexistence will reflect whether the selective herbivore acts as a density-independent *limiting* agent, with herbivore numbers set by mechanisms other than food availability or, instead, as a density-dependent *regulatory* agent.

COMPETITION

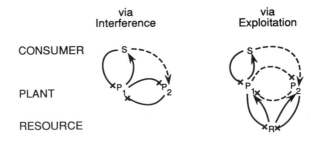

APPARENT COMPETITION

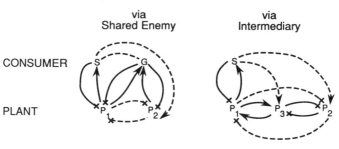

Figure 2 Incorporation of selective, differential herbivory into the conceptual model in Fig. 1. We divide the consumer trophic level into generalized, shared herbivores (G) and selective, more specialized herbivores (S). Other symbols follow Fig. 1. Selective consumers (S) feeding on competitor P_1 should indirectly benefit competitor P_2, independent of whether the competitive mechanism involves interference, exploitation, or apparent competition.

Most of the large body of theory on the potential role of herbivory on plant species coexistence (see Abrams, 1977; Holt, 1985) relies on the following structure for plant dynamics:

$$dN_i/dt = N_i[f_i(N_1, N_2, \ldots, N_j) - m_i(H, N_1, N_2, \ldots, N_j)] \qquad (1)$$

where H is herbivore density, and N_i is a measure of population size, such as density or biomass, for plant species i. The function f_i implicitly encapsulates both intra- and interspecific interactions among plants and, in particular, the negative density dependence inherent in competitive interactions (i.e., $\partial f_i/\partial N_j < 0$). The function m_i describes how herbivores depress the growth rate of population i, by increasing mortality or decreasing individual growth or reproduction. We use this basic theory and we assume both that the underlying competitive interaction between plants matches one of the models of Fig. 1, and that one competitor dominates the other, i.e., there are no priority effects.

In Fig. 3, we plot several possible density-dependent growth functions for the competitive dominant, species 1. The strength of density dependence is given by the slope of the per capita growth function, evaluated at density N_1. In fact, the absolute value of the slope is the marginal effect of a small change in density on per capita growth rate. With logistic population growth (Fig. 3A) the strength of density dependence is independent of population density. Alternatively, with nonlogistic growth (e.g., Fig. 3B), the strength of density dependence may diminish (curve a) or intensify (curve b) with increasing density. Herbivory may thus reduce plant population size or growth rate in either a density-independent (Fig. 3A,B) or density-dependent manner (Fig. 3C). The plant population will be in balance when its per capita growth just matches the reduction in size or growth rate caused by herbivory.

This graphical model illustrates several important potential effects of selective herbivory. First, increasing the level of selective herbivory on a dominant plant will clearly lower its equilibrial population size (Fig. 3A–C), altering interactions with subdominant competing plants.

Second, however, the magnitude of the herbivore-caused reduction will depend on several factors. For density-independent herbivory on a plant population with logistic growth, lowering r will increase the impact of a given level of herbivory (e.g., from $N_1 = 1$ to $N_1 = 2$: Fig. 3A). With nonlogistic growth and weak density dependence near K (Fig. 3B: curve a), even low levels of herbivory can severely depress host plant abundance. In contrast, however, with strong density dependence near K (Fig. 3B, curve b), the plant population can compensate and is not as severely depressed, for even quite high levels of herbivory.

Third, density-independent herbivory can also reduce a population's rate of increase at low plant densities (Fig. 3A,B). High levels of her-

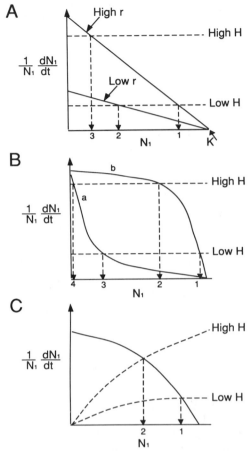

Figure 3 Effects of intraspecific density dependence and alternate population growth responses to herbivory. The solid lines in each part depict per capita growth rate as a function of density for the dominant competitor (N_1); the density dependence may include both direct interference and the effects of exploitative competition for a limiting resource, but it does not reflect effects of herbivores. In adding either density-independent (A, B) or density-dependent (C) herbivory, we assume for now that the impact of herbivores on plant populations is additive, decreasing population growth rate without modifying the underlying dynamics of intraspecific and interspecific competition. The difference between a given solid curve and the dashed line measures the realized intrinsic growth rate for the plant. Equilibrial density is determined by the density at which a solid curve and dashed line intersect. In A, the plant grows logistically and herbivory is density independent, as would be the case when herbivore dynamics are limited by factors other than plant density. With a low level of herbivory, a plant species with high r is depressed in abundance from K to the density denoted by 1, and a species with low r is depressed to the lower density denoted by 2. With a high level of herbivory, the high-r species is depressed to level 3, whereas the low-r species is completely eliminated. In B, the plant population grows nonlogistically, and herbivory remains density independent. Since the slope of the growth curve represents the

bivory may eliminate the plant from the local community (Fig. 3A), or reduce the relative growth rate of the competitive dominant significantly (Fig. 3B), potentially allowing it to be excluded competitively by the nonconsumed, subdominant plant species. Thus, density-independent herbivory is as likely to preclude coexistence as to allow it.

In contrast, if herbivory is strongly density dependent (Fig. 3C), it will only have a small effect on the capacity of the consumed species to increase when that species is rare. In this case, species with a density-dependent herbivore can be pushed to a lower density by consumption, permitting the coexistence of a competing species, but it should not itself tend to be excluded from the community by the herbivore.

In summary, this model suggests that selective herbivory on the competitive dominant will have its strongest effect when the plant has either a low r or weak density dependence near K, i.e., when the species has constraints on its capacity to compensate for reductions in abundance below its environmental carrying capacity. In these cases, the competitive dominant will be greatly reduced in density by herbivory, and its impact on competing, subordinate plant species will be correspondingly less. Conversely, a high level of compensatory recruitment by the consumed plant population should reduce the influence of selective herbivory on competitive interactions among co-occurring plants. Finally, the overall effect of selective herbivory on species coexistence will be sensitive to whether herbivory is density dependent or not.

The field evidence surveyed in our review (below) leads to similar predictions and, thus, appears consistent with these models. However, there are, as yet, few field experiments which simultaneously analyze herbivory and competition. Moreover, empirical studies of herbivore effects show that herbivores often modify individual plant traits in ways not yet incorporated into theoretical models (Louda and Keeler, 1989).

III. Herbivore Impact on Individual Plants

In many cases, chronic herbivory modifies the growth rate, the form, or the developmental timing of plants, and so changes plant traits that

---

degree of density dependence at any particular plant density, line **a** represents the case of weak density dependence at high N; in this case, low levels of herbivory will severely depress abundance for a plant population (3) and high levels of herbivory will cause near extinction (4). Line **b** illustrates the case of strong intraspecific density dependence at high N; in this case, high levels of herbivory are required to depress equilibrial abundance at all. In C, plant growth and herbivory are both density dependent. Here, if herbivory increases with population size, it may greatly depress the size of a plant population without affecting the rate at which the plant population can increase when its density is low.

appear critical to the acquisition of resources. In fact, the characteristics that determine competitive ability, either to deplete resources or to tolerate low resource levels (see Goldberg, this volume), are the traits most often modified by herbivores (Louda and Keeler, 1989). These observations lead to the hypothesis that herbivory can change the ability of a plant to acquire limited resources by altering key morphological traits.

Herbivores often slow growth, reduce biomass, or decrease plant stature (e.g., Gradwell, 1974; Morrow and LaMarche, 1979; Louda, 1984; Crawley, 1988a). Foliage consumption frequently modifies leaf traits, such as density, age structure, metabolism, and canopy shape (e.g., Mattson, 1977; Coley, 1983; Louda, 1984; Clark and Clark, 1985; Whitham and Mopper, 1985). Herbivory may also change (1) internal allocation of resources (e.g., Detling *et al.*, 1979; Chapin, 1980; Caldwell *et al.*, 1981; Mooney and Gulmon, 1982; Bazzaz *et al.*, 1987), (2) root:shoot ratios (e.g., Richards, 1984), (3) nutrient turnover rates (Mattson and Addy, 1975), and (4) litter accumulation rates (Belsky, 1986). All of these herbivore-induced modifications will influence plant performance, and thus potential competitive interactions among plants.

Plants with higher nutrient requirements (Berendse, 1985) or growth rates (Taylor and Bardner, 1968) were vulnerable to foliage losses. The quantitative impact of herbivory was then related to factors that determined the individual plant's ability to compensate for herbivore-caused foliage losses, such as its nutritional status (e.g., Karban and Courtney, 1987; Polley and Detling, 1989) or the specific growing conditions (e.g., Louda, 1982b, 1983; Maschinski and Whitham, 1989). Thus, generally, for plants with high resource requirements, herbivory should be expected to influence resource-mediated interactions when the limiting resource is scarce.

Factorial experiments designed to evaluate the effects of the impact of chronic herbivory on competitive interactions support the hypothesis that herbivores often modify individual plant growth and, in doing so, influence plant position in competitive hierarchies. Included in laboratory studies that are relevant here, is a 1964 study by Sibma *et al.*, summarized by Harper (1977). In this study, oats contributed disproportionately more to seed yield than did barley in a replacement series experiment in the greenhouse. However, the introduction of a root-feeding nematode (*Heterodera avenae*) depressed the competitive advantage of oats over barley and reversed the outcome of the interaction. Windle and Franz (1979) found similar results testing the effect of greenbugs on barley in a growth chamber experiment. Also, defoliation of *Holcus lanatus* had different effects on yield, depending on whether it was growing with versus without its competitor, *Lolium perenne* (Watt and Hagger, 1980). And, using clipping to simulate herbivory on competitive mixes of plants in greenhouse pots, Fowler and Rausher (1985) observed that defoliation modified the competitive outcome for *Aristolochia reticulata* and a dominant grass.

Field experiments that we have found in the literature include a recent evaluation of herbivore impact on competitive interactions of *Rumex*. A specialized leaf beetle, *Gastrophysa viridula* (Chrysomelidae), changed the relative growth and success of *Rumex obtusifolius* and *Rumex crispus*: alone, with each other, and with co-occurring grasses (Bentley and Whittaker, 1979; Bentley *et al.*, 1980; Cottam *et al.*, 1986). Cottam *et al.* (1986) concluded that (1) herbivory made the critical difference in the outcome of competition between *Rumex* species and grasses, and (2) the magnitude of the effect of herbivores depended on "other factors which impinge on the growth of the plant." Similarly, another experiment demonstrated that the suppression of *Dactylis glomerata* (Poaceae) by larger-statured *Trifolium* species (Fabaceae) was reversed by preferential grazing of slugs on *T. repens* (Cottam, 1986). Insects or slugs are also known to influence the growth or relative competitive performance of pasture species (Dirzo and Harper, 1980, 1982), a prairie forb (Louda *et al.*, 1989), forest sedges (Handel, 1976), and a desert shrub (Parker and Salzman, 1985).

In summary, preferential consumption by herbivores among co-occurring plants is common and it can affect individual plant characteristics (see Louda and Keeler, 1989). As predicted theoretically, such modifications can lead to differential performance among co-occurring competitors, especially when the environment or resource constraints prevent full regrowth and complete compensation for losses. However, most of the extensive evidence supporting this prediction is observational. More studies that directly measure the influence of herbivory and its interaction with individual competitive performance are needed, especially for native plants under field conditions.

IV. Herbivore Alteration of Population Dynamics and Resource Demand

Clearly, herbivory that kills plants decreases plant density, affecting density-dependent interspecific interactions. Thus, when herbivory changes the density or distribution of a host population, it should also alter the competitive pressure exerted by that population on its competitors. Mortality caused by selective herbivory contributes to local spatial variation in density. Such variation reflects both central place foraging by herbivores (Reichman, 1979; Huntly, 1987) and changes in herbivore pressure correlated with environmental heterogeneity (e.g., Burdon and Chilvers,

1974; Halligan, 1974; Rausher and Feeny, 1980; Parker and Root, 1981; Louda, 1982b, 1983).

Herbivory can also have a differential effect on reproduction among co-occurring plants (see Janzen, 1971; Crawley, 1983, 1989a; Hendrix, 1988; Louda, 1989). Although compensatory flowering occurs (e.g., Hendrix, 1979; Paige and Whitham, 1987), flower and seed consumption generally decreases seed numbers or quality and lowers the potential recruitment component of fitness (Janzen, 1971; Louda, 1982b, 1983, 1989; Hendrix, 1988; Louda *et al.*, 1989; Maschinski and Whitham, 1989).

The evidence that herbivory reduces density or recruitment, leaving space unoccupied and resources unsequestered, comes from several bodies of literature, as discussed in the rest of this section.

A. Rangeland Management

Exclosures of grazers, both native and domesticated, usually lead to shifts in species composition (see Scott *et al.*, 1979; Dyer *et al.*, 1982; Crawley, 1983, 1989a; Huntly and Inouye, 1988; Naiman, 1988; Naiman *et al.*, 1988). Selective use of plants by vertebrates provides a mechanism for such shifts (Harper, 1977), either by causing differential mortality or by changing competitive ability of the preferred plant. The plants that decrease under moderate grazing tend to be (1) relatively short-lived, often colonizing, perennials; (2) relatively palatable and nutritious to the large grazers; and (3) intermediate in growth rate and other life history traits between weedy annuals (ruderals) and long-lived, competitively dominant perennials. Interestingly, dominant grasses tend to respond positively to moderate-to-light grazing or clipping, *if* sufficient resources are available (e.g., Jameson, 1963; Norton-Griffiths, 1979; McNaughton, 1985; Painter and Detling, 1981; Archer and Detling, 1984; Seastedt, 1985). However, these responses also depend on the timing and severity of losses and the availability of resources for regrowth (Lee and Bazzaz, 1980; Wallace *et al.*, 1984; Coughenour *et al.*, 1985; Polley and Detling, 1989). Alternatively, nondominant grasses and forbs usually respond negatively to removal of leaf area (e.g., Bell, 1970; Crawley, 1983; Coughenour *et al.*, 1985).

Insect and nematode feeding is also selective, and its effects on plant morphology, growth, and reproduction could influence competitive interactions in rangelands. Examples of invertebrates using and differentially affecting grassland plants include foliage-feeding grasshoppers (e.g., Chandra and Williams, 1983; Landa and Rabinowitz, 1983; Joern, 1989), flower- and seed-feeding insects (e.g., Henderson and Clements, 1979; Kinsman and Platt, 1984; Louda *et al.*, 1987b), and root-feeding invertebrates (Ueckert, 1979; Ingham and Detling, 1986; Seastedt *et al.*, 1987). Exclusions of insects, such as fruitfly (*Oscinella* sp.), have led to changes in plant species composition, with decreases in the weedier species in British grasslands (Clements and Henderson, 1979). Consumers and plant competition also interact to determine the recruitment of a native, monocarpic thistle (*Cirsium canescens*) in Sandhills Prairie (Louda *et al.*, 1989). In this case, predispersal seed predation by insects restricted both the number of viable seeds and the number of seedlings established (Table 1). Additionally, seedling recruitment was 16 times higher in the open than in grass (8 versus 0.5%). Seedling survivorship was reduced in proximity to switchgrass (*Panicum virgatum*; Table 2). In this case, insect seed destruction and subsequent interaction of thistle seedlings with established grasses should jointly determine population density and growth rate (Louda and Potvin, in preparation). Many other studies also suggest that both herbivory and competition may be important to population dynamics (see Janzen, 1971; Hendrix, 1988; Louda, 1989).

Table 1 Effects of Flower and Seed Herbivory on Seed Production, Average Seedling Establishment, and Subsequent Survival of Platte Thistle (*Cirsium canescens*) in Disturbed Blowouts at Arapaho Prairie, in the Sandhills Prairie Ecosystem of Arthur County, Nebraska[a,b]

	Control, Water Only		Insecticide in Water	
	$\overline{X}$	SE	$\overline{X}$	SE
1984 Exclusion Experiment ($N = 6,5$)				
Total seeds	487	89	651	101
Viable seeds	27	6.3	105	20.1
1985 seedlings	0.6	0.37	3.3	0.71
1988 adults	0.05	0.03	0.33	0.16
1985 Exclusion Experiment ($N = 6,6$)				
Total seeds	652	76	783	121
Viable seeds	53	5.6	105	12.3
1986 seedlings	0.35	0.20	2.7	0.69
1988 adults	0.07	0.05	0.41	0.16

[a] The data given here were taken from Louda and Potvin (in prep.) and were calculated per plant by blowout for each treatment, *N* being the number of blowouts per treatment; establishment and survival were recorded in the third week of May (1985–1988); the same blowouts were used in 1985, with treatments reversed. By May 1988, we had accounted for the fate of 96.4 and 78.0% of all seedlings established in the 1984 and 1985 experiments, respectively.

[b] Two-way ANOVA model explained a significant portion of the variance ($p < 0.01$); there was a significant treatment effect for all stages (all $p < 0.03$) subsequent to the initial one (total seeds); years were not significantly different.

Table 2   Survival of Thistle (*Cirsium canescens*) Seedlings Transplanted Into Open Areas of a Grass (*Panicum virgatum*) Clone and into the Open Area Adjacent to the Clone[a]

	Number Surviving[b]		Percent Surviving
	$\bar{X}$	SE	
After 9 weeks			
In grass	0.14	0.14	4.8
In open	1.29	0.36	42.9
After 2 years			
In grass	0.00	—	0.0
In open	0.29	0.18	9.5

[a] From Louda *et al.* (1989).
[b] $\bar{X}$, Average for the 7 replicates of 3 seedlings per treatment replicate, with two treatments: (a) in open areas within a clone of grass (*Panicum virgatum*), and (b) in open areas adjacent to the clone; univariate ANOVA on arcsine-transformed proportions, $p < 0.02$ after 9 weeks. The study was carried out from May 22, 1986 to May 20, 1988 at Arapaho Prairie, Arthur County, Nebraska.

Vertebrate and invertebrate herbivores may have different, and at times conflicting, effects on plant dynamics and thus on the competitive interactions among plants (e.g., Inouye *et al.*, 1980; Davidson *et al.*, 1985; Gibson *et al.*, 1987; Crawley, 1989b). However, the actual role of chronic, differential insect herbivory, as well as its interactions with vertebrate herbivory in plant competitive interactions, remains to be assessed directly under field conditions for the majority of plants in most natural communities.

B. Outbreaks

Significant temporal and spatial patterns of damage and influence are correlated with eruptions of herbivores. Outbreaks usually reduce the growth and resource acquisition of dominant plants (see Mattson, 1977; Barbosa and Schultz, 1987; Mattson and Haack, 1987; Joern, 1989), as well as accelerate nutrient turnover and regeneration in the community (e.g., Mattson and Addy, 1975).

For example, the interaction of heather, heather beetles, and grasses provides a case where experimental data suggest a coupling between cycles of insect abundance and of vegetation change (see Berdowski and Zeilinga, 1983, 1987; Heil and Diemont, 1983). In the heathlands of The Netherlands declines in abundance of the dominant heather (*Calluna vulgaris,* Ericaceae) and simultaneous expansions of subdominant grasses are correlated with the opening of the canopy by outbreaks of the monophagous heather beetle. Field experiments showed that (1) removal of heather led to increased grass cover, (2) increases were greatest in heather removal plots that were fertilized, and (3) feeding and growth of heather beetles increased most on fertilized plants (Brunsting and Heil, 1985). Such experimental data are especially intriguing since in England cycles in abundance and dominance of heather have been explained as cycles of plant senescence (e.g., Watt, 1985).

Similarly, outbreaks of beetles (*Trirhabda*) on goldenrods (*Solidago*) are often observed in prairie grasslands. McBrien *et al.* (1983) found that, after an outbreak of beetles, grasses invaded large areas previously monopolized by goldenrods. Medium intensity, early defoliation retarded clonal development and reduced flowering, and heavy defoliation caused clonal mortality. Beetle herbivory also eliminated goldenrod transplants in the drier portion of a soil moisture gradient in prairie (Werner, 1989). Such observations suggest that periodic episodes of intense herbivory interact with high levels of competition for soil resources in the prairie in the determination of goldenrod establishment and persistence. However, a direct test of this hypothesis for the interaction of herbivory with competition in relation to changing environmental conditions is still needed for goldenrod dynamics.

C. Biological Control of Weeds

Successful biological control projects show that specialized insect herbivores can lead to dramatic reductions in host plant densities, and they also suggest properties that may promote strong herbivore effect (Goeden, 1978; Julien, 1984). Less successful projects can provide insight into those factors that limit herbivore impact (Goeden and Louda, 1976; Murdoch *et al.,* 1984). We summarize two cases to make these points clear.

In the first case, the degree of control of *Hypericum perforatum* (Saint-John's-wort, Clusiaceae) by the introduced insects varied geographically. In northern California, an area with dry summers, the introduction of a specialized leaf beetle led to drastically reduced total plant density and to a severe compression of its realized distribution (Huffaker and Kennett, 1959; Harper, 1969). The exact mechanism may be more complicated than originally thought. Damage by herbivores appears to have interacted with soil moisture deficits in determining this result. Growth was reduced and plants eliminated only in the open sun and not in the shade (Huffaker, 1951; Harris, 1980). Also, in areas where control was incomplete, such as British Columbia (Goeden, 1978) and New Zealand (Cameron, 1935), soil moisture remained high throughout the growing season, allowing compensatory regrowth and maintenance of competitive position even after beetle herbivory (Cameron, 1935; Clark, 1953; Williams, 1985). Such results provide indirect evidence for the simultaneous interaction of competition and herbivory in the net outcome.

In the second case, the degree of control of tansy ragwort (*Senecio jacobaea*, Asteraceae) by a moth (*Tyria jacobaeae*, Arctiidae) also varied geographically and it was also correlated with growing conditions. Defoliation usually does not kill individuals, but it does reduce plant growth and size. These losses to herbivores limited tansy ragwort populations, but only where the physical conditions, particularly dry soils, prevented sufficient compensatory regrowth (van der Meijden, 1979; Myers, 1980; Dempster and Pollard, 1981; Cox and McEvoy, 1983; Islam and Crawley, 1983). Thus, the persistence of tansy ragwort, relative to its potential competitors, was inversely correlated with the environment and the plant's regrowth capacity after herbivory.

D. Experimental Studies

Most tests of herbivory on native plants have evaluated the direct effects of consumption on plant growth or fitness. These studies show, in general, that consumption (1) is variable in both space and time and selective among species (see Crawley, 1983, 1989a,b; Hendrix, 1988; Louda, 1989), (2) often alters plant growth and reproduction (Waloff and Richards, 1977; Louda, 1984; Marquis, 1984; Parker, 1985; Paige and Whitham, 1987), (3) can change abundance (e.g., Cantlon, 1969; Rausher and Feeny, 1980; Kinsman and Platt, 1984; Stamp, 1984; Parker, 1985), or (4) modify distribution (Parker and Root, 1981; Louda, 1982a,b, 1983). However, very few studies (see below) explicitly evaluate the joint effects of herbivore consumption and plant competitive interactions over a relevant spectrum of field conditions.

V. Spatial Variation in Herbivore Effect

If herbivore impact has a spatial dimension, then the influence of herbivory on plant competition should also vary in space. Our model for selective herbivory (Fig. 3) predicts that the impact of loss will depend inversely on the plant population's capacity to compensate for losses. Clearly, if the ability to compensate for consumption is related to environmental conditions, then variation in impact along environmental gradients is to be expected. Interactions for which data exist from several locations support the suggestion that the contribution of herbivory to plant dynamics and potential interactions shifts as growing conditions shift.

A. Herbivore Impact along Environmental Gradients

Consumer effects can, in fact, often be ordered along gradients: geographic gradients (Levin, 1976), productivity gradients (Fretwell, 1977; Oksanan *et al.*, 1981), elevational gradients (Louda, 1982a, 1983), disturbance gradients (Huntly, 1987; Coley, 1987), and local gradients (Handel, 1976; Louda *et al.*, 1987a–c; Louda, 1988).

On the geographic scale, several examples demonstrate spatial variation in herbivore influence. The effectiveness of control of Saint-John's-wort increased along a wet-to-dry regional gradient in the Pacific northwest (e.g., Harris, 1980). Herbivore pressure on native ginger (*Asarum caudatum*, Aristolocaceae) changed along the Pacific Coast: plants from moist northern areas with high exposure to slugs had slower growth, higher investment in defense, and lower seed output than plants in drier southern areas with less herbivore pressure (Cates, 1975). Frequency of cyanogenic morphs of *Lotus corniculatus* (Fabaceae) in Europe was directly correlated with herbivore pressure and inversely correlated with time available for regrowth (Jones, 1973). Such observations suggest that herbivory could be a major pressure affecting relative growth and resource exploitation, and thus the balance between competition and predation effects, along geographic gradients (Menge and Sutherland, 1976).

On the regional scale, plant species often replace each other along elevational gradients. One hypothesis for such replacement is that physiologically superior species competitively displace each other as the environment changes (Cody, 1978; Bunce *et al.*, 1979). However, herbivore abundances and damage also change along elevational gradients (Janzen and Schoener, 1967; Janzen *et al.*, 1976; Louda, 1982b, 1983; Randall, 1982), and thus the effect of herbivory also can change along the gradient. For example, control of gorse (*Ulex europaeus*, Fabaceae) by introduced seed predators shifted along an elevational gradient in Hawaii (Goeden, 1978). Thus, an alternative hypothesis is that differential consumption along such gradients may also lead to the observed replacement of plant species.

In a test of this hypothesis, Louda (1982a,b, 1983) excluded predispersal seed predators from two native goldenbushes (*Haplopappus squarrosus* and *Haplopappus venetus*, Asteraceae) at sites along an ocean-to-mountains gradient over which those species replace each other (Fig. 4A). Insect seed predation limited the production and the release of viable seed and, subsequently, the establishment of seedlings by both species (Louda, 1982a,b, 1983). The limitation was disproportionately strong near the coast for *H. squarrosus*, limiting its recruitment coastally and restricting it to the inland part of the gradient (Fig. 4C; Louda, 1982b). In addition, for the other, more coastal species (*H. venetus*), more intense

A

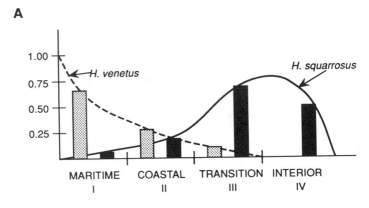

B *H. venetus*   **C** *H. squarrosus*

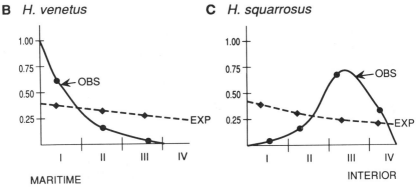

ZONES OF SAN DIEGO COUNTY

Figure 4 (A) Replacement of two goldenbushes, *Haplopappus venetus* by *Haplopappus squarrosus,* along the complex environmental gradient from maritime coast to interior mountains in San Diego County, California. Bars represent observed relative frequency of each species in a zone [stipled, *H. venetus,* adapted from Louda (1983); solid, *H. squarrosus,* adapted from Louda (1982b)], N = 15,250. (B, C) Comparison of observed frequency of each species after herbivory (solid line) against the potential distribution without herbivory (dashed line). The expected (= potential) distribution in the absence of consumption was projected on the basis of several measures of individual performance of control plants, especially flower and seed production, when insects were excluded (details in Louda, 1982b, 1983). These experiments demonstrated that consumption played a significant role in determining in the coastal abundance of *H. venetus* and the more inland distribution of *H. squarrosus* along the gradient.

seedling consumption inland on the gradient followed the seed preda-tion (Louda, 1983). The higher mortality inland compressed the realized adult distribution toward the coast (Fig. 4B). So, seed predation reduced density, and differential levels of total consumption by herbivores along the gradient dramatically changed the subsequent spatial pattern of both species along the gradient. Using only the observational data (Fig. 4A), the replacement of these species along the gradient might have been interpreted as driven by competition. However, the experiments showed that the replacement was actually caused by herbivore consumption that caused disproportionate reductions in densities of *H. squarrosus* near the coast and of *H. venetus* inland. These decreases in density presumably released resources for other coastal scrub and chaparral plants in specific portions of the gradient, potentially affecting other, competitive-mediated interactions.

On the local scale, herbivory also varies along topographic gradients. Handel (1976) found that significant differential feeding by slugs on *Carex platyphylla* along a soil moisture gradient in northeastern deciduous forest altered *Carex* distribution. Insect consumption of *Cleome serrulata* (Capparidaceae), a fugitive annual of shortgrass prairie, shifted the population's mode for most successful seed set from the wet to the medium-wet portion of a 30-m soil moisture gradient (Louda *et al.*, 1987b; Fig. 5A). Plants grew larger and flowered more, but both foliage and seed losses to insects were much higher, in the wettest part of the gradient.

To summarize, such cases illustrate that herbivory can dramatically alter plant population dynamics along environmental gradients, thereby changing the intensity, character, and possibly outcome of competitive interactions among plants along these gradients. More studies are required to identify the characteristics, or groups, of species vulnerable or resistant to such herbivore interference in plant–plant interactions along various environmental gradients.

B. Herbivore Impact across Habitat Discontinuities

We predicted, and found, that the *net* impact of consumption generally changed in relation to conditions for regrowth. Growing conditions usually vary substantially between adjacent habitats. Herbivore pressure also often varies between adjacent habitats, including between sun versus shade sites (e.g., Huffaker and Kennett, 1959; Lincoln and Langenheim, 1979; Lincoln and Mooney, 1983; Louda and Rodman, 1983; Louda *et al.*, 1987a,c). Thus, consistently different levels of herbivory could determine comparative plant performance and relative competitive ability in adjacent habitats.

The best-known example of a case where herbivory appears to determine habitat-specific competitive ability is that of Saint-John's-wort (Clark, 1953). Insect herbivory was greater in the sun (Huffaker and Kennett, 1959). In addition, water deficits limited compensatory regrowth in the sun, presumably lowering the ability of attacked plants to compete with unconsumed neighbors (Clark, 1953; Williams, 1985).

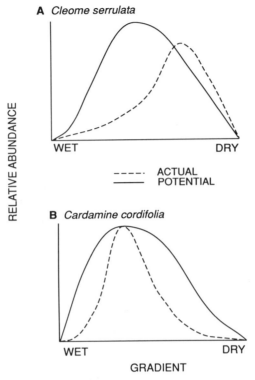

Figure 5 Schematic representation of herbivore-caused shifts in plant performance along small-scale topographic soil gradients: (A) Rocky Mountain bee plant, *Cleome serrulata,* in dry, shortgrass prairie of eastern Colorado (adapted from Louda *et al.,* 1987b); and (B) bittercress, *Cardamine cordifolia,* within wet, willow-shaded habitats in a montane area of the Rocky Mountains, central Colorado (see Louda and Rodman, 1983; Louda, 1988).

Other cases in which herbivory appears significant and varies in magnitude between adjacent habitats include *Lotus corniculatus* (Fabaceae) off versus on minetailings (Jones, 1966); *Machaeranthera canescens* (Asteraceae) in open versus in more closed vegetation in cool desert (Parker and Root, 1981); and *Cardamine cordifolia* (Brassicaceae) in dry, sunny habitats versus in adjacent moist, shaded habitats (Louda and Rodman, 1983; Collinge and Louda, 1988a,b).

For the latter species, field experiments have shown (1) consistently higher levels of herbivory in sun than in natural shade (Louda and Rodman, 1983; Collinge and Louda, 1988a,b); (2) increased vulnerability of shade plants to insects after removal of overhanging willow branches (Table 3); (3) growth and survival in the sun when insects were excluded (Table 3); and (4) decreased vertical growth, reduced leaf initi-

Table 3 Insect Herbivory and Bittercress (*Cardamine cordifolia*) Survival and Growth in a Transplant Experiment[a,b,c]

	Control (Water Only)		No Insects (Insecticide in Water)	
	$\overline{X}$	SE	$\overline{X}$	SE
Increase in leaf area removed (mm²/leaf)[d]				
Sun	68.2	1.93	5.3	1.94
Shade	9.5	0.5	3.2	0.55
Ramet height (cm)				
Sun	13.1	1.7	18.5	1.9
Shade	20.6	3.6	24.2	2.6
Number of fruits per plant				
Sun	0.2	0.1	0.6	0.2
Shade	0.8	0.2	0.7	0.2

[a] From Louda (in review).

[b] Half of the ramets were transplanted into the sun (one portion was used as controls, and the other portion was treated with pyrethrum to reduce foliage-feeding by insects) and half were transplanted back into the shade (one portion used as controls, the other treated with pyrethreum to reduce foliage-feeding by insects) at First Ford Meadow, Copper Creek, Gothic area, Gunnison County, Colorado (1980).

[c] MANOVA, Wilks-Barr Trace, $p < 0.01$.

[d] Damage evaluated using square-root transformed data.

ation, accelerated leaf senescence, lowered total leaf area, and reduced seed set caused by chronic insect herbivory (Louda, 1984). The more intense herbivory imposed on plants in the sun and in drier areas restricts the local population to moist and shaded areas (Fig. 5B) and thus releases resources for other, co-occurring plants in the adjacent sunny habitats.

Most work on herbivory has assessed the contribution of losses to the performance of established plants, i.e., their growth, survival, and seed production. However, in some cases herbivory on seedlings may be even more important in determining establishment, abundance, and subsequent competitive potential (e.g., Cantlon, 1969; Louda, 1983; Mills, 1983, 1984; Clark and Clark, 1985). Further direct tests of this hypothesis, especially along gradients, appear warranted.

Controlled field experiments that actually evaluate herbivory as a mechanism altering competitive ability and outcome are rare. However, the experimental studies that we did find do show that herbivory does not have to cause mortality in order to affect relative competitive performance. Herbivore defoliation in these cases removed a demonstrated competitive advantage or changed the expected outcome of competition (Bentley and Whittaker, 1979; Windle and Franz, 1979; Bentley *et al.*, 1980; Watt and Hagger, 1980; McBrien *et al.*, 1983; Fowler and Rausher, 1985; Parker and Salzman, 1985; Cottam, 1986; Cottam *et al.*, 1986). However, such tests need to be extended to other species and to a range of environmental conditions.

Herbivory could also be an important aspect of temporal variation in plant interactions. Differential herbivory on plants occurring at each stage of succession could either augment or inhibit competitively driven changes in species composition (Ellison, 1960; Cates and Orians, 1975; Reader and Southwood, 1981; Brown, 1985; Mills, 1986). Foraging by vertebrate herbivores, including deer, elk, and rabbits, has been observed to cause changes in species composition (e.g., Leopold, 1956; Watt, 1957; Crawley, 1983). Also, the activities and feeding of fossorial herbivores disturb plants, influencing plant densities and interactions. These herbivores have an effect both by consumption and by indirect facilitation of soil turnover and nutrient cycling (e.g., Reichman and Smith, 1981; Tilman, 1983; Huntly, 1987; Inouye *et al.*, 1987). Such herbivory could be a particularly important modifier of competitive interactions during succession.

VI. Discussion

A. Predictions of the Effect of Selective Herbivory

Inclusion of selective herbivory into Holt's (1977, 1984) conceptual model (Fig. 1) suggests how herbivore consumption could be imposed and influence the various types of competitive interactions among plants (Fig. 2). The graphical analysis of selective herbivory on a competitive dominant (Fig. 3) shows that the net effect of selective herbivory should depend on (1) the plant's intrinsic rate of increase (Fig. 3A), a key aspect of compensatory ability; (2) the strength of the plant's density dependence near K (Fig. 3B); and (3) the degree of density dependence in the herbivory (Fig. 3C). The empirical evidence, especially the few experimental studies, broadly supports these expectations. The impact of selective herbivory varied in relation to (1) regrowth capability (a dimension of compensatory ability), and (2) competitive rank in the absence of herbivory. Our review also suggests that by inducing shifts in plant shape, phenology, resource allocation, recruitment patterns, etc., herbivores could frequently alter competitive interactions between plants in complex, nonadditive ways.

B. Environmental Context and Net Herbivore Effect

In most cases, the actual impact of herbivory was scaled by abiotic conditions. Losses had greater demographic and competitive repercussions under harsher physical conditions or under more limited resource conditions. Thus, herbivore-caused decreases in density or plant growth generally changed along environmental gradients, i.e., with elevation, habitat, gap occurrence, or soil moisture. Spatial variation in insect herbivory along gradients is correlated with differences in insect activity and abundance (Clark, 1953); Janzen and Schoener, 1967; Williams, 1985), or with plant physiological condition and foliage quality for insect feeding and growth (Lincoln and Langenheim, 1979; Lincoln and Mooney, 1983; Louda, 1986; Louda *et al.*, 1987b,c; Collinge and Louda, 1988a). However, this prediction and the relative importance of such mechanisms in creating spatial variation in herbivory, and thus in plant–plant interactions, along gradients still needs to be evaluated experimentally for a broader range of plant species and communities.

C. Exploitation versus Compensation in Net Herbivore Effect

Logically and empirically it seems clear that the relationship between herbivory and ultimate plant performance is determined by the opportunity for compensatory responses at both individual and population levels (see above, and Maschinski and Whitham, 1989). In theory, factors that determine the opportunity and capacity for compensatory responses include (1) physical constraints on plant growth, (2) resource constraints on plant growth, (3) flexibility in internal resource allocation, and (4) type of intraspecific density dependence. Thus, prediction of plant responses to herbivory requires analysis of environmental controls on those responses, such as (a) resource distributions, (b) resource levels, (c) resource renewal rates, as well as (d) spatial and temporal variation in these variables. Both spatial and temporal variation in resources may limit plant capacity to make up for losses and, thus, influence net herbivore impact and plant–plant interactions. Relevant temporal variables include (i) length of growing period, (ii) fluctuations in resource regeneration rates, and (iii) frequency of external disturbance. Theory and observation suggest the importance of each of these constraints on regrowth, and thus on the ability of herbivores to influence relative competitive ability; however, all of these ideas require more experimental analysis under field conditions.

Plants with significant, selective insect herbivores did not grow as well, nor succeed in maintaining themselves as long, when *both* herbivory and potential resource limitation occurred. Given time and resources for regrowth, compensation for loss of leaf tissue or seeds often occurred. However, compensatory physiological responses and growth depend on physical conditions and require resources (Belsky, 1986). When abiotic environmental conditions or time precluded full recovery, or when individual compensatory responses required diversion of limited resources and thus delayed relative growth and reproduction, damaged plants appeared to be at a significant competitive disadvantage. As a result, we expect compensatory responses to become more restricted as the species' distributional boundaries are approached, or as limiting resources become scarce. We predict that the importance of chronic herbivory, for the establishment of seedlings and for maintenance of relative competitive position of established plants, will increase *in general*: (1) as environmental conditions decrease plant capacity for compensatory regrowth, and (2) as herbivore pressure on competitors declines. When net losses (the excess of exploitative loss over compensatory regrowth) are differential among competitors, herbivory should make a substantial contribution to the character and outcome of interactions among competing plants.

A controversy exists over whether herbivory generally harms or benefits plants (e.g., Seastedt, 1985; Belsky, 1986; McNaughton, 1986; Paige and Whitham, 1987; Maschinski and Whitham, 1989). However, our arguments do not depend on whether herbivory, in isolation from its community context, is injurious or beneficial to the plant eaten. The evidence suggests that, if competition for limited resources exists and if levels of herbivory and plant compensatory responses vary in either direction among co-occurring plants, then herbivores could be critical in the determination of relative competitive ability. In such cases, herbivory leads to patterns in the plant community that would be unlikely in the absence of herbivory.

D. Detection of Interaction between Herbivory and Competition

The abundant evidence for the *potential* impact of herbivores suggests that relative resistance to herbivory is an important part of the suite of competition-mediating characteristics of plants. Few analyses have been done to assess this expectation. Herbivory needs to be assessed more routinely in studies of plant competition, and, vice versa, examination of the competitive milieu would benefit studies of herbivory. Such assessments could be made by first comparing spatial and temporal differences in plant traits, resource use, and herbivore damage (e.g., Tingey, 1986), and then by doing the appropriate factorial experiments in the field.

There are several ways to facilitate the detection and evaluation of herbivory in competition studies. Obviously, both herbivory and competition need to be examined. Competition needs to be quantified in the absence of natural enemies (e.g., in exclosures), and response to limiting resources needs to be identified. The effect of herbivory on competing plants should be characterized experimentally if possible. Additionally, since physical conditions influence compensatory growth and thus actual herbivore impact, the experiments should evaluate competition and consumption over the range of abiotic conditions under which the plants occur. The models presented here suggest that intraspecific density dependence of the dominant species needs to be studied along with interspecific competition and herbivory.

Patterns in damage among competitors provide clues to potential effects, suggesting the relevant spatial or temporal dimensions. However, observed losses of leaf area or of seeds tend to underestimate actual herbivore impact (e.g., Coley, 1983; Louda, 1984; Fox and Morrow, 1986; Anderson, 1988). Consistent, differential patterns of herbivory can be important, often independent of the absolute amount of loss (Fox and Morrow, 1986). So, variation in herbivory among populations, sites, habitats, regions, and so forth, should be measured in relation to plant abundance, resource availability, and total plant biomass. An inverse relationship between damage and plant occurrence along a particular gradient, for example, suggests that differential herbivory cannot be ruled out as an explanation. In addition, we recommend simple exclusion tests and bioassays, done by placing vulnerable plants in different habitats. These allow spatial, temporal, and species-specific variation in damage to be categorized and quantified throughout the plant's range of habitats.

The obvious subsequent step is to evaluate the interaction of herbivory and resource competition experimentally in the field, modifying both herbivore load (Harper, 1969, 1977) and plant growing conditions (Ellenberg, 1954). The basic experimental design should include decreases and enhancements of herbivory, resources, and competitors, singly and in combination. The specific system and particular subsidiary questions will determine the details of the design.

Given the evidence for variation along gradients in both plant abundance (Austin, this volume) and herbivory (above), the most important experiments will clearly be those that are done in more than one portion of a biologically interesting gradient and repeated along replicate gradients. We also need long-term, multiyear experiments. The contribution of herbivory to plant growth, density, or competitive ability may well shift with population age or size structure or with changes in the background vegetation.

In summary, herbivory can be a critical aspect in plant competitive interactions. We recommend an approach for the further analysis of this hypothesis that integrates both observational and experimental tests of theoretical and empirically based predictions for the role of both physical and biological variation in such interactions. We predict integrated studies of competition and predation in plant communities will yield a more robust, synthetic picture of the functioning of such assemblages.

VII. Summary

Herbivory can influence plant competitive interactions in two ways. First, herbivory often modifies plant growth or morphology, changing access to resources. Thus, when resources are limited, differential losses to herbivores could be critical, qualitatively altering the interaction among competing plants. Second, herbivory affects population distribution and abundance of some species, thereby changing resources for competitors. Generalist herbivores often modify total resource demand by plants. Selective herbivores in some cases shift relative competitive abilities. Either can cause predator-mediated patterns that mimic those generated by resource competition.

We incorporate selective herbivory into conceptual and graphical models of competition. The models suggest that the impact of selective herbivory on plant competition will reflect both the plant's compensatory responses to herbivory, including intraspecific density dependence, and the strength of density dependence of the herbivory. Our literature review leads to the same hypotheses. The net effect of herbivory generally depended on the balance between loss and compensatory regrowth. Also, the relative influence of herbivory on plant performance, and thus competitive potentials, often changed along environmental gradients. Herbivory was particularly important where constraints in resources, growing season, or growth strategies limited plant compensation for losses, and diminished the species's capacity to maintain itself against competitors.

We conclude that herbivory, by modifying individual traits and affecting population dynamics, represents a potentially significant dimension of competitive interactions among plants. However, few studies have as yet directly tested the interaction of herbivory and competition. So, we end by suggesting methods for doing so in future research.

Acknowledgments

Our heartiest thanks go to all of the people who have helped with these studies, especially to those who have discussed, challenged, argued, edited, and otherwise forced us to try to clarify the ideas presented here. These people include Gail Baker, Paulette Bierzychudek, Diane Campbell, Lissy Coley, Sharon Collinge, Mick Crawley, Laurel Fox, Sally Gaines, Tony Joern, Hank Howe, Trice Morrow, Annette Olson, Wayne Polley, Claudia Tyler, and several anonymous reviewers. Joe Connell's influence was strong, helpful and must be obvious. Discussions at the workshop were also instructive and are greatly appreciated. Our work could not have been done without the generous financial support provided by the National Science Foundation (DEB80-11106, DEB82-07955, BSR84-05625, BSR85-16515, BSR87-04705, BSR87-18088), University of Nebraska Research Council, General Research Fund of the University of Kansas, Sigma Xi Scientific Research Society, and Organization of Tropical Studies, as well as logistical support from the staffs of the Rocky Mountain Biological Laboratory and Cedar Point Biological Station.

References

Abrams, P. A. (1977). Density-independent mortality and interspecific competition: A test of Pianka's niche-overlap hypothesis. *Am. Nat.* **11**, 539–552.

Anderson, A. N. (1988). Insect seed predators may cause far greater losses than they appear to. *Oikos* **52**, 337–340.

Archer, S., and Detling, J. K. (1984). The effects of defoliation and competition on regrowth of tillers of two North American mixed-grass prairie graminoids. *Oikos* **43**, 351–357.

Barbosa, P., and Schultz, J. C. (eds.) (1987). "Insect Outbreaks." Academic Press, Orlando, Florida.

Bazzaz, F. A., Chiariello, N. R., Coley, P. D., and Pitelka, L. F. (1987). Allocating resources to reproduction and defense. *BioScience* **37**, 58–67.

Bell, R. H. V. (1970). The use of the herb layer by grazing ungulates in the Serengeti. *In* "Animal Populations in Relation to Their Food Resources" (A. Watson, ed.), pp. 111–124. Blackwell Scientific Publications, Oxford.

Belsky, A. J. (1986). Does herbivory benefit plants? A review of the evidence. *Am. Nat.* **127**, 870–892.

Bentley, S., and Whittaker, J. B. (1979). Effects of grazing by a chrysomelid beetle, *Gastrophysa viridula*, on competition between *Rumex obtusifolius* and *Rumex crispus*. *J. Ecol.* **67**, 79–90.

Bentley, S., Whittaker, J. B., and Malloch, A. J. C. (1980). Field experiments on the effects of grazing by a chrysomelid beetle (*Gastrophysa viridula*) on seed production and quality in *Rumex obtusifolius* and *Rumex crispus*. *J. Ecol.* **68**, 671–674.

Berdowski, J. J. M., and Zeilinga, R. (1983). The effect of the heather beetle (*Lochmaea suturalis* Thomson) on heather (*Calluna vulgaris* (L.) Hull) as a cause of mosaic patterns in heathlands. *Acta Bot. Neerl.* **32**, 250–251.

Berdowski, J. J. M., and Zeilinga, R. (1987). Transition from heathland to grassland: Damaging effects of the heather beetle. *J. Ecol.* **75**, 59–175.

Berendse, F. (1985). The effect of grazing on the outcome of competition between plant species with different nutrient requirements. *Oikos* **44**, 35–39.

Brown, V. K. (1985). Insect herbivores and succession. *Oikos* **44**, 17–22.

Brunsting, A. M. H., and Heil, G. W. (1985). The role of nutrients in the interactions between a herbivorous beetle and some competing plant species in heathlands. *Oikos* **44**, 23–26.

Bunce, J. A., Chabot, B. F., and Miller, L. N. (1979). Role of annual leaf carbon balance in the distribution of plant species along an elevational gradient. *Bot. Gaz.* **140**, 288–294.

Burdon, J. J., and Chilvers, G. A. (1974). Fungal and insect parasites contributing to niche differentiation in mixed species stands of eucalypt saplings. *Aust. J. Bot.* **22**, 103–114.

Caldwell, M. M., Richards, J. H., Johnson, D. A. Nowak, R. S., and Dzurec, R. S. (1981). Coping with herbivory: Photosynthetic capacity and resource allocation in two semiarid *Agropyron* bunchgrasses. *Oecologia* **50**, 14–24.

Cameron, E. (1935). A study of the natural control of ragwort (*Senecio jacobaea* L.). *J. Ecol.* **23**, 265–322.

Cantlon, J. E. (1969). The stability of animal populations and their sensitivity to technology. *Brookhaven Symp. Biol.* **22**, 197–203.

Cates, R. G. (1975). The interface between slugs and wild ginger: Some evolutionary aspects. *Ecology* **56**, 391–400.

Cates, R. G., and Orians, G. H. (1975). Successional status and the palatability of plants to generalist herbivores. *Ecology* **56**, 410–418.

Chandra, S., and Williams, G. (1983). Frequency dependent selection in grazing behavior of the desert locust *Schistocerca gregarii*. *Ecol. Entomol.* **8**, 13–21.

Chapin, F. S., III (1980). Nutrient allocation and responses to defoliation in tundra plants. *Arct. Alp. Res.* **12**, 553–563.

Clark, D. A., and Clark, D. B. (1985). Seedling dynamics of tropical trees: Impacts of herbivory and meristem damage. *Ecology* **66**, 1884–1892.

Clark, L. R. (1953). The ecology of *Chrysomela gemellata* Rossi and *C. hyperici* Forst. and their effect on St. John's wort in the Bright District, Victoria. *Aust. J. Zool.* **1**, 1–69.

Clements, R. O., and Henderson, I. F. (1979). Insects as a cause of botanical change in swards. *J. Br. Grassl. Soc.* **10**, 157–160.

Cody, M. L. (1978). Distribution ecology of *Happlopappus* and *Chrysothamnus* in the Mohave Desert. I. Niche position and niche shifts on north-facing granitic slopes. *Am. J. Bot.* **65**, 1107–1116.

Coley, P. D. (1983). Herbivory and defense characteristics of tree species in a lowland tropical forest. *Ecol. Monogr.* **53**, 209–233.

Coley, P. D. (1987). Interspecific variation in plant anti-herbivore properties: The role of habitat quality and rate of disturbance. *New Phytol.* **106**, 251–263.

Collinge, S. K., and Louda, S. M. (1988a). Herbivory by leaf miners in response to experimental shading of a native crucifer. *Oecologia* **75**, 559–566.

Collinge, S. K., and Louda, S. M. (1988b). Patterns of resource use by a dropsophilid (Dipteran) leaf minder on a native crucifer. *Ann. Entomol. Soc. Am.* **81**, 733–741.

Cottam, D. A. (1986). The effects of slug-grazing on *Trifolium repens* and *Dactylis glomerata* in monoculture and mixed sward. *Oikos* **47**, 275–279.

Cottam, D. A., Whittaker, J. B., and Malloch, A. J. C. (1986). The effects of chrysomelid beetle grazing and plant competition on the growth of *Rumex obtusifolius*. *Oecologia* **70**, 452–456.

Coughenour, M. B., McNaughton, S. J., and Wallace, L. L. (1985). Responses of an African graminoid (*Thermeda triandra* Forsk.) to frequent defoliation, nitrogen and water: A limit of adaptation to herbivory. *Oecologia* **68**, 105–111.

Cox, C. S., and McEvoy, P. B. (1983). Effect of summer moisture stress on the capacity of tansy ragwort (*Senecio jacobaea*) to compensate for defoliation by cinnabar moth (*Tyria jacobaea*). *J. Appl. Ecol.* **20**, 225–234.

Crawley, M. J. (1983). "Herbivory: The Dynamics of Animal–Plant Interactions." Univ. of California Press, Berkeley, California.

Crawley, M. J. (1989a). Herbivores and plant population dynamics. *In* "Plant Population Biology" (A. J. Davy, M. J. Hutchings, and A. R. Watkinson, eds.), in press. Blackwell, Oxford, England.

Crawley, M. J. (1989b). The relative importance of vertebrate and invertebrate herbivores in plant population dynamics. *In* "Insect–Plant Interactions" (E. A. Bernays, ed.), in press. CRC. Press, Boca Raton, Florida.

Davidson, D. W., Samson, D. A., and Inouye, R. S. (1985). Granivory in the Chihuahuan Desert: Interactions within and between trophic levels. *Ecology* **66**, 486–502.

Dempster, J. P., and Pollard, E. (1981). Fluctuations in resource availability and insect populations. *Oecologia* **50**, 412–416.

Denno, R. F., and McClure, M. S. (eds.) (1983). "Variable Plants and Herbivores in Natural and Managed Systems." Academic Press, New York.

Detling, J. K., Dyer, M. I., and Winn, D. T. (1979). Net photosynthesis, root respiration, and regrowth of *Bouteloua gracilis* following simulated grazing. *Oecologia* **41,** 127–134.

Dirzo, R. (1985). The role of the grazing animal. *In* "Studies on Plant Demography: A Festschrift for John L. Harper" (J. White, ed.), pp. 343–355. Academic Press, New York.

Dirzo, R., and Harper, J. L. (1980). Experimental studies on slug–plant interactions. II. The effect of grazing by slugs on high density monocultures of *Capsella bursa-pastoris* and *Poa annua. J. Ecol.* **68,** 999–1011.

Dirzo, R., and Harper, J. L. (1982). Experimental studies of slug–plant interactions. IV. The performance of cyanogenic and acyanogenic morphs of *Trifolim repens* in the field. *J. Ecol.* **70,** 119–138.

Dyer, M. I., Detling, J. K., Coleman, D. C., and Hilbert, D. W. (1982). The role of herbivores in grasslands. *In* "Grasses and Grasslands: Systematics and Ecology" (J. R. Estes, R. J. Tyrl, and J. N. Brunken, eds.), pp. 255–295. Univ. of Oklahoma Press, Norman, Oklahoma.

Edwards, P. J., and Wratten, S. D. (1980). "Ecology of Insect–Plant Interactions." Arnold, London

Ellenberg, H. (1954). Physiologisches und okologishes Verhalten derselben Pilanzanzenarten. *Ber. Dtsch. Bot. Ges.* **65,** 350–361.

Ellison, L. (1960). Influence of grazing on plant succession of rangelands. *Bot. Rev.* **26,** 1–78.

Fowler, N. L. (1986). The role of competition in plant communities in arid and semi-arid regions. *Annu. Rev. Ecol. Syst.* **17,** 89–110.

Fowler, N. L., and Rausher, M. D. (1985). Joint effects of competitors and herbivores on growth and reproduction in *Aristolochia reticulata. Ecology* **66,** 1580–1587.

Fox, L. R., and Morrow, P. A. (1986). On comparing herbivore damage in Australian and north temperate systems. *Aust. J. Ecol.* **11,** 387–393.

Fretwell, S. D. (1977). The regulation of plant communities by food chains exploiting them. *Perspect. Biol. Med.* **20,** 169–185.

Gibson, C. W. D., Brown, V. K., and Jepsen, M. (1987). Relationships between the effects of insect herbivory and sheep grazing on seasonal changes in an early successional plant community. *Oecologia* **71,** 245–253.

Goeden, R. D. (1978). Biological control of weeds. *In* "Introduced Parasites and Predators of Arthropod Pests and Weeds: A World Review." (C. P. Clausen, ed.), pp. 357–414. USDA Handb. **480,** 545 pp.

Goeden, R. D., and Louda, S. M. (1976). Biotic interference with insects imported for weed control. *Annu. Rev. Entomol.* **21,** 325–342.

Gradwell, G. R. (1974). The effect of defoliation on tree growth. *In* "The British Oak" (M. G. Morris and F. H. Perring, eds.), pp. 182–193. Classey, Farringdon, England.

Halligan, J. P. (1974). Relationship between animal activity and bare areas associated with California sagebrush in annual grassland. *J. Range Manage.* **27,** 358–362.

Handel, S. N. (1976). "Population Ecology of Three Woodland *Carex* Species," Ph.D. thesis. Cornell Univ., Ithaca, New York.

Harper, J. L. (1969). The role of predation in vegetational diversity. *Brookhaven Symp. Biol.* **22,** 48–62.

Harper, J. L. (1977). "Population Biology of Plants." Academic Press, New York.

Harris, P. (1980). Stress as a strategy in the biological control of weeds. *BARC Symp.* **5,** 333–340.

Heil, G. W., and Diemont, W. H. (1983). Raised nutrient levels change heathland to grassland. *Vegetatio* **53,** 113–120.

Henderson, I. F., and Clements, R. O. (1979). Differential susceptibility to pest damage in agricultural grasses. *J. Agric. Sci.* **93**, 465–472.

Hendrix, S. D. (1979). Compensatory reproduction in a biennial herb following insect defloration. *Oecologia* **42**, 107–118.

Hendrix, S. D. (1988). Herbivory and its impact on plant reproduction. *In* "Plant Reproductive Ecology" (J. Lovett Doust and L. Lovett Doust, eds.). Oxford University Press, Oxford.

Hodgkinson, I. D., and Hughes, M. K. (1982). "Insect Herbivory." Chapman and Hall, New York.

Holt, R. D. (1977). Predation, apparent competition and the structure of prey communities. *Theor. Pop. Biol.* **12**, 197–229.

Holt, R. D. (1984). Spatial heterogeneity, indirect interactions, and the coexistence of prey species. *Am. Nat.* **124**, 377–406.

Holt, R. D. (1985). Density-independent mortality, non-linear competitive interactions, and species coexistence. *J. Theor. Biol.* **116**, 479–493.

Huffaker, C. B. (1951). The return of native perennial bunchgrass following the removal of Klamath weed (*Hypericum perforatum* L.) by imported beetles. *Ecology* **32**, 443–458.

Huffaker, C. B., and Kennett, C. E. (1959). A ten-year study of vegetational changes associated with the biological control of Klamath weed. *J. Range Manage.* **12**, 69–82.

Huntly, N. J. (1987). Effects of refuging herbivores (pikas, *Ochotona princeps*) on subalpine meadow vegetation. *Ecology* **68**, 274–283.

Huntly, N. J., and Inouye, R. S. (1988). Pocket gophers in ecosystems: Patterns and mechanisms. *BioScience* **38**, 786–793.

Ingham, R. E., and Detling, J. K. (1986). Effects of defoliation and nematode consumption on growth and leaf gas exchange in *Bouteloua curtipendula*. *Oikos* **46**, 23–28.

Inouye, R. S., Byers, G. S., and Brown, J. H. (1980). Effects of predation and competition on survivorship, fecundity, and community structure of desert annuals. *Ecology* **61**, 1344–1351.

Inouye, R. S., Huntly, N. J., Tilman, D., and Tester, J. R. (1987). Pocket gophers *Geomys bursarius*), vegetation, and soil nitrogen along a successional sere in east central Minnesota. *Oecologia* **68**, 178–184.

Islam, Z., and Crawley, M. J. (1983). Compensation and regrowth in ragwort (*Senecio jacobaea*) attacked by Cinnabar moth, *Tyria jacobaeae*. *J. Ecol.* **71**, 829–843.

Jameson, D. A. (1963). Responses of individual plants to harvesting. *Bot. Rev.* **29**, 532–594.

Janzen, D. H. (1971). Seed predation by animals. *Annu. Rev. Ecol. Syst.* **2**, 465–492.

Janzen, D. H., and Schoener, T. W. (1967). Differences in insect abundance and diversity between wetter and drier sites during a tropical dry season. *Ecology* **39**, 5–16.

Janzen, D. H., Ataroff, M., Farinas, M., Reyes, S., Soler, A., Sorino, P., and Vera, M. (1976). Changes in the arthropod community along an elevational gradient in the Venezuelan Andes. *Biotropica* **8**, 193–203.

Jeffries, M. J., and Lawton, J. H. (1985). Enemy free space and the structure of ecological communities. *Biol. J. Linn. Soc.* **23**, 269–286.

Joern, A. (1989). Insect herbivory in the transition to California annual grasslands: Did grasshoppers deliver the coup de grass? *In* "Grassland Structure and Function: California Annual Grassland" (L. F. Huenneke and H. S. Mooney, eds.), pp. 117–134. Kluwer Academic Publishers, Dordrecht.

Jones, D. A. (1966). On the polymorphism of cyanogenesis in *Lotus corniculatus*. Selection by animals. *Can. J. Genet. Cytol.* **8**, 556–567.

Jones, D. A. (1973). Coevolution and cyanogenesis. *In* "Taxonomy and Ecology" (H. Heywood, ed.), pp. 213–242. Academic Press, New York.

Julien, M. H. (1984). Biological control of weeds: An evaluation. *Prot. Ecol.* **7**, 3–25.

Karban, R., and Courtney, S. (1987). Intraspecific host plant choice: Lack of consequences for *Streptanthus tortuosus* (Cruciferae) and *Euchloe hyantis* (Lepidoptera : Pieridae). *Oikos* **48,** 243–248.

Kinsman, S., and Platt, W. J. (1984). The impact of herbivores (*Heliodines nyctaginella* : Lepidoptera) upon *Mirabilis hirsuta,* a fugitive prairie plant. *Oecologia* **65,** 2–6.

Landa, K., and Rabinowitz, D. (1983). Relative preference of *Arphia sulphurea* (Orthoptera : Acrididae) for sparse and common prairie grasses. *Ecology* **64,** 392–395.

Lee, T. D., and Bazzaz, F. A. (1980). Effects of defoliation and competition on growth and reproduction of the annual plant, *Abutilon theophasti. J. Ecol.* **68,** 813–821.

Leopold, A. S. (1956). Deer in relation to plant succession. *Trans. North Am. Wildl. Conf.* **21,** 159–172.

Levin, D. A. (1976). Alkaloid-bearing plants: An ecogeographic perspective. *Am. Nat.* **110,** 261–284.

Lincoln, D., and Langenheim, J. H. (1979). Variation in *Satureja douglasii* monoterpenoids in relation to light intensity and herbivory. *Biochem. Syst. Ecol.* **7,** 289–298.

Lincoln, D. E., and Mooney, H. A. (1983). Herbivory on *Diplacus aurantiacus* shrubs in sun and shade. *Oecologia* **64,** 173–176.

Louda, S. M. (1982a). Limitation of the recruitment of the shrub *Haplopappus squarrosus* (Asteraceae) by flower- and seed-feeding insects. *J. Ecol.* **70,** 43–53.

Louda, S. M. (1982b). Distribution ecology: Variation in plant recruitment over a gradient in relation to insect seed predation. *Ecol. Monogr.* **52,** 25–41.

Louda, S. M. (1983). Seed predation and seedling mortality in the recruitment of a shrub, *Haplopappus venetus* (Asteraceae), along a climatic gradient. *Ecology* **64,** 511–521.

Louda, S. M. (1984). Herbivore effect on stature, fruiting and leaf dynamics of a native crucifer. *Ecology* **65,** 1379–1386.

Louda, S. M. (1986). Insect herbivory in response to root-cutting and flooding stress on a native crucifer under field conditions. *Acta Oecol. Oecol. Gen.* **7,** 37–53.

Louda, S. M. (1988). Insect pests and plant stress as considerations for revegetation of disturbed ecosystems. *In* "Rehabilitating Ecosystems, Vol. II," pp. 51–67. CRC Press, Boca Raton, Florida.

Louda, S. M. (1989). Predation in the dynamics of seed regeneration. *In* "Ecology of Soil Seed Banks" (M. A. Leck, V. T. Parker, and R. L. Simpson, eds.), pp. 25–51. Academic Press, San Diego, California.

Louda, S. M., and Keeler, H. H. (1989). Plant traits determining relative competitive ability: Influence of endemic levels of chronic herbivory. *Vegetatio,* in review.

Louda, S. M., and Potvin, M. A. (in prep.). Alteration of regeneration probability and lifetime fitness of a monocarpic perennial by insect seed predation. *Ecology,* in review.

Louda, S. M., and Rodman, J. E. (1983). Concentration of glucosinolates in relation to habitat and insect herbivory for the native crucifer *Cardamine cordifolia. Biochem. Syst. Ecol.* **11,** 199–207.

Louda, S. M., Dixon, P., and Huntly, N. J. (1987a). Herbivory in sun versus shade at a natural meadow–woodland ecotone in the Rocky Mountains. *Vegetatio* **72,** 141–149.

Louda, S. M., Farris, M. A., and Blua, M. J. (1987b). Variation in methylglucosinolate and insect damage to *Cleome serrulata* (Capparaceae) along a natural soil moisture gradient. *J. Chem. Ecol.* **13,** 569–581.

Louda, S. M., Huntly, N. J., and Dixon, P. (1987c). Insect herbivory across a sun/shade gradient: Response to experimentally-induced *in situ* plant stress. *Acta Oecol. Oecol. Gen.* **8,** 357–363.

Louda, S. M., Potvin, M. A., and Collinge, S. K. (1989). Biotic factors influencing seedling recruitment by Platte thistle in Sandhills prairie. *Amer. Midl. Nat.,* in press.

Marquis, R. J. (1984). Leaf herbivores decrease fitness of a tropical plant. *Science* **226,** 537–539.

Maschinski, J., and Whitham, T. G. (1989). The continuum of plant responses to herbivory: The influence of plant association, nutrient availability and timing. *Am. Nat.* **134,** 1–19.

Mattson, W. J. (ed.) (1977). "The Role of Arthropods in Forest Ecosystems." Springer-Verlag, New York.

Mattson, W. J., and Addy, N. D. (1975). Phytophagous insects as regulators of forest primary production. *Science* **190,** 515–522.

Mattson, W. J., and Haack, R. A. (1987). The role of drought in outbreaks of plant-eating insects. *BioScience* **37,** 110–118.

McBrien, H., Harmsen, R., and Crowder, A. (1983). A case of insect grazing affecting plant succession. *Ecology* **64,** 1035–1039.

McNaughton, S. J. (1985). Ecology of a grazing ecosystem: The Serengeti. *Ecol. Monogr.* **55,** 259–294.

McNaughton, S. J. (1986). On plants and herbivores. *Am. Nat.* **128,** 765–770.

McNaughton, S. J., Ruess, R. W., and Seagle, S. W. (1988). Large mammals and process dynamics in African ecosystems. *BioScience* **38,** 794–800.

Menge, B. A., and Sutherland, J. P. (1976). Species diversity gradients: Synthesis of the roles of predation, competition and temporal heterogeneity. *Am. Nat.* **110,** 351–369.

Mills, J. N. (1983). Herbivory and seedling establishment in post-fire southern California chaparral. *Oecologia* **60,** 267–270.

Mills, J. N. (1984). Effects of feeding by mealy bugs (*Planococcus citri*, Homoptera : Pseudococcidae) on the growth of *Colliguaga odorifera* seedlings. *Oecologia* **64,** 142–144.

Mills, J. N. (1986). Herbivores and early postfire succession in southern California chaparral. *Ecology* **67,** 1637–1649.

Mooney, H. A., and Gulmon, S. (1982). Constraints on leaf structure and function in reference to herbivory. *BioScience* **332,** 198–206.

Morrow, P. A. (1977). The significance of phytophagous insects in the *Eucalyptus* forests of Australia. *In* "The Role of Arthropods in Forest Ecosystems" (W. Mattson, ed.), pp. 19–30. Springer-Verlag, Berlin.

Morrow, P. A., and LaMarche, V. C. (1977). Tree ring evidence for chronic insect suppression of productivity in subalpine *Eucalyptus*. *Science* **201,** 1244–1245.

Murdoch, W. W., Reeve, J. D., Huffaker, C. B., and Kennett, C. E. (1984). Biological control of olive scale and its reference to ecological theory. *Am. Nat.* **123,** 371–392.

Myers, J. H. (1980). Is the insect or the plant the driving force in the cinnabar moth–tansy ragwort system? *Oecologia* **42,** 307–323.

Naiman, R. J. (1988). Animal influences on ecosystem dynamics. *BioScience* **38,** 750–752.

Naiman, R. J., Johnston, C. A., and Kelley, J. C. (1988). Alteration of North American streams by beaver. *BioScience* **38,** 753–763.

Norton-Griffiths, M. (1979). The influence of grazing, browsing, and fire on vegetation dynamics of the Serengeti. *In* "Serengeti: Dynamics of an Ecosystem" (A. R. E Sinclair and M. Norton-Griffiths, eds.), pp. 310–352. Univ. of Chicago Press, Chicago.

Oksanen, L., Fretwell, S. D., Arruda, J., and Niemelä, P. (1981). Exploitation ecosystems in gradients of primary productivity. *Am. Nat.* **118,** 240–261.

Paige, K. N., and Whitham, T. G. (1987). Overcompensation in response to mammalian herbivory: The advantage of being eaten. *Am. Nat.* **129,** 407–416.

Painter, E. L., and Detling, J. K. (1981). Effects of defoliation on net photosynthesis and regrowth of western wheatgrass. *J. Range Management* **34,** 68–71.

Parker, M. A. (1985). Size-dependent herbivore attack and the demography of an arid grassland shrub. *Ecology* **66,** 850–860.

Parker, M. A., and Root, R. B. (1981). Insect herbivores limit habitat distribution of a native composite, *Machaeranthera canescens*. *Ecology* **62,** 1390–1392.

Parker, M. A., and Salzman, A. G. (1985). Herbivore exclosure and competitor removal: Effects on juvenile survivorship and growth in the shrub, *Gutierrezia microcephala. J. Ecol.* **73,** 903–913.

Polley, H. W., and Detling, J. K. (1989). Defoliation, nitrogen, and competition: Effects on plant growth and nitrogen nutrition. *Ecology* **70,** 721–727.

Randall, M. G. M. (1982). The dynamics of an insect population throughout its altitudinal range: *Coleophora alticolella* (Lepidoptera) in Northern England. *J. Anim. Ecol.* **50,** 993–1016.

Rausher, M. D., and Feeny, P. P. (1980). Herbivory, plant density and plant reproductive success: The effect of *Battus philenor* on *Aristolochia reticulata. Ecology* **61,** 905–917.

Reader, P. M., and Southwood, T. R. E. (1981). The relationship between palatability to invertebrates and the successional status of a plant. *Oecologia* **51,** 271–275.

Reichman, O. J. (1979). Desert granivore foraging and its impact on seed densities and distributions. *Ecology* **60,** 1085–1092.

Reichman, O. J., and Smith, C. C. (1981). Impact of pocket gopher burrows on overlying vegetation. *J. Mammal.* **66,** 720–725.

Richards, J. H. (1984). Root growth response to defoliation in *Agropyron* bunchgrasses: Field observations with an improved root periscope. *Oecologia* **64,** 21–25.

Scott, J. A., French, N. R., and Leetham, J. W. (1979). Patterns of consumption in grasslands. *In* "Perspectives in Grassland Ecology" (N. R. French, ed.), pp. 89–105. Springer-Verlag, New York.

Seastedt, T. R. (1985). Maximization of primary and secondary productivity by grazers. *Am. Nat.* **126,** 559–564.

Seastedt, T. R., Todd, T. C., and James, S. W. (1987). Experimental manipulations of arthropod, nematode and earthworm communities in a North American tallgrass prairie. *Pedobiologia* **30,** 9–18.

Sih, A., Crowley, P., McPeek, M., Petranka, J., and Strohmeier, K. (1985). Predation, competition and prey communities: A review of field experiments. *Annu. Rev. Ecol. Syst.* **16,** 269–311.

Stamp, N. E. (1984). Effect of defoliation by checkerspot caterpillars (*Euphydryas phaeton*) and sawfly larvae (*Macrophya nigra* and *Tenthredo grandis*) on their host plants (*Chelone* spp.). *Oecologia* **63,** 275–280.

Taylor, W. E., and Bardner, R. (1968). Effects of feeding by larvae of *Phaedon cochlearieae* (F.) and *Plutella maculipennis* (Curt.) on the yield of radish and turnip plants. *Ann. Appl. Biol.* **62,** 249–254.

Tilman, D. (1983). Plant succession and gopher disturbance along an experimental gradient. *Oecologia* **60,** 285–292.

Tingey, W. M. (1986). Techniques for evaluating plant resistance to insects. *In* "Insect–Plant Interactions" (J. R. Miller and T. A. Miller, eds.), pp. 251–284. Springer-Verlag, Berlin.

Ueckert, D. N. (1979). Impact of a white grub (*Phyllophaga crinita*) on a shortgrass community and evaluation of selected rehabilitation practices. *J. Range Manage.* **32,** 445–448.

van der Meijden, E. (1979). Herbivore exploitation of a fugitive plant species: Local survival and extinction of the cinnabar moth and ragwort in a heterogeneous environment. *Oecologia* **42,** 307–323.

Wallace, L. L., McNaughton, S. J., and Coughenour, M. B. (1984). Compensatory photosynthetic responses of three African graminoids to different fertilization, watering and clipping regimes. *Bot. Gaz.* **145,** 151–156.

Waloff, N., and Richards, O. W. (1977). The effect of insect fauna on growth, mortality and natality of broom, *Sarothamnus scoparius. J. Appl. Ecol.* **14,** 787–798.

Watt, A. S. (1955). Bracken versus heather: A study in plant sociology. *J. Ecol.* **40**, 1–10.

Watt, A. S. (1957). The effect of excluding rabbits from grassland B (*Mesobrometum*) in Breckland. *J. Ecol.* **45**, 861–878.

Watt, T. A., and Hagger, R. J. (1980). The effect of defoliation upon yield, flowering and vegetative spread of *Holcus lanatus* growing with and without *Lolium perenne*. *Grass Forage Sci.* **35**, 227–234.

Werner, P. A. (1989). Goldenrods (*Solidago* spp.) on edaphic gradients: A field experiment of competition vs. tolerance using reciprocal transplants. *Ecol. Monogr.*, in press.

Whitham, T. G., and Mopper, S. (1985). Chronic herbivory: Impacts on architecture and sex expression of pinyon pine. *Science* **228**, 1089–1091.

Whittaker, J. B. (1979). Invertebrate grazing, competition and plant dynamics. *In* "Population Dynamics" (R. M. Anderson, B. D. Turner, and L. R. Taylor, eds.), pp. 207–222. Blackwell, Oxford, England.

Williams, K. S. (1985). Climatic influences on weeds and their herbivores: Biological control of St. John's wort in British Columbia. *Proc. Int. Symp. Biol. Control Weeds, 6th*, pp. 127–132.

Windle, P. N., and Franz, E. H. (1979). The effects of insect parasitism on plant competition: Greenbugs and barley. *Ecology* **60**, 521–529.

20

Predation, Herbivory, and Plant Strategies Along Gradients of Primary Productivity

Lauri Oksanen

I. Vegetation Processes in Benign and Stressful Environments: Variations on the Same Theme?

Plant ecologists have divergent ideas on the impact of stress on vegetation processes. One view can be traced to Darwin's (1859) statement that, for organisms inhabiting extreme environments, the struggle for existence is almost exclusively against the "elements," not between different organisms. This view has a strong foothold in British animal and plant ecology (Southwood, 1977; Grime, 1977, 1979; Callaghan and Emman-

Perspectives on Plant Competition. Copyright © 1990 by Academic Press, Inc. All rights of reproduction in any form reserved.

uelsson, 1985). In Central Europe, struggle between organisms is often regarded as important, even in extreme environments. Environmental stresses are chiefly seen in the role of external constraints that determine which characteristics of plants are most competitive in a given environment (Walter, 1964, 1968; Ellenberg and Mueller-Dombois, 1974; Ellenberg, 1978), although the possibility that environmental stresses directly exclude superior competitors is accepted. A still less compromising view of competition as the organizing principle of plant communities prevails in northern Europe, dominated by Cajander's (1909) view of site-specific community types that consist of those plants that are maximally competitive in the habitat and utilize resources in sufficiently different ways to permit competitive coexistence. For Cajander, distributional limits in both local and global scale are always products of changes in the competitive balance between different species. Major vegetational transitions represent similar changes in the competitive balance between different types of plants. A similar view of competition and community structure has spread to American plant ecology, when researchers with roots in the zoological tradition of Gause (1934), Lack (1954), and MacArthur (1972) have become interested in plants (Cody, 1986; Tilman, 1982, 1984, 1985, 1987, 1988a).

In the inevitable clash between Grime's school and what could be called the Cajander–Tilman school (Thompson, 1987; Tilman, 1988b), the logical position of the latter school is stronger. Grime's (1979) idea that a single plant type would be a superior competitor under all circumstances is in conflict with the principle of allocation (Tilman, 1988a), with well-known ecophysiological facts (e.g., photosynthetic responses of different plants to moisture, light, and temperature; see Kershaw, 1975; Lechowicz, 1978; Bazzaz, 1979) and with the existence of a tradeoff between short-term and long-term competitiveness (Smith, 1976). It is also difficult to see how stress could reduce the importance of competition. In every habitat, there is some limit for the plant biomass that the resource basis can support. When this level is approached, competition seems inevitable.

Tilman (1988a) showed that a large part of global vegetational patterns can be explained on the basis of changing terms of competition. In arid areas where water is a minimum factor and canopy never gets dense enough to create substantial competition for light, the size and shape of the root system largely determines the competitiveness of the plant. Plants allocating many resources to roots and only little to stems can thus outcompete plants with an opposite allocation pattern. Also, in early successional habitats and in cold areas, where mineral nutrients are often in short supply, competition can mainly take place in the soil, favoring plants with low shoot/root ratios (Chapin and Shaver, 1985; Tilman,

1987). In cold and windy habitats with lots of bare ground, thermal conditions can further increase the advantages of low stature (Billings and Mooney, 1968; Walter, 1968; Ellenberg, 1978).

Although many types of environmental stresses may favor high allocation of resources to roots, it is difficult to explain the transition from forests to steppes and arctic or alpine tundras solely as a consequence of changing terms of competition. These transitions occur in areas where the vegetation is still closed and competition for light has by no means ceased to exist. Yet, typical tundra and steppe plants do not differ from trees just by having different allocation patterns. Typically, these plants are herbaceous and have leaf-producing meristems close to the ground, and even the woody ones are characterized by weak apical dominance, so that the above-ground shoot system remains much shorter than would be possible with the prevailing level of allocation to shoots (see Walter, 1964, 1968; Knapp, 1965; Ellenberg, 1978). If competition were equally important in stressful and benign environments, it would seem more advantageous to make the allocation to shoots in the form of an erect woody stem. In a few years, such a stem would overtop the shoots of herbaceous plants and prostrate dwarf shrubs. Winter ecology can explain a part of this enigma: plants extending above the snow cover must cope with extremely hostile thermal and moisture conditions in late winter and early spring (Tranquillini, 1957, 1970). However, most tundra plants are either herbaceous or have much lower stature than could be explained by winter ecology alone. Thermal advantages of staying close to the ground do not explain prostrateness, either: robust cushion form (i.e., tight packing of erect shoots) is an equivalent, possibly even superior way of maximizing the warming effect of direct solar radiation (Gauslaa, 1984). Also, explanations based on the superiority of the fibrous root system of graminoids (Walter, 1964, pp. 280–292) or their superior ability to recycle nutrients (Jonasson and Chapin, 1985) are problematic, because they presuppose that erect woody plants could not evolve functionally similar root systems or recycling mechanisms.

Although the transition from forests to steppes and tundras does not quite fit the implications of the Cajander–Tilman competition approach, it does not fit Grime's theory either. If arid and arctic timberlines represented cases where superior competitors become directly excluded by environmental stresses, the timberlines should be abrupt. As pointed out by Ellenberg (1978, p. 522), where one tree can grow, so could another one too. Thus, in the absence of disturbance, the site should become a closed forest. However, parklands with scattered trees are typical for transitions from forests to steppes and tundras from equatorial to (McNaughton, 1985) subarctic (Norin, 1961) areas, with just two notable exceptions. One consists of subarid temperate areas without fire-resistant trees, where fire prevents the development of savanna-like formations (Walter, 1968). This does not qualify as an example of an undisturbed timberline. The other one consists of islands without native herbivores (e.g., Hawaii and New Zealand; see Cockayne, 1958; Connor, 1965; Knapp, 1965; Mark, 1955; Burrell, 1965; Mueller-Dombois, 1967; Walter, 1968; Williams, 1975). On these islands, timberlines in the normal sense do not exist at all. In their natural state, the gradients of increasing aridity on Hawaii and New Zealand are characterized by forests grading into scrublands with successively lower stature, whereas graminoid-dominated vegetation has only occurred on disturbed sites. (Even there, the graminoids are functionally more like shrubs than like northern hemisphere grasses: leaves are extremely robust and sit on 1–2 m tall perennial pedestals.) On the mountains of Hawaii, rain forests directly grade into alpine barrens with rosette trees. On New Zealand, undisturbed altitudinal gradients consist of forests grading to successively lower scrublands and, finally, to barren crests dominated by robust cushion plants. Consequently, Ellenberg's point implies that undisturbed timberlines are practically nonexistent: either there is some form of disturbance or the transition from the dominance of upright woody plants to the prevalence of prostrate or semiprostrate plants is replaced by some continuous gradient where plants change their allocation patterns but retain an erect growth form.

In conclusion, the Cajander–Tilman approach is logically sound but its premises seem to be inapplicable to the rapid or sharp transitions from forests and closed scrublands to typical steppe and tundra communities. Grime, in turn, may be right in proposing that increasing stress sometimes implies lower intensity competition but cannot explain why this should happen. The enigma can be solved by assuming that stress is inevitably accompanied by disturbance (Tilman, 1988a, pp. 140–145; Tilman, 1988b, pp. 313–318). The aberrant vegetational patterns of grazer-free islands (Oksanen, 1988) suggest that grazing vertebrates are a central source for this disturbance, although fire can sometimes be an important contributing source. What remains to be explained is why a certain intensity of stress and intense grazing by vertebrates inevitably go together.

II. Trophic Dynamics and Primary Productivity

A theory which connects stress and grazing was outlined by Fretwell (1977). He found the hypothesis of Hairston *et al.* (1960) on trophic dynamics and population regulation to be appealing, but could not apply it to grassland ecosystems. According to Hairston *et al.*, the world is green (i.e., plant biomass is abundant) because predators regulate herbivores at a level where only a small fraction of primary production is consumed by grazers. However, this did not seem to apply to North American short-grass plains. There, plant biomass is not strikingly plentiful, and histori-cal evidence suggests that this is not a man-made situation. Studies from similar but still undisturbed ecosystems suggest that grazers are re-source-limited (Sinclair, 1977, 1985) and have strong impact on the veg-etation (McNaughton, 1979, 1985). Moreover, the native plants of the plains are clearly adapted to intense grazing pressure (Stebbins, 1981). Fretwell's solution was to accept the general view of Hairston *et al.*, that consumers can regulate their resources, but to reject the specific point that carnivores always limit herbivore populations and prevent depletion of forage. Fretwell suggested that trophic dynamics depend on primary productivity: the hypothesis of Hairston *et al.* applies to relatively pro-ductive habitats, whereas in more barren areas, grazers are resource-limited and the vegetation is subjected to intense natural herbivory.

Fretwell's (1977) idea was formally analyzed by Oksanen *et al.* (1981). They assumed that the equation for the growth of plant biomass contains a term representing the potential gross primary productivity of the habi-tat, and that both herbivores and carnivores are capable of regulating the abundances of their resources (Sih *et al.*, 1985). On both trophic levels, interactions between consumers are chiefly indirect, taking place via shared resources. The potential primary productivity of the habitat de-termines how much plant biomass it can maximally sustain and the den-sity of grazers needed to keep the plant biomass at any fixed level smaller than the maximum. With these assumptions, one can construct a phase space with biomasses of plants, grazers, and carnivores as its axes and find zero isoclines (actually: isosurfaces) where consumption and growth are equal for each of the three components. As three-dimensional iso-cline models are somewhat cumbersome to work with, it is fortunate that the ecologically relevant conclusions for plants can be presented as a two-dimensional projection, corresponding to the simple models of Ro-senzweig and MacArthur (1963).

Figure 1 represents a set of such two-dimensional isocline graphs for habitats with vast differences in primary productivity. In all habitats, the plant isocline is arch-shaped, but the archs differ in size and curvature. The isoclines corresponding to extremely barren environments are very close to the origin, whereas those for more productive habitats have higher and wider archs. As consumers are not directly affected by the factors which regulate the growth of plants, their isoclines are assumed to be identical for all habitats. In the absence of carnivores, the herbivore isocline is a pure consumer isocline, i.e., a straight vertical line: the short-term well-being of the grazers depends only on the amount of forage

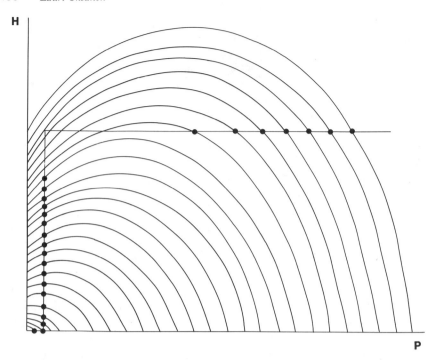

Figure 1   An isocline model of a herbivore (H)–plant (P) exploitation system with the impact of carnivores implicitly included. Archs, plant isoclines for habitats with different primary productivities; rectangle, herbivore isocline; dots, equilibrium points for each herbivore–plant system.

and is independent of the density of grazers (see Noy Meir, 1975; Rosenzweig, 1977; Caughley and Lawton, 1981). In the presence of carnivores, the herbivore isocline becomes complicated (Rosenzweig, 1973; Oksanen *et al.*, 1981), but we can get around this problem by noticing that, in laissez-faire exploitation systems, a top consumer limits the density of its resources to some constant level. Consequently, the effect of carnivores on grazer–plant dynamics can be mimicked by making the herbivore isocline consist of a vertical piece (the herbivore isocline in the absence of carnivores) and a horizontal piece which actually represents the carnivore isocline.

The isocline analysis of Fig. 1 shows that increasing primary productivity is accompanied by qualitative changes in the dynamics of the grazing chain. In extremely barren habitats, plant and herbivore isoclines do not meet at all. Consequently, the equilibrial community is predicted to be free of grazers and increasing productivity is predicted to increase the equilibrial plant biomass. In somewhat more productive habitats, plant and herbivore isoclines meet in the vertical section of the herbivore isocline. Consequently, the equilibrial community consists of plants and grazers. In this productivity interval, the herbivore–plant system exhibits a phenomenon dubbed as the "Paradox of Enrichment" by Rosenzweig (1971): increased primary productivity only increases the equilibrial densities of grazers, whereas accessible (i.e., above-ground) plant biomass remains constant. The next qualitative shift occurs when the environment becomes so productive that the plant isocline meets the horizontal section of the herbivore isocline (actually: the carnivore isocline). From this point onward, Paradox of Enrichment shifts to the interaction between carnivores and herbivores, and plant biomass starts to increase again in response to increasing primary productivity.

In order to have labels for ecosystems with different dynamics in the grazing chain, Fretwell (1977) called the three productivity zones where the model predicts qualitatively different trophic dynamics "one link," "two link," and "three link" ecosystems. This terminology creates the impression that food chains are predicted to be shorter in less productive habitats, which is debatable (Pimm, 1982; Pimm and Kitching, 1987). Excluding transients and those predators which exploit temporary outbreaks generated by a locally unstable grazer–plant equilibrium, this impression is in a way correct. Notice, however, that what has been modeled is only a part of the grazing chain. The model is restricted to animals which move on the land in their search of food and are active throughout the year or have high costs of dormancy, i.e., to vertebrates. Typical invertebrate grazing systems have many features (low mobility of herbivores, predictable association between the herbivore and the food plant, low costs of dormancy) which make it unlikely that their part of the grazing chain reacts to changes in primary productivity. Moreover, the carnivores that isocline models deal with are only a part of the carnivore trophic level—those capable of killing healthy prey. There are also carnivores adapted to search for weak prey (and carrion) and the presence of such carnivores (e.g., wolverines, jackals) in two-link ecosystems is not in conflict with the isocline model (see Oksanen and Ericson, 1987a).

How the predicted changes in trophic dynamics influence the life of the plants can be visualized by plotting the predicted equilibrial phytomass against the potential primary productivity of the habitat (Fig. 2a). The distance between the phytomass in the absence of grazers (dashed line) and the predicted equilibrium (solid line) represents the intensity of natural grazing. The plant ecological implications become still more tangible when the information is rearranged by counting the predicted percent difference between maximum and equilibrial biomass. This represents the intensity of natural grazing pressure from the point of view

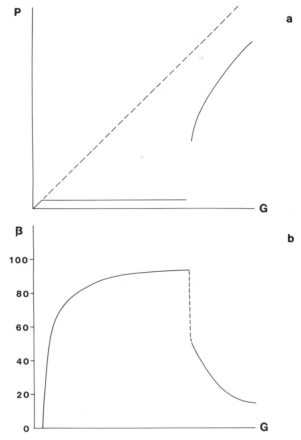

Figure 2 Predictions on phytomass and grazing pressure generated by Fig. 1. (a) Predicted relation between gross primary productivity (G) and above-ground phytomass (P) in the absence (dashed line) and presence (solid line) of grazers. (b) Predicted percent reduction of above-ground phytomass due to natural grazing pressure (β) as a function of gross primary productivity (G).

of plants. Focusing on this measure (β, Fig. 2b), we see that the productivity interval of two-link ecosystems stands out as a distinct zone with much higher natural grazing pressure than in other parts of the productivity cline.

The predicted relationship between primary productivity and natural grazing pressure is not especially sensitive to deviations from the assumption of strictly laissez-faire consumer–resource dynamics. Interference among herbivores would tilt the plateau part of Fig. 2a upward, but the impact on Fig. 2b would be very slight. Interference among carnivores would have still less influence: the increase in equilibrial phytomass in the right-hand part of Fig. 2a would become a bit more gradual, but only extreme levels of interference would produce notable changes in Fig. 2b. However, the plant predictions of the isocline model are sensitive to deviations from two assumptions, which are not yet stated. The isoclines are drawn so that grazers are assumed to be able to survive on very scanty vegetation, whereas carnivores are assumed to require relatively high grazer densities in order to break even. Converse assumptions would reduce two-link ecosystems to a marginal phenomenon at the transition from one-link to three-link ecosystems, and if both herbivores and carnivores were assumed to survive on a very scanty resource basis, the predictions would change so that practically all ecosystems would have three-link dynamics (which I think is the case in arthropod grazing chains). With regard to vertebrate grazers, my assumption is based on observations of semidomesticated reindeer, which are able to survive on ranges with very low biomasses of food plants. (A similar observation could be obtained, for example, by looking at goats and camels in the Middle East semideserts.) With regard to carnivores, I rely on studies showing that efficient pursuers of vastly different body sizes indeed require quite high prey densities (Schaller, 1972; Erlinge, 1974).

The central message of Fig. 2 is that what Grime (1977, 1979) proposes to be the stress-tolerant strategy might actually represent adaptations to intense grazing pressure. Facts and logics seem to support this reinterpretation (Tilman, 1988a). British heathlands play a central role in Grime's (1979) discussion on "stress tolerators," and this type of vegetation is indeed a product of long-lasting grazing (Gimingham, 1972). Moreover, Grime's (1979; Table 6) list of characteristics of supposed stress tolerators is easily reinterpreted as a catalogue of grazing-tolerant features. Low palatability is a clear antigrazer adaptation; small, leathery and needlelike leaves are more grazing-resistant than broad and mesomorphic ones; prevalence of vegetative reproduction is adaptive in heavily grazed habitats where inflorescences tend to be consumed (McNaughton, 1979; Oksanen and Ericson, 1987b; Tihomirov, 1959). Long life span of leaves makes heavy investments in defense more feasible, and low maximum growth rate is an inevitable consequence of such investments.

There is a straightforward way to test the relative merits of the two interpretations of Grime's stress-tolerant strategy. If the strategy really represented adaptations to environmental stresses, there should be a monotonous trend of increasing amount of stress-tolerant features along a gradient of increasing environmental stress. If, in turn, natural grazing pressure is the crucial factor, the so-called stress tolerators should be absent from the extremely stressful habitats (Fig. 2b). These contrasting predictions can be tested with ordination data of Oksanen and Ranta

(1989) on the vegetation of a mountain chain in Norwegian Lapland (Oksanen, 1980), rising from birch forests to eternal snow. In Table 1, 15 species with extremely high, middle, and extremely low scores on the first ordination axis (i.e., altitude) are compared to those morphological, reproductive (see Söyrinki, 1939), and chemical characteristics with which Grime (1979) predicts unambiguous differences between his competitive and stress-tolerant strategies (Table 1). The supposedly stress-tolerant features are relatively common at middle altitudes, but the high-alpine element is again characterized by features which Grime regards as competitive. This is not a peculiarity of this particular mountain chain: *Ranunculus glacialis* holds the altitudinal record throughout the Scandinavian mountains (Gjaerevoll and Jörgensen, 1952) and the European Alps (Ellenberg, 1978) and is the dominating constituent of Scandinavian high-alpine vegetation (Nordhagen, 1927, 1943; Gjaerevoll, 1956). *Oxyria digyna* has wide circumpolar distribution in high-alpine habitats

Table 1 Occurrence of Grime's Competitive (C) and Stress-Tolerant (S) Features in the High-, Middle-, and Low-Altitude Plants of Iddonjárga Mountains[a]

	1	2	3	4	5	6	7	8	Score C-S
High-altitude plants									
Cardamine bellidifolia	C	C	C	C	C	C	C	C	8-0
Cerastium cerastiodes	S	S	S	C	S	C	S	C	3-5
Carex lachenalii	—	S	S	S	C	C	S	C	3-4
Luzula confusa	—	S	S	C	S	C	S	C	3-4
Oxyria digyna	C	C	C	C	C	C	C	C	8-0
Poa alpina	—	S	S	C	C	C	S	C	4-3
Ranunculus glacialis	C	C	C	C	C	C	C	C	8-0
Ranunculus nivalis	C	C	C	C	C	C	C	C	8-0
Saxifraga aizoides	C	S	S	C	S	C	S	S	3-5
Saxifraga caespitosa	S	S	S	C	S	C	S	C	3-5
Saxifraga cernua	C	C	C	C	C	C	C	S	7-1
Saxifraga oppositifolia	S	S	S	S	S	S	S	C	1-7
Saxifraga rivularis	C	C	C	C	C	C	C	C	8-0
Saxifraga tenuis	S	S	C	C	S	C	S	C	4-4
Trisetum spicatum	—	S	S	C	C	C	S	C	4-3
Score for S features	4	9	8	2	6	1	9	2	75-41
Middle-altitude plants									
Alchemilla alpina	C	C	C	S	C	C	C	C	7-1
Anthoxanthum odoratum	—	S	S	C	C	C	S	C	4-3
Carex bigelowii	—	S	S	S	C	C	S	S	2-5
Calamagrostis lapponica	—	S	S	S	C	C	S	S	2-5
Diapensia lapponica	S	S	S	S	S	S	S	C	1-7
Euprasia frigida	C	C	S	C	C	C	C	C	7-1

(continued)

Table 1   *(Continued)*

	1	2	3	4	5	6	7	8	Score C-S
Festuca ovina	—	S	S	S	S	C	S	C	2-5
Hieracium alpinum	S	S	C	C	C	C	C	C	6-2
Juncus trifidus	—	S	S	S	S	C	S	C	2-5
Luzula frigida	—	S	S	S	S	C	S	C	2-5
Loiseleuria procumbens	A	S	S	S	S	S	S	C	1-7
Phyllodoce coerulea	S	S	S	S	S	S	S	C	1-7
Pyrola minor	S	S	C	S	S	S	S	S	1-7
Ranunculus acris	C	S	C	C	C	C	C	C	7-1
Silene acaulis	S	S	S	S	S	S	S	C	1-7
Score for S features	6	13	11	11	8	5	11	3	46-68
Low-altitude plants									
Betula pubescens	C	C	C	C	C	C	C	C*	8-0
Cornus suecica	C	C	C	C	C	C	C	S	7-1
Deschampsia flexuosa	—	S	S	C	S	C	S	S	2-5
Gymnocarpium dryopteris	C	C	C	C	C	C	C	S	7-1
Equisetum arvense	—	S	S	S	C	C	S	S	2-5
Equisetum pratense	—	S	S	S	C	C	S	S	2-5
Juniperus communis	C	S	S	S	S	S	S	C*	2-6
Linnaea borealis	S	S	S	S	S	C	S	S	1-7
Luzula pilosa	—	S	C	C	S	C	S	C*	4-3
Lycopodium annotinum	S	S	S	S	S	S	S	S	0-8
Pedicularis lapponica	C	S	S	C	C	C	C	S	5-3
Salix caprea	C	C	C	C	C	C	C	C*	8-0
Salix myrsinifolia	C	C	C	C	C	C	C	C*	8-0
Salix phylicifolia	C	C	C	C	C	C	C	C*	8-0
Trientalis europaea	C	C	C	C	C	C	C	S	7-1
Score for S features	2	8	7	5	5	2	7	9	71-45
χ^2-tests[b]									
High vs. low altitude	—	—	—	—	—	—	—	*1	—
High vs. middle altitude	—	—	—	*2	—	—	—	—	*3
Middle vs. low altitude	o4	—	—	o5	—	—	—	o6	*7

[a] (1) Life form: C, tree, upright shrub, or ordinary herb; S, cushion plant, rosette plant, or trailing dwarf shrub; —, graminoid or *Equisetum* (relation to Grime's categories unclear). (2) Shoot structure: C, leaves well differentiated and apically attached to an upright stem or elevated by a stem-like petiole; S, photosynthesizing organs with wide vertical spreading or entirely basal position. (3) Leaf shape: C, robust and mesomorphic; S, small or narrow and needle-like. (4) Leaf texture: C, soft; S, tough and leathery. (5) Leaf longevity: C, short (deciduous); S, long (evergreen). (6) Palatability: C, high, moderate, or low due to small amounts of acute toxins; S, very low, due to large concentrations of secondary compounds. (7) Perennation: C, specialized buds or seeds; S, stress-tolerant leaves or roots. (8) Reproduction: C, generative reproduction has a significant role in the life cycle; S, reproduction overwhelmingly vegetative; C*, little or no generative reproduction in Söyrinki's (1939) material but substantial generative reproduction in more southern areas (i.e., generative reproduction belongs to the strategy of the plants but cannot be executed under subarctic conditions), combined with C in the calculation of the C-S score. Life form has been ignored in the tests of total C-S scores because of its positive correlation with shoot structure.

[b] *, $p < 0.05$; o, $p < 0.1$; —, not significant. (1) $\chi^2 = 5.17$, (2) $\chi^2 = 8.69$, (3) $\chi^2 = 13.90$, (4) $\chi^2 = 3.31$, (5) $\chi^2 = 3.34$, (6) $\chi^2 = 3.47$, (7) $\chi^2 = 8.41$.

(see Mooney and Billings, 1961; Billings and Mooney, 1968). Also, the high-alpine saxifrages of Table 1 (and their morphologically similar congeners) are a dominating element in the boulderfield flora of arctic, boreal, and temperate mountains (see Nordhagen, 1927, 1943; Böcher, 1933, 1954; Braun-Blanquet, 1948–1950; Gjaerevoll, 1956; Reisgl and Pitschmann, 1958; Komarková, 1978).

However, just changing the label of Grime's stress-tolerant strategy to "grazing-tolerant strategy" is not sufficient for relating broad vegetational patterns to the predictions of the isocline model. Grime's catalog is not derived from first principles but represents summary of observations, and comparing observation-based generalizations to further observations creates a risk of circular reasoning. With regard to such a key characteristic as stature, Grime refrains from making any clear statements. Moreover, graminoids, which are a prominent constituent of many tundra and steppe communities, show less than perfect fit to the list of features that Grime regards as stress-tolerant (see Table 1).

An attempt to connect Fig. 2b to broad vegetational patterns thus requires that two questions be answered. First, the impact of grazing on stature must be analyzed. Second, one must find out when grazing will favor plants that Grime regards as stress tolerators—with slow growth rate and a large quantity of resources allocated to the production of secondary chemicals—and when will graminoid-type plants with low defensive investment and an ability to recover rapidly be favored. In both issues, my approach is to apply the general theory of Evolutionarily Stable Strategies (ESS; see Maynard Smith, 1974) and to turn attention to patterns in nature only after predictions have been derived from first principles.

III. Grazing and the ESS Foliage Height of Plants

The main reason for plants to grow tall is competition for light, which makes it advantageous to divert resources from leaf surfaces to erect stems. If there is enough competition for light to favor any vertical growth at all, woody plants should normally have an advantage over herbs and grasses (see above). Conversely, if there is no selective pressure for tall stature in herbs, there is hardly any point in producing erect, woody stems. Thus, an analysis of the ESS foliage height in herbs also tells a good deal about the relative advantages of woody and herbaceous habits.

In a purely competitive situation, the foliage height ESS of herbs represents a balance between two factors. On one hand, it is always advantageous to be slightly taller than the neighbors. On the other hand, the taller the herb, the greater fraction of available resources must be allocated to support structures. This balance summarized below, has been analyzed by Givnish (1982) in an ingenious way. Let $f(h)$ represent the amount of resources available to leaves as a function of foliage height and $g(\Delta h)$ be the photosynthetic rate per unit of leaf area as a function of the difference between the foliage height of the plant and the average foliage height in the vegetation. If the function $f(h)$ and the value and first derivative of $g(\Delta h)$ at $\Delta h = 0$ are known, the ESS foliage height can be found as the value of h that satisfies the equation

$$-f'(h)/f(h) = g'(0)/g(0) \tag{1}$$

The reason is that the photosynthetic performance of a plant depends on the product $f(h)g(\Delta h)$. A product increases as long as the relative decrease rate of one factor $[f(h)]$ is slower than the relative increase rate of the other factor $[g(\Delta h)]$. The ratio of the first derivative to the function represents these relative rates; the minus sign on the left is needed because we compare a decrease to an increase [i.e., $f'(h) < 0$].

Givnish (1982) argued that $g(\Delta h)$ is a sigmoid function, steepest at $h = 0$. The denser the vegetation, the further down is the lower asymptote of the $g(\Delta h)$ curve which represents the net photosynthetic rate of below-canopy leaves. With regard to the $f(h)$ function, Givnish was a bit inconsistent. He assumed linearly decreasing $f(h)$, which was compatible with his data (where scatter prevents firm conclusions about the form of $f(h)$ from being drawn). However, his equations showed that $f(h)$ must be strongly upward convex, as h appears in fourth power (Givnish, 1982; Eq. A9), and also common sense says this. Because of mechanical constraints, height increments of a given magnitude must require greater investments of raw materials if the shoot is tall to begin with. This point has an important corollary: even if the plant cover is sparse and, consequently, $g'(0)/g(0)$ has a low value, prostrateness will not easily be an ESS for herbs in a purely competitive situation, because of the low marginal costs of raising leaves slightly above the ground (see Fig. 3). As an erect, woody stem in a few years lifts leaves higher up than a herbaceous stem produced by corresponding annual investment, we can conclude that undisturbed competition favors erect, woody plants even in habitats where competition for light is only a minor factor and optimally allocating plants invest most of their resources in roots.

The model of Givnish (1982) is based on the tacit assumption that tissue losses either do not occur or they are independent of shoot height. In grazed systems, these premises are unrealistic. Leaf tissues have some positive mortality rate (m) due to grazing, and this rate is a function of foliage height: pieces of foliage which are so close to the ground that it is difficult for the grazer to get them will have a much lower mortality rate

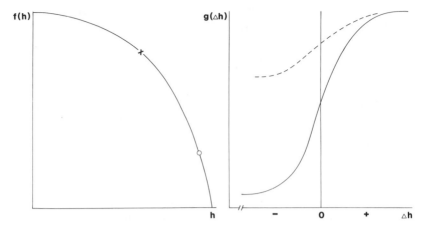

Figure 3 Amount of resources available for foliage (f) as a function of foliage height (h) and photosynthetic rate per unit of foliage (g) as a function of the difference (Δh) between the foliage height of the plant and the average foliage height of the vegetation. Dashed line, sparse vegetation; solid line, dense vegetation. Corresponding ESS foliage heights are marked by × and ○, respectively.

than more elevated pieces of foliage. The simplest plausible form for $m(h)$ is a sigmoidally increasing function of h (Fig. 4). However, other forms are also possible. Small grazers especially can have problems with handling tall leaves of herbaceous plants. In that case, $m(h)$ will be hump-backed, with a negative slope at high values of h.

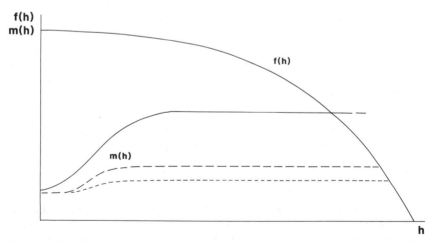

Figure 4 Amount of resources available to foliage (f) and rate of tissue loss (m) as functions of foliage height. Solid m curve, intense grazing; dashed line, moderate grazing; dotted line, light grazing.

With a constant instantaneous rate of grazing, leaf tissues produced at the time $t = 0$ disappear in a negative exponential manner, with the amount of remaining leaf tissues (L) after time t being

$$L = e^{-mt} \tag{2}$$

The contribution of leaves produced at $t = 0$ to the energy balance of the plant can be obtained by multiplying the instantaneous contribution rate, i.e., $f(h)g(\Delta h)$, by the area between Eq. (2) and the time axis, which represents the effective lifetime (T) of the leaves. Integrating Eq. (2) from $t = 0$ to $t = \infty$ yields the simple result that $T = 1/m(h)$. By noting $f(h)/m(h) = z(h)$, we can directly apply the results of Givnish (1982): a given foliage height is an ESS if it satisfies the equation

$$-z'(h)/z(h) = g'(0)/g(0) \tag{3}$$

There are two ways to proceed further: to construct the $z(h)$ curve dividing $f(h)$ by $m(h)$ point by point and to perform a graphical analysis comparable to Fig. 2. The result (Fig. 5) shows that, in most cases, weak or moderate grazing pressure has no impact on the foliage height ESS. When grazing pressure becomes high enough to have any impact at all, the impact is drastic at once: the ESS foliage height "jumps" from high to very low, and is thereafter only little affected by further increases in the intensity of grazing. The grazing pressure which is intense enough to cause this shift depends on the density of the vegetation, but the abrupt

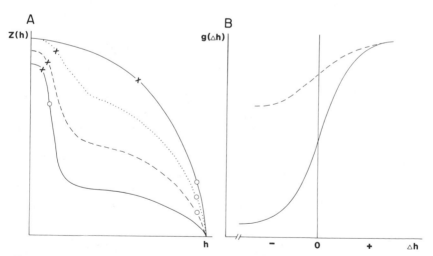

Figure 5 An ESS analysis corresponding to Fig. 3, but with $f(h)$ replaced by $z(h)$, obtained by dividing $f(h)$ for the appropriate $m(h)$ values. (A) solid line (top), $z(h)$ for no grazing [$= f(h)$]; dotted line, light grazing; dashed line, moderate grazing; solid line (bottom), intense grazing. For each $z(h)$ curve, ESS heights for sparse and dense vegetation have been marked by $\times$ and $\bigcirc$ as in Fig. 1. (B) As described in Fig. 3.

nature of the shift does not. A hump-backed $m(h)$ curve would further aggrevate the abruptness of this shift. In that case, the impact of weak grazing pressure is to favor plants that are taller than the foliage height ESS for the purely competitive situation provided that this ESS is greater than the foliage height for the hump of the $m(h)$ curve; sufficiently high grazing pressure again favors prostrate plants.

An analytical approach clarifies the situation. Substituting $z(h) = f(h)/m(h)$ and doing some rearrangements yields

$$-f'(h)/f(h) + m'(h)/m(h) = g'(0)/(g(0)) \qquad (4)$$

If the $m(h)$ function is sigmoid, $m'(h) \approx 0$ for a wide range of h values (Fig. 4), including the neighborhood of the purely competitive ESS for practically all conceivable $g(\Delta h)$ curves (see Fig. 2). In the neighborhood of the competitive ESS, Eq. (4) thus degenerates to Eq. (1). If the $m(h)$ curve is hump-backed and the vegetation is at least moderately dense, $m'(h)$ is negative in the neighborhood of the competitive ESS, thus low grazing intensities will indeed increase foliage height ESS. Whether the $m(h)$ function is sigmoid or hump-backed, the foliage height ESS will shift to a very low value when the rising part of the $m(h)$ curve becomes steep enough to satisfy Eq. (4).

The above analysis presupposes that all plants have the same photosynthetic response to changed light intensities. Under this premise, foliage heights below the ESS level are always suboptimal and probably even lethal if the vegetation is dense. However, a pronounced shade plant could readily invade plant communities consisting of tall plants. For an invading shade plant, even weak grazing pressure would represent strong selection for low stature. Being well below the average foliage height of the dominants, a shade plant is in a situation where $g'(\Delta h)/g(\Delta h)$ is very small; consequently, fairly low maximum values of $m'(h)/m(h)$ can suffice to make $m'(h)/m(h)$ alone greater than $g'(\Delta h)/g(\Delta h)$. Plants with an intermediate shoot height between the canopy and the handling threshold of grazers get the worst of both worlds by being too low to compete for light but yet too tall to escape grazing.

The all-or-none impact of grazing on the foliage height ESS gets especially interesting in systems with substantial fluctuations in the intensity of grazing which, indeed, is more the rule than the exception in areas grazed by mobile herds of ungulates (McNaughton, 1979, 1985), fluctuating populations of microtine rodents (Fuller et al., 1977; Černjavskij and Tkačev, 1982; Andersson and Jonasson, 1986) or both (Batzli et al., 1980). Even if the average grazing pressure were high enough to favor prostrate plants, they might go extinct in years of low grazing intensity. Conversely, even if erect woody plants were favored on the average, they might never manage to reach safe size before becoming grazed and trampled by an ungulate herd or bark-gnawed to death by rodents. As long-term fitness in a fluctuating environment depends on the product of annual fitnesses (Levins, 1968; Schaffer, 1974), it is imperative for a plant to hedge its bet and to make sure that its fitness in the worst year will not be zero (see Stearns, 1976). Typical basal-leaved graminoids can be regarded as masters of the trade of coping with fluctuating intensity of grazing. They can flexibly increase leaf height if grazers remain absent and competition for light becomes intense. Yet, they hold the most valuable tissues out of the reach of grazers. After grazing they can first produce leaves that lie flat on the ground. Later on, these pieces of foliage can be lifted up by producing vertical leaf segments which start to function like petioles.

IV. Grazing and the ESS Level of Plant Defenses

Natural selection does not reward a plant for damaging or killing grazers but for directing herbivory from oneself to one's neighbors (Moran and Hamilton, 1980). Thus, the level of defense that represents an ESS can be studied with the same method that has above been applied to the study of foliage height ESS. Assuming that a plant has a fixed amount of reduced carbon available and an allocation pattern between different plant organs which is independent of the allocation of carbon to defense, the amount of photosynthetically active leaf tissues (f) that a plant can construct is a decreasing function of their concentration of defensive compounds (d). As structural tissues and defensive compounds are two alternative ways of allocating reduced carbon, the relation can be specified as $f(d) = 1/(d + 1)$ (see Fagerström, 1989). The rate of net energy accumulation, $p(d)$, must be closely related to $f(d)$. If plants did not respire and if reduced carbon were a limiting resource for the construction of photosynthetically active tissues, $p(d)$ would be directly proportional to $f(d)$, so that the two would be interchangeable with appropriate scaling of units. However, respiration will inevitably consume some reduced carbon. A simple way to model this is to assume that respiratory costs are independent of the allocation between productive tissues and defensive compounds, in which case, carbon-limitation implies

$$p(d) = f(d) - c = 1/(1 + d) - c \tag{5}$$

where c stands for respiratory costs. The assumption of constancy can indeed be criticized, but the deduction given below is quite robust with respect to deviations from a constant c. The main thing is that c is positive for all values of d, as it indeed has to be.

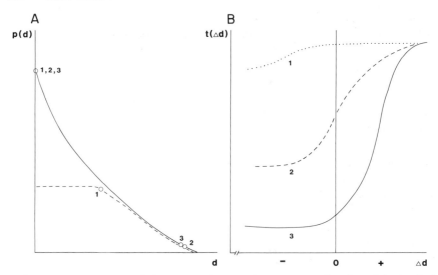

Figure 6 (A) Amount of photosynthetic parenchyma (p) as a function of the level of defense (d) in a nutrient-rich (solid line) and nutrient-poor (dashed line) habitat. (B) The effective production time of the leaves (t) as a function of the difference between the ambient level of defense and the level of defense in the plant. (Δd) for lightly grazed (1, dotted line), moderately grazed (2, dashed line), and intensely grazed (3, solid line) habitat. The corresponding ESS levels of defense are marked by circles that are numbered as the $t(\Delta d)$ curves.

Equation (5) generates a downward convex $p(d)$ curve (Fig. 6, solid line); which should be realistic for nutrient-rich sites. In nutrient-poor sites, the values for $p(d)$ for low levels of d will be less than predicted by Eq. (5): even if plants do not allocate anything to defense, nutrient shortage prevents them from using all available reduced carbon for making photosynthetically active leaf tissues. Consequently, the leftmost part of the $p(d)$ curve will be almost flat (see Tuomi *et al.*, 1984), and Eq. (5) becomes relevant first when all surplus carbon has been channeled to defensive compounds. The resulting $p(d)$ curve (Fig. 6, dashed line) will thus be sigmoidly decreasing.

The rate of tissue loss (m), in turn, will be a sigmoidally decreasing function of the difference between the level of defense in the plant and the average level of defense in the vegetation [$m(\Delta d)$]: plants that are much less palatable than average are rejected, those which are more palatable are preferred. Small differences in the level of defense matter most for plants which are at the limit of being included in the diet. The rate of tissue loss for rejected plants will be low regardless of the intensity of grazing, whereas the loss rate of preferred plants must be an increasing function of the intensity of grazing. According to the theory of optimal foraging (Krebs, 1978), the level of defense where m is maximally sensitive to small changes in Δd depends on the availability of high-quality forage. In lightly grazed environments where high-quality items (e.g., reproductive organs; see Tast and Kalela, 1971) abound, only the best will be good enough. When the grazing pressure is intense and there is acute shortage of subsistence food, only the worst food items will be rejected (see Thomas and Edmonds, 1983). Consequently, increasing grazing pressure will both increase the steepness of the $m(\Delta d)$ curve by raising its upper asymptote and shift the steepest part rightward (from $\Delta d < 0$ to $\Delta d > 0$). With a rate of tissue mortality m, the effective production time (t) of the foliage will be $1/m$ (see previous section). As the inverse of a sigmoidally decreasing function, $t(\Delta d)$ will be a sigmoidally increasing one (Fig. 6).

The contribution of the leaves to the energy balance of the plant is their production rate, $p(d)$, times their effective lifetime, $t(\Delta d)$. Thus, the results of Givnish (1982) can be directly applied. The level of defense is an ESS if it satisfies

$$-p'(d)/p(d) = t'(0)/t(0) \tag{6}$$

For situations where $p(d)$ and $f(d)$ curves match (i.e., plants are carbon-limited), we can substitute $p(d) = 1/(d + 1) - c$. After rearrangements, this gives

$$1/[1 - c + (1 - 2c)d - cd^2] = t'(0)/t(0) \tag{7}$$

Three predictions immediately emerge from Fig. 6 and Eq. (7). First, zero level of carbon-based chemical defense is practically never optimal in nutrient-poor habitats. As long as mineral nutrients limit the construction of photosynthetically active leaf tissues, defense is so cheap that even slight grazing pressure favors increasing levels of defense, up to the point where reduced carbon becomes a limiting resource. Second, if grazing will ever favor allocation of "costly" carbon to defense, moderate intensities will do it already. At high grazing pressures, $t'(0)$ starts to be a decreasing function of grazing intensity, because grazers cannot be choosy. Although this is to some extent compensated by decreasing $t(0)$, the result is still that $t'(0)/t(0)$ reaches a maximum at moderate grazing intensity. Third, provided that respiratory costs are not overwhelmingly high ($c < 0.5$), the marginal costs of defense are a decreasing function of d for low and moderate levels of defense. Consequently, it is unlikely that an ESS could be found at low allocations of costly carbon to defense. If the left-hand side of Eq. (7) is smaller than the right-hand side at $d = 0$ or when all "surplus carbon" has become allocated to defense (i.e., at the lowest value of d where Eq. (7) becomes relevant), there will be runaway selection to higher and higher levels of defense, until the second-order term in the denominator of the left-hand expression starts to dominate. This has an interesting corollary. Because nutrient limitation pushes the region of applicability of Eq. (7) rightward where marginal costs of allocating costly carbon to defense are not as forbiddingly high as at $d = 0$, moderate nutrient shortage increases the likelihood of runaway selection to high levels of defense. However, this does not apply to extreme levels of nutrient shortage: then the plateau of the $p(d)$ curve extends to high values of d and, consequently, the second-order term dominates immediately when Eq. (7) becomes relevant, making runaway selection impossible.

In less technical terms, the above predictions can be summarized as follows. Allocation of surplus carbon to defense is part of a defense ESS almost regardless of the intensity of grazing. In relatively nutrient-poor habitats, it is likely that moderate intensities of grazing shift the ESS to high levels of carbon-based chemical defense, and it is possible that extremely high grazing intensity shifts the ESS back to the use of surplus carbon only. In nutrient-rich habitats, the ESS level of purely carbon-based chemical defense is likely to remain zero regardless of the intensity of grazing. Instead, plants will opt for rapid recovery or forms of defense which yield better marginal gains at $d = 0$ (nitrogen-containing toxins, spines). These predictions are not especially novel: rather similar ideas have been deduced from observations by, for example, Bryant *et al.* (1983), Coley *et al.* (1985), and van der Mejden *et al.* (1988). However, the model shows that these ideas follow from first principles, and that reference to the somewhat confusing concept of compensatory growth (see Belsky, 1986) is not needed. The essential issue is the relation between marginal gains and marginal costs of defense, not the direct reaction of plants to grazing.

V. Graminoid, Ericoid, and *Dryas* Strategies

From Sections III and IV we can conclude that, depending on conditions, grazing will favor one out of three broad adaptational syndromes. If the habitat is nutrient-rich and grazing is both intense and frequent, prostrate plants will be favored. These should have morphological and reproductive features which improve grazing tolerance (small, narrow, or finely lobed leaves, either chiefly vegetative reproduction or many inflorescences, each with a small number of small seeds which ripen quickly), and they should have structures which decrease their attractiveness to grazers (hairs, thick cuticle). However, there should not be accumulations of defensive chemicals above the level of carbon surplus. This combination of characteristics is found in many short grasses, but also in many arctic–alpine dicots. To emphasize that this syndrome is by no means restricted to monocots, I have chosen to call it the *Dryas* strategy, after mountain avens. In nutrient-poor habitats, constant and relatively high intensity of grazing will favor plants which differ from the *Dryas* strategy by having high levels of strictly carbon-based chemical defense. This combination of traits should usually go together with evergreen leaves, which make it more easy to achieve high concentrations of defensive chemicals with a feasible annual investment. Thus, the syndrome becomes practically identical to what Grime (1977, 1979) called the stress-tolerant strategy. I call the syndrome the ericoid strategy after the dicot group for which this combination of traits is especially typical.

Fluctuating intensity of grazing will favor plants that share most characteristics of the *Dryas* strategy but differ from it by having a capacity for rapid vertical growth by means of basal intercalary meristems or an apical bud at ground level. I call this combination of traits the graminoid strategy, as the majority of plants with these characteristics are graminoids (members of families Poaceae, Cyperaceae, and Juncaceae). However, dicots with rosettes of erect, finely lobed leaves (e.g., *Geum rossii* of Rocky Mountains) represent this strategy, whereas tall grasses, tussock grasses, and obligately prostrate graminoids do not.

Data on global and continental vegetation patterns (Walter, 1964, 1968; Knapp, 1965; Ellenberg, 1978) and on the vegetation of well-studied mountain, tundra, and steppe–desert areas (Nordhagen, 1927, 1943; Böcher, 1933, 1954; Kalliola, 1939; Braun-Blanquet, 1948–1950; Whittaker and Niering, 1965; Bliss, 1975; Komarková, 1978; Olsvig-Whittaker *et al.*, 1983) suggest that moderately barren ecosystems are consistently dominated by one of the three variants of the grazing-tolerant strategy. The *Dryas* strategy is especially typical for dry but at least moderately nutrient-rich tundras. The ericoid strategy is characteristic for nutrient-poor oceanic tundras (continuously chilly weather, strongly leached soils) and arid regions with Mediterranean-type climate, where favorable thermal and moisture conditions do not coincide and soils get leached by winter rains. There are some peculiar types of grasslands in habitats which are even more nutrient-poor than the areas where typical representatives of the ericoid strategy prevail [e.g., the tussock tundra of nonglaciated parts of Alaska and northern USSR (see Wein and Bliss, 1974; Chapin and Shaver, 1985) and grasslands in interior Australia (see Winkworth, 1967)]. However, these formations are dominated by tussock graminoids which have little to do with the graminoid strategy as defined above but seem rather to represent cases where high levels of defense arise automatically, as a consequence of the extreme nutrient shortage of the habitat. Genuine representatives of the graminoid strategy—herbaceous plants with basal growth points and narrow or finely lobed leaves—prevail in at least moderately nutrient-rich arid, alpine, and arctic habitats with strongly pulsatory primary production which inevitably makes grazing intermittent.

If the vegetational patterns of tundra, steppe, and semidesert environments were caused by changes in environmental constraints on classical competition, the vegetation of extremely barren environments should be either basically similar or represent a continuation of the same trend. However, extreme deserts are dominated by tall shrubs or low trees and by annuals with relatively "mesomorphic" looks (Walter, 1964, 1968). Extreme high-altitude barrens of temperate and boreal mountains are dominated by mesomorphic perennial herbs like *Ranunculus glacialis* (see Table 1). The dominant of polar deserts, *Papaver radicatum*, has basically similar morphology, and the distinctive feature of polar deserts (as opposed to semideserts) is the rarity of graminoids and trailing dicots (Bliss *et al.*, 1984). On tropical mountains, rosette trees represent the ultimate high-alpine life form: they occur above alpine grasslands (Troll, 1941; Smith and Young, 1987) and seem to be limited downward by herbivory (Kofford, 1957; Mulkey *et al.*, 1984).

Further evidence can be obtained from herbivore exclosure experiments. A large-scale experiment with temperate steppe vegetation was inadvertently performed when a remnant of the Ukrainian steppe was protected against grazers. The immediate result was an expansion of shrubs and tall grasses at the cost of the original dominants (*Stipa*, bunch grasses). Later on, the area was invaded by Scots pines (Walter, 1968, p. 602). A controlled long-term exclosure experiment has been run on the lowland tundra at Barrow, Alaska, since the 1950s. The general result has been a pronounced decline in the abundance of typical tundra graminoids. Tall grasses have flourished in favorable habitats and a gradual build-up of moss banks has taken place in less favorable ones (Batzli *et al.*, 1980). The exclosure experiments in arid plains of the Serengeti, Tanzania, have resulted in an even quicker and more complete replacement of short grasses by tall grasses (McNaughton, 1979). In my tundra exclosures on Finnmarksvidda, Norwegian Lapland, changes in plant cover have been equally dramatical (Oksanen, 1988). During the first 3 years, there was a pronounced expansion of the blueberry (*Vaccinium myrtillus*) on lichen–moss tundra and almost total disappearance of mosses (Oksanen and Oksanen, 1981). In the other two habitats (snowbed and low herb meadow), changes proceeded at a more uneven pace. After 8 years, however, the results were clear in all habitats: the plants that were closest to the *Dryas* strategy were suffering heavy losses, whereas the tallest and most broad-leaved ones were favored (Fig. 7). Age structure differences between blueberry twigs in exclosures and on open plots (Fig. 8) support the interpretation that

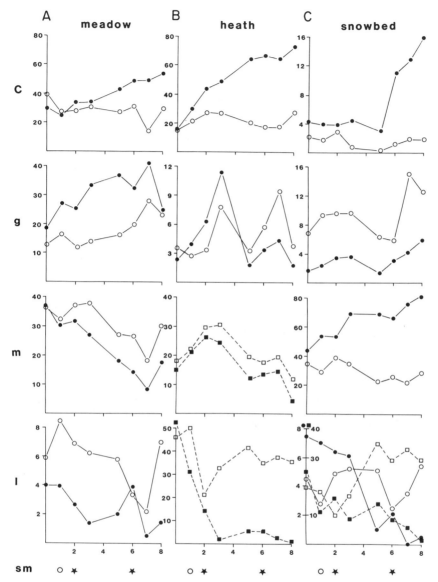

Figure 7 Changes in the relative abundance of pronounced competitors (c), typical graminoids (g), medium-sized plants (m), and low plants (l) in exclosures (solid symbols) and on open plots (open symbols) in the following different tundra habitats: (A) shrubby meadow (c are broad-leaved shrubs and dwarf shrubs and tall grasses, m are herbs with foliage heights from 5 to 15 cm, l are smaller herbs), (B) lichen–moss heath (c are broad-leaved shrubs and dwarf shrubs, m are fruticose lichens, l are low mosses and hepatics), and (C) dry snowbed (c are the tallest herbs and grasses, m are dwarf willows, goldenrods, and dandelions, l are prostrate herbs, low mosses, and hepatics). Squares and dashed lines refer to cryptogams, dots/circles and entire lines to vascular plants.

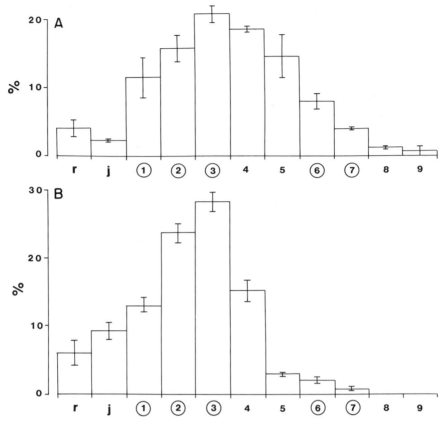

Figure 8 Age structures of (A) exclosure and (B) open plot populations of *Vacinium myrtillus* on the lichen–moss heath. r is a horizontal runner without an above-ground part; j is a juvenile shoot without overwintered above-ground parts; numbers 1 through 9 refer to the number of overwintered annual above-ground segments, which are circled if the oldest segment grew up under high or moderate vole density. Vertical bars refer to standard errors.

differences in survival rates during vole peaks account for the observed vegetational changes.

VI. Concluding Remarks

My chapter has focused on presenting a view of steppes, arctic–alpine tundras, and tropical alpine grasslands as the rangelands of the nature, where grazing and grazer-mediated indirect interactions (apparent competition; see Holt, 1977) are overridingly important for structuring plant communities. This must, however, be seen in proper perspective. The model of Oksanen *et al.* (1981) implies that natural herbivory is *not* important in ecosystems with primary productivity greater than 700 g/m²/ yr dry matter, except for such parts of the vegetation which receive very little resources and disproportionately much herbivory, and is even less important in extremely barren habitats such as polar deserts, alpine boulderfields, and arid deserts. When it comes to explaining the structure of these plant communities which cover most of our globe, I am a firm advocate of the Cajander–Tilman approach. In the present chapter, I have focused on plant communities and vegetation processes which I regard as exceptional in a global perspective, because that is where I have found new challenges and unexplained patterns.

VII. Summary

A model of population dynamics in the grazing chain predicts that the importance of natural herbivory is small in productive habitats because predators prevent the persistence of excessively high grazer densities, and in extremely barren areas, where grazers only occur at transients. In areas with intermediate productivity, natural grazing pressure is predicted to be intense, because moderate herbivore densities are sufficient to consume the relatively meager production of the vegetation. Consequently, classical resource competition should be an overridingly important vegetation-structuring process in productive and in extremely barren areas, whereas, in areas with intermediate primary productivity, i.e., arctic and alpine tundras, steppes, arid savannas, and the grassland belts of tropical mountains, grazing and grazer-mediated indirect interactions between plants should play a correspondingly central role. Traits of morphology and life history in dominating plants of these habitats fit the idea that the native grazers play a central role in structuring their vegetation. This hypothesis is also supported by the results of experimental exclusions of grazers and by the aberrant vegetational patterns of grazer-free islands.

Acknowledgments

In the fieldwork on the tundra, the help of Tarja Oksanen and Aslak Lukkari has been indispensable. Section IV on plant defenses was substantially revised in response to the comments of Torbjörn Fagerström, who showed that marginal cost of defense can be a decreasing function of d, and David Tilman, who pointed out the need to incorporate respiratory costs in the model. Useful comments were also provided by Don Alstad and two anonymous referees. The work has been supported by a grant from NFR (Swedish Council For Natural Sciences).

References

Andersson, M., and Jonasson, S. (1986). Rodent cycles in relation to food resources on an Alpine heath. *Oikos* **46**, 93–106.

Batzli, G. O., White, R. G., McLean, S. F., Jr., Pitelka, F. A., and Collier, B. D. (1980). The herbivore-based trophic level. *In* "An Arctic Ecosystem: The Coastal Tundra at Barrow, Alaska" (J. Brown, L. Tieszen, and F. Bunnell, eds.), pp. 335–410. Dowden, Hutchinson & Ross, Stroudsburg, Pennsylvania.

Bazzaz, F. A. (1979). The physiological ecology of plant succession. *Annu. Rev. Ecol. Syst.* **10**, 351–371.

Belsky, J. (1986). Does herbivory benefit plants? A review of the evidence. *Am. Nat.* **127**, 870–892.

Billings, W. D., and Mooney, H. A. (1968). The ecology of Arctic and Alpine plants. *Biol. Rev.* **43**, 481–529.

Bliss, L. C. (1975). Tundra grasslands, herblands and shrublands and the role of herbivores. *Geosci. Man* **10**, 51–79.

Bliss, L. C., Svoboda, J., and Bliss, D. I. (1984). Polar deserts, their plant cover and plant production in the Canadian high Arctic. *Holarct. Ecol.* **7**, 305–324.

Böcher, T. (1933). Studies on the vegetation of the east coast of Greenland. *Medd. Groenl.* **104**, 1–132.

Böcher, T. (1954). Oceanic and continental vegetation complexes in southwestern Greenland. *Medd. Groenl.* **148**, 1–336.

Braun-Blanquet, J. (1948–1950). Übersicht der Pflanzengesellschaften Rätiens. *Vegetatio* **1**, 29–41, 129–146, 258–316; **2**, 20–37, 214–237, 341–360.

Bryant, J. P., Chapin, F. S., and Klein, D. J. (1983). Carbon/nutrient balance of boreal plants in relation to vertebrate herbivory. *Oikos* **40**, 357–368.

Burrell, J. (1965). Ecology of *Leptospermum* in Otago. *N.Z. J. Bot.* **3**, 3–16.

Cajander, A. K. (1909). Über die Waldtypen. *Acta For. Fenn.* **1**, 1–175.

Callaghan, T. V., and Emmanuelsson, U. (1985). Population structure processes of tundra plants and vegetation. *In* "The Population Structure of Vegetation" (J. White, ed.), pp. 399–439. Junk, Dordrecht, The Netherlands.

Caughley, G., and Lawton, J. H. (1981). Plant–herbivore systems. *In* "Theoretical Ecology: Principles and Applications" (R. May, ed.), pp. 132–166. Blackwell, Oxford, England.

Černjavskij, F. B., and Tkačev, A. V. (1982). "Populjacionnye Cikli Lemmingov v Arktike: Ekologisčeskie i Endokrinnye Aspekty." Nauka, Moscow.

Chapin, S. F., and Shaver, G. R. (1985). Individualistic growth response of tundra plant species to environmental manipulations in the field. *Ecology* **66**, 564–576.

Cockayne, L. (1958). "The Vegetation of New Zealand," 3rd ed. Engelmann, Leipzig, German Democratic Republic.

Cody, M. L. (1986). Structural niches in plant communities. *In* "Community Ecology" (J. Diamond and T. J. Case, eds.), pp. 381–405. Harper & Row, New York.

Coley, P. D., Bryant, J. P., and Chapin, S. F. (1985). Resource availability and plant antiherbivory defense. *Science* **230**, 895–899.

Connor, H. E. (1965). Tussock grasslands in the middle Rakaia Valley, Canterbury, New Zealand. *N.Z. J. Bot.* **3**, 261–276.

Darwin, C. (1859). "The Origin of Species by Means of Natural Selection or the Preservation of Favoured Races in the Struggle for Life." Murray, London.

Ellenberg, H. (1978). "Vegetation Mitteleuropas mit den Alpen in Ökologischer Sicht." Ulmer, Stuttgart, Federal Republic of Germany.

Ellenberg, H., and Mueller-Dombois, D. (1974). "Aims and Methods of Vegetation Study." Wiley, New York.

Erlinge, S. (1974). Distribution, territoriality and numbers of the weasel *Mustela nivalis* in relation to prey abundance. *Oikos* **25**, 308–314.

Fagerström, T. (1988). Antiherbivory chemical defense in plants: A note on the concept of cost. *Am. Nat.* **133**, 281–283.

Fretwell, S. D. (1977). The regulation of plant communities by food chains exploiting them. *Perspect. Biol. Med.* **20**, 169–185.

Fuller, W. A., Martell, A. M., Smith, R. F. C., and Speller, S. W. (1977). *In* "True Cone Lowland, Denom Island, Canada, an Arctic Ecosystem" (L. C. Bliss, ed.), pp. 437–466. Univ. of Alberta Press, Edmonton, Alberta.

Gause, G. F. (1934). "The Struggle for Existence." Hafner, New York.

Gauslaa, Y. (1984). Heat resistance and energy budget in different Scandinavian plants. *Holarct. Ecol.* **7**, 1–78,

Gimingham, C. H. (1972). "Ecology of Heathlands." Chapman and Hall, London.

Givnish, T. J. (1982). On the adaptive significance of leaf height in forest herbs. *Am. Nat.* **120**, 353–381.

Gjaerevoll, O. (1956). The plant communities of Scandinavian Alpine snow-beds. *Skr., K. Nor. Vidensk. Selsk.* **1956**, 1–405.

Gjaerevoll, O., and Jörgensen, R. (1952). "Fjellflora." Universitetsforlaget, Oslo.

Grime, J. P. (1977). Evidence for the existence of three primary strategies in plants and its relevance to ecological and evolutionary theory. *Am. Nat.* **111**, 1169–1194.

Grime, J. P. (1979). "Plant Strategies and Vegetation Processes." Wiley, Chichester, England.

Hairston, N. G., Smith, F. E., and Slobodkin, L. B. (1960). Community structure, population control and competition. *Am. Nat.* **94**, 421–425.

Holt, R. D. (1977). Predation, apparent competition, and the structure of prey communities. *Theor. Pop. Biol.* **12**, 197–229.

Jonasson, S., and Chapin, F. S. (1985). Significance of sequential leaf development for nutrient balance of the cotton sedge, *Eriophorum vaginatum* L. *Oecologia* **67**, 511–518.

Kalliola, R. (1939). Pflanzensoziologische Untersuchungen in der alpinen Stufe Finnisch-Lapplands. *Ann. Bot. Soc. "Vanamo"* **13**, 1–321.

Kershaw, K. A. (1975). Studies on lichen-dominated systems. XIV: The comparative ecology of *Alectoria nitidula* and *Cladonia alpestris*. *Can. J. Bot.* **53**, 2608–2613.

Knapp, R. (1965). "Die Vegetation von Nord- und Mittelamerika und der Hawaii-Inseln." Fischer, Stuttgart, Federal Republic of Germany.

Kofford, C. (1957). The vicuna and the puna. *Ecol. Monogr.* **27**, 153–219.

Komarková, V. (1978). "Alpine Vegetation of the Indian Peaks Area, Front Range, Colorado." Cramer, Vaduz, Liechtenstein.

Krebs, J. R. (1978). Optimal foraging: Decision rules for predators. *In* "Behavioral Ecology: An Evolutionary Approach" (J. Krebs and N. Davies, eds.), pp. 23–63. Blackwell, Oxford, England.

Lack, D. (1954). "The Natural Regulation of Animal Numbers." Oxford Univ. Press, Oxford, England.

Lechowicz, M. J. (1978). Carbon dioxide exchange in *Cladina* lichens from subarctic and temperate habitats. *Oecologia* **32**, 225–237.

Levins, R. (1968). "Evolution in Changing Environments." Princeton Univ. Press, Princeton, New Jersey.

MacArthur, R. H. (1972). "Geographical Ecology." Harper & Row, New York.

Mark, A. F. (1955). Grassland and shrubland on Mangataua, Otago. *N.Z. J. Sci. Technol.* **37**, 349–366.

Maynard Smith, J. (1974). "Models in Ecology." Cambridge Univ. Press, Cambridge, England.

McNaughton, S. J. (1979). Grazing as an optimization process: Grass–ungulate relationships in the Serengeti. *Am. Nat.* **113,** 691–703.

McNaughton, S. J. (1985). Ecology of a grazing system: The Serengeti. *Ecol. Monogr.* **55,** 259–294.

Mooney, H. A., and Billings, W. D. (1961). Comparative ecological physiology of Arctic and Alpine populations of *Oxyria digyna*. *Ecol. Monogr.* **31,** 1–29.

Moran, N., and Hamilton, W. D. (1980). Low nutritive quality as defense against herbivores. *J. Theor. Biol.* **86,** 247–254.

Mueller-Dombois, D. (1967). Ecological relations in the Alpine and subalpine vegetation of Mauna Loa, Hawaii. *J. Indian Bot. Soc.* **46,** 403–411.

Mulkey, S. S., Smith, A. P., and Young, T. P. (1984). Predation by elephants on *Senecio keniodendron* in the Alpine zone of Mount Kenya. *Biotropica* **16,** 246–248.

Nordhagen, R. (1927). Die Vegetation und Flora des Sylenegebietes. I. Die Vegetation. *Skr. Nor. Vidensk.-Akad., 1: Mat.–Naturvidensk. Kl.* **1927,** 1–612.

Nordhagen, R. (1943). Sikilsdalen og Norges fjellbeiter. *Bergens Mus. Skr.* **22,** 1–607.

Norin, B. N. (1961). Čto takoe lesotundra? *Bot. Zhurn. (Leningrad)* **46,** 21–28.

Noy Meir, I. (1975). Stability of grazing systems. An application of predator–prey graphs. *J. Ecol.* **63,** 459–481.

Oksanen, L. (1980). Abundance relationships between competitive and grazing-tolerant plants in productivity gradients of Fennoscandian mountains. *Ann. Bot. Fenn.* **17,** 410–429.

Oksanen, L. (1988). Ecosystem organization: Mutualism and cybernetics or plain Darwinian struggle for existence? *Am. Nat.* **131,** 424–444.

Oksanen, L., and Ericson, L. (1987a). Concluding remarks: Trophic exploitation and community structure. *Oikos* **50,** 417–422.

Oksanen, L., and Ericson, L. (1987b). Dynamics of tundra and taiga populations of herbaceous plants in relation to the Tihomirov–Fretwell and Kalela–Tast hypotheses. *Oikos* **50,** 381–388.

Oksanen, L., and Oksanen, T. (1981). Lemmings (*Lemmus lemmus*) and grey-sided voles (*Clethrionomys rufocanus*) in interaction with their resources and predators on Finnmarksvidda, northern Norway. *Rep. Kevo Subarct. Res. Stn.* **17,** 7–31.

Oksanen, L., and Ranta, E. (1989). Plant strategies along vegetational gradients on the mountains of Iddonjárga: a test of two theories. MS thesis, submitted to *Evol. Ecol.*

Oksanen, L., Fretwell, S. D., Arruda, J., and Niemelä, P. (1981). Exploitation ecosystems in gradients of primary productivity. *Am. Nat.* **118,** 240–261.

Olsvig-Whittaker, L., Schack, M., and Yair, A. (1983). Vegetation pattern related to environmental factors in the Negev watershed. *Vegetatio* **54,** 153–165.

Pimm, S. L. (1982). "Food Webs." Chapman and Hall, London.

Pimm, S. L., and Kitching, R. L. (1987). The determinants of food chain lengths. *Oikos* **50,** 336–346.

Reisgl, H., and Pitschmann, H. (1958). Oberen Grenzen von Flora und Vegetation in der Nivalstufe der zentralen Ötztaler Alpen (Tirol). *Vegetatio* **8,** 93–129.

Rosenzweig, M. L. (1971). Paradox of enrichment: Destabilization of exploitation ecosystems in ecological time. *Science* **171,** 385–387.

Rosenzweig, M. L. (1973). Exploitation in three trophic levels. *Am. Nat.* **107,** 275–294.

Rosenzweig, M. L. (1977). Aspects of biological exploitation. *Q. Rev. Biol.* **52,** 371–380.

Rosenzweig, M. L., and MacArthur, R. H. (1963). Graphical presentation and stability conditions of predator–prey interactions. *Am. Nat.* **97,** 209–223.

Schaffer, W. F. (1974). Optimal reproductive effort in fluctuating environments. *Am. Nat.* **108**, 783–790.

Schaller, G. B. (1972). "The Serengeti Lion." Univ. of Chicago Press, Chicago.

Sih, A., Crowley, P., McPeek, M., Petranka, J., and Strohmeier, K. (1985). Predation, competition and communities: A review of field experiments. *Annu. Rev. Ecol. Syst.* **16**, 269–311.

Sinclair, A. R. E. (1977). "The African Buffalo: A Study of Resource Limitation of Populations." Univ. of Chicago Press, Chicago.

Sinclair, A. R. E. (1985). Population regulation of the Serengeti wildebeest: A test of the food limitation hypothesis. *Oecologia* **65**, 266–268.

Smith, A. P., and Young, T. P. (1987). Tropical Alpine plant ecology. *Annu. Rev. Ecol. Syst.* **18**, 137–158.

Smith, C. C. (1976). When and how much to reproduce: The trade-off between power and efficiency. *Am. Zool.* **16**, 763–774.

Southwood, T. R. E. (1977). Habitat, the templet of species. *J. Anim. Ecol.* **46**, 337–365.

Söyrinki, N. (1939). Vermehrung der Samenpfanzen in der alpinen Vegetation Petsamo-Lapplands. II. Spezieller Teil. *Ann. Bot. Soc. "Vanamo"* **14**, 1–404.

Stearns, S. C. (1976). Life history tactics: A review of ideas. *Q. Rev. Biol.* **51**, 3–47.

Stebbins, G. L. (1981). Coevolution of grasses and herbivores. *Ann. Mo. Bot. Gard.* **68**, 75–86.

Tast, J., and Kalela, O. (1971). Comparisons between rodent cycles and plant production in Finnish Lapland. *Ann. Acad. Sci. Fenn., Ser. A4* **186**.

Thomas, D. C., and Edmonds, J. (1983). Rumen contents and habitat selection of Peary caribou in winter. *Arct. Alp. Res.* **15**, 97–105.

Thompson, K. (1987). The resource ratio hypothesis and the meaning of competition. *Funct. Ecol.* **1**, 297–303.

Tihomirov, B. A. (1959). "Vzaijmosvjazi Životnogo Mira i Rastitel'nogo Pokrova Tundry." Akad. Nauk SSSR, Bot. Inst. "Komarova," Moscow.

Tilman, D. (1982). "Resource Competition and Community Structure." Princeton Univ. Press, Princeton, New Jersey.

Tilman, D. (1984). Plant dominance along an experimental nutrient gradient. *Ecology* **65**, 1445–1453.

Tilman, D. (1985). The resource ratio hypothesis of plant succession. *Am. Nat.* **125**, 827–852.

Tilman, D. (1987). Secondary succession and the pattern of plant dominance along experimental nitrogen gradients. *Ecol. Monogr.* **57**, 189–214.

Tilman, D. (1988a). "Plant Strategies and the Structure and Dynamics of Plant Communities." Princeton Univ. Press, Princeton, New Jersey.

Tilman, D. (1988b). On the meaning of competition and the mechanisms of competitive superiority. *Funct. Ecol.* **1**, 304–315.

Tranquillini, W. (1957). Standortsklima, Wasserbilanz und CO_2-Gaswechsel junger Zirben (*Pinus cembra* L.) an der alpinen Waldgrenze. *Planta* **49**, 612–661.

Tranquillini, W. (1970). Einfluss des Windes auf den Gaswechsel der Pflanzen. *Umschau* **1970**, 860–861.

Troll, C. (1941). Studien zur vergleichenden Geographie der Hochgebirgen der Erde. *Bonn. Mitt.* **21**, 1–50.

Tuomi, J., Neuvonen, S., Haukioja, E., and Sirén, S. (1984). Nutrient stress: An explanation for plant anti-herbivory responses to defoliation. *Oecologia* **61**, 208–210.

van der Mejden, E., Wijn, M., and Verkaar, H. J. (1988). Defense and regrowth, alternative plant strategies in the struggle against herbivores. *Oikos* **51**, 355–363.

Walter, H. (1964). Vegetation der Erde in öko-physiologischer Betrachtung. I. Die tropischen und subtropischen Zonen. Fischer, Jena, German Democratic Republic.

Walter, H. (1968). "Vegetation der Erde in öko-physiologischer Betrachtung. II. Die gemässigten und arktischen Zonen." Fischer, Jena, German Democratic Republic.

Wein, R. W., and Bliss, L. C. (1974). Primary production in Arctic cottongrass tussock communities. *Arct. Alp. Res.* **6,** 261–274.

Whittaker, R. H., and Niering, W. A. (1965). Vegetation of Santa Catalina mountains of Arizona: A gradient analysis of the south slope. *Ecology* **46,** 429–452.

Williams, J. W. (1975). Studies on the tall tussock (*Chionochloa*) vegetation/soil systems of the southern Tararua Range, New Zealand. 2: The vegetation/soil relationships. *N.Z. J. Bot.* **13,** 269–303.

Winkworth, R. E. (1967). The composition of several arid spinifex grasslands of central Australia in relation to rainfall, soil water relations, and nutrients. *Aust. J. Bot.* **15,** 107–130.

Index